COLLECTING EVOLUTION

Collecting Evolution

THE GALÁPAGOS EXPEDITION THAT

VINDICATED DARWIN

Matthew J. James

OXFORD
UNIVERSITY PRESS

OXFORD

UNIVERSITY PRESS

Oxford University Press is a department of the University of Oxford. It furthers
the University's objective of excellence in research, scholarship, and education
by publishing worldwide. Oxford is a registered trade mark of Oxford University
Press in the UK and certain other countries.

Published in the United States of America by Oxford University Press
198 Madison Avenue, New York, NY 10016, United States of America.

CIP data is on file at the Library of Congress
ISBN 978–0–19–935459–7

9 8 7 6 5 4 3 2 1
Printed by Sheridan Books, Inc., United States of America

To my colleague and mentor
Robert I. Bowman
1925–2006
The Dean of Galápagos Science

The *Academy*

By F. M. Dickie

⌒⸻

POEM READ AT the christening and renaming of the schooner *Earnest* as the schooner *Academy* on June 23, 1905. Miss Anna Dickie formally christened and dedicated the schooner, a bottle of champagne being broken over the bows. Poem recited by Miss Mary E. Hyde, *Academy* librarian.

OH it's HO for the Galápagos
    Far southward o'er the sea,
Whose lonely caves' low booming roar
    Breathes forth a mystery

For there the fishes build their nests
    On high in aged trees,
While the cats and mice all play at tag
    Under the long green seas;

Where monstrous turtles crawl about
    All sizzling in the sun,
Until the cook in strident tones
    Cries, "Fetch him in, he's done."

Insects and beetles, so I've heard,
    Darken the noon-day sky;
This is a blessed isle for them,
    They never hear "Shoo fly"

To you most gallant mariners
    That boldly brave the main
Your daring in this enterprise
    Will in our hearts remain

To cast aside all friendships ties,
    O'er the blue sea to roam,
Ah, may your welcome be sincere,
    When this good ship comes home.

In all your loneliness, brace hearts,
    One comfort may be shown,
These decks will never be hen-pecked
    For all the wild oats sown.

Then let us here With right good cheer,
    All toast our much loved host.
May every move Of this trip prove
    Fit theme for honest boast.

I christen thee "Academy,"
    And away the wine goes splashing;
Its creamy foam Is fit alone
    For a vessel brave and dashing.

Our hopes for thee Will ever be
    Right manfully confirmed
May'st glorify unto the sky
    The name which thou hast earned

Contents

Preface

⌒

IN 1962, WHEN I was seven years old, my parents subscribed to a series of Time-Life books on the natural world, the Life Nature Library. This collection contained several well-illustrated volumes with enough color to hold the attention of three boys who had an excess of the natural world just outside their oceanfront back door in Hawaii. The books filled a foot of bookshelf space and allowed us to vicariously visit places far away from our volcanic island home. With simple titles such as *The Earth, The Desert, Ecology, The Fishes, The Plants,* and *The Mountains,* we learned through text and photos about places and organisms "out there" across the blue ocean. Despite the beauty and allure of these distant places, we never really understood why someone would want to actually visit them. Perhaps smugly, we were convinced that my parents had landed us in paradise.

But one book in particular in the series caught my attention, even back then. On its cover, two bright red crabs and two black marine iguanas clung to a wave-splashed volcanic rock. The photo could almost have been taken on the shore near our house; we had the red crabs, but no iguanas. The title was elegant in its simplicity and consistent with the others in the series: *Evolution.* All of what I saw and experienced growing up on the island of Oahu paralleled the geology and flora and fauna of the Galápagos Islands. These were two separate, yet connected volcanic archipelagoes, where the dominant scientific paradigm was Darwin and evolution and natural selection.

Years later on the mainland, after I had completed my doctorate and several years of university teaching and Galápagos research, I learned who took that photo. The photographer turns out to have been a man who had become my colleague, friend, and mentor, a man who was recently honored with the venerable label "the dean of Galápagos science": the late Dr. Robert I. Bowman.

Bob Bowman began working in Galápagos, particularly on Darwin's finches, several years before I was born, right around the time my physician father got out of Texas so that he could pursue his specialty in southern California. My path crossed Bob's, as seemingly happens with all Galápagos researchers, at a scientific meeting in San Francisco. Bob taught at San Francisco State University for more than thirty years before a productive retirement. In addition to sharing his encyclopedic knowledge of all things Galápagos with me, Bob was my link to the past.

Soon after I met him, I began exploring the story of the California Academy of Sciences' 1905–1906 mission to the Galápagos, the story told in this book. Fortunately for me, Bob knew several of the "eight young men" who appear here as sailor-scientists, and I got to know those men better and more fully through him. As I researched and wrote this book, I interviewed Bob at his home in the Berkeley hills and on the phone several times. Often unexpectedly he would come up with an anecdote about one of the eight young men. These anecdotes breathed life into men whose lives I have followed from cradle to grave for the past seventeen years and whom I will never meet.

With Bob's encouragement, I began exploring the topics I was drawn to outside of my specialty of paleontology. I began this writing project after becoming aware of the untapped wells of Galápagos archival material housed at the California Academy of Sciences in San Francisco. The Academy's archival collection from the Galápagos expedition consists of manuscripts, correspondence, photographs, field notes, and personal diaries. I found abundant material untouched in any academic sense, even down to opened letters still folded in their envelopes, read perhaps only once by the addressee, and not yet accessioned into the archival collections. Having walked in the footsteps of the expedition paleontologist, Washington Henry Ochsner, when I first visited Galápagos with colleagues in 1982, I now had a chance to see the islands again through the eyes of Ochsner and his shipmates.

What I found piqued my interest, both about the life and work of Rollo Beck, the leader of the 1905–1906 expedition, and the maritime history of the schooner *Earnest/Academy*, the vessel of discovery and adventure. Soon enough—and after many long conversations with Bob and the Academy archivist, Michele Wellck—my interest expanded to the lives of the eight young men aboard the ship and the expedition itself. It seemed to have a trajectory similar to that of a novel, with a beginning, middle, and end. The overall picture of the expedition is richer and fuller than that of just one person or of the vessel.

Nowadays one hears about the Galápagos mainly when the islands are threatened by an oil spill or when a volcano erupts and slow-moving lava threatens even slower-moving tortoises— or when the Galápagos are touted online or in a glossy brochure as an expensive ecotourism destination, promising a once-in-a-lifetime experience "before it is too late," a promotional gimmick to entice potential visitors to see islands where Darwin's ideas were set in motion. Little do they know the same call to action was issued in the late nineteenth century by men who were the predecessors of today's conservationists.

There have been many expeditions to the Galápagos. Even today, zoologists, geologists, botanists, and park wardens undertake expeditions to remote islands in the archipelago or to remote parts of islands. But save Darwin's visit in 1835, no expedition to the Galápagos has been more important for understanding the islands' natural history than the 1905–1906 expedition of the California Academy of Sciences. The timing of the expedition also mattered tremendously for the Academy, which lost virtually all of its collections and its museum building in the 1906 earthquake and fires that devastated San Francisco. In what follows, I have focused on the details, the intricate interconnections, and the broad implications of the expedition, believing that details reveal the true nature of something and allow us to place it into a wider scientific and historical context.

NOTE

Page xi: *One book in particular*: R. Moore, *Evolution* (New York: Time Nature Library).

Acknowledgments

THROUGHOUT THE PROCESS of research and writing, I approached archivists and librarians on bent knee. Some of them kindly allowed me unprecedented access to the original documents on which this story is based. With some I interacted in person, others answered my questions by email or post, or sent photocopies of documents. Their collective generosity and professionalism made my writing project possible, both the broader story of the schooner *Earnest/Academy* and the specific story of the 1905–1906 Galápagos Islands scientific collecting expedition. For their kind assistance, I offer my sincere gratitude. They are the keepers of the good stuff.

I first visited the Galápagos in 1982 as a graduate student on a paleontology collecting expedition, where we walked in the footsteps of and collected at many of the same fossil localities as did Washington Henry Ochsner in 1905–1906, using the field notes of Joseph Slayton Hunter. For their collegiality and camaraderie on that first trip, I thank William D. Pitt, Lois J. Pitt, Carole S. Hickman, and Jere H. Lipps. I traveled in the islands over the years with several Galápagos National Park guides and lecturers, including Greg Estes, John Madunich, Josephina (Pepy) Arevalo De Madunich, Juan Carlos Avila, John Woram, Maggie Fusari, and Carolina Larrea Angermeyer

For general assistance with Galápagos, Darwin, and topics on evolution, I thank Robert I. Bowman, John Woram, Joel W. Hedgpeth, Heidi Snell, Howard Snell, Marc L. Miller, Frank Sulloway, Leo F. Laporte, Linda Cayot, Roberta L. Millstein, Roslyn Cameron, Gayle Davis Merlen, Godfrey Merlen, Greg Estes, Thalia Grant, Jack Nelson, Peter Pritchard, Ricardo Crespo Plaza, Elizabeth Hennessey, William H. Durham, Johannah Barry, Swen Lorenz, Dennis Geist, Chantal Blanton, Edward Larson, Martin Wikelski, Rocio Rueda, Diane Keech, Alfredo Carrasco, Sylvia Harcourt, Jack Stein Grove, Günther Reck, Ole Hamann, Carlos Valle, Peter Kramer, Friedemann Koester, and Rodrigo Bustamante. For help with specimens and archival material at the Museum of Vertebrate Zoology, University of California, Berkeley, I thank Seth Benson, Ward Russell, George Bartholomew, Frank Pitelka, James L. Patton. In the California Academy of Sciences Library and Archives, I owe particular thanks to Michele Wellck, former archivist, for nurturing the early years of this project, having numerous productive conversations, and allowing very generous access to the collections; to Barbara West, volunteer at the Academy, for her superb work of editing the field notes from the 1905–1906 expedition; and to the very helpful Academy library staff: Karren Elsbernd, Larry Currie, Danielle Castronovo, Becky Morin, Heather Yager, Yolanda Bustos, and Kelly Jensen (especially with photograph assistance). In the research departments and specimen collections of the California Academy of Sciences, I thank Jean Farrington (Library), Al Leviton, Jens Vindum, and Angelica Arenas Rodríguez (Herpetology), John P. Dumbacher, Maureen Moe Flannery, Louis Baptista, and Betsy Cutler (Birds and Mammals), John E. McCosker (Ichthyology), Thomas F. Daniel, Debra Trock (Botany), Edward Ross, David H. Kavanaugh, Vick Lee (Entomology), Meg Burke (Education), Christina Piotrowski, Terrence M. Gosliner, Gary C. Williams, Elizabeth Kools, Robert Van Syoc, and Jean DeMouthe (Invertebrates and Geology), and George E. Lindsay (former director). At the Pacific Grove Museum of Natural History, Paul Finnegan and Esther Trosow, for access to Rollo Beck correspondence. At the J. Porter Shaw Library, San Francisco Maritime Museum, Bill Kooiman, Debbie Grace, Melani Van Petten, David Hull, Ted Miles, Gina Bardi, and Steve Hyman. I also thank for their generous help and access when I visited their collections Rebecca Livingston (National Archives, Washington, DC), Margery Charlante (Archives II, College Park, MD), Bill Cox (Smithsonian Archives, Washington, DC), Albert "Skip" Theberger (NOAA Central Library, Silver Spring, MD), Heather Haggstrom (Maryland Historical Society, Baltimore), Mary LeCroy (American Museum of Natural History, New York), and Molly Hagemann (Bishop Museum, Honolulu).

For assistance and insights into the lives of the sailor-scientists and crewmembers of the schooner *Academy*, I thank Frank Pitelka, Don Fellers, Polly Fellers, Gibbe Parsons, Kerry Parsons, Florence C. Haimes, Rollo Beck Parsons, Gary Fellers, Phyllis E. Slattery, Maureen S. Frederickson, Richard Shutler, Jr., Walter Slayton Hunter, John H. Ochsner, Deanna Ochsner, Cheryl Ochsner Swenson, Winifred J. "Grammy" Ochsner Clark, Gordon Clark, Judy Tucker, Bruce Clark, Susan Clark, Craig Ochsner, Vera Kirachenko, Al Leviton, Robert I. Bowman, Paul Brown (during a visit to the grave of F. X. Williams in Chula Vista, California), the son and grandson of the expedition botanist Alban Stewart—both of whom are also named Alban Stewart, Mark R. Jennings, Howard Mel, Amelie Mel de Fontenay, Lena Nelson Duffin, and Lois Schneider.

My faculty writing group colleagues over multiple years at Sonoma State University provided detailed written and verbal feedback of chapter drafts: Kathy Charmaz (stalwart director of the writing program), Noel Byrne, Melinda Milligan, Sheila Katz, Myrna Goodman, Andy Roth (Sociology); Jeannette Koshar (Nursing); Julie Allen, Mira-Lisa Katz (English); Judith Abbott, Mary Halavais (History); Rebecca Bryan, Lauren Morimoto (Kinesiology); Lena McQuade (Women's and Gender Studies); Nicole Myers (Geology); Carmen Works (Chemistry); Tom Rosen, Margie Purser, Richard Senghas (Anthropology); Diana Grant, Tryon Woods (Criminology and Criminal Justice Studies); Carlos Ayala (Education); Jeff Baldwin, Dolly Friedel (Geography); Daniel Malpica (Chicano and Latino Studies); Liz Thatch (Business Administration).

Mary Beth Glisson performed saintly work, reading the entire manuscript and making many helpful changes. Linda Cayot read and commented on the entire manuscript; any remaining errors are my own. Audra Wolfe masterfully helped shape the final manuscript, and her enthusiasm for this project is deeply appreciated.

My family, Cynthia Leung, Emma James, and Malia James, lived with this project, supporting me in the process and nudging me to completion. At Oxford University Press I have worked closely with Jeremy Lewis, Erik Hane, and all through the editing and production process with Anna Langley. I also thank Rajakumari Ganessin for overseeing and coordinating the numerous final production steps as project manager at Newgen Knowledge Works, Mary Becker for insightful copyediting, and Leslie Johnson as production editor.

Note to the Reader

⌒

THROUGHOUT THE TEXT, in order to maintain the historical connection to the 1905–1906 expedition and to maintain fidelity to the original field notes, correspondence, and the contemporary published scientific papers and monographs, I have used the English names of the individual islands in the Galápagos Islands. Several islands have multiple names, and consistency of usage is documented in John Woram's *Charles Darwin Slept Here*, as well as online at galapagos.to.

Prologue

EARLY ON THE morning of April 2, 1906, Rollo Beck, Joseph Hunter, and Frederick Nelson set out in a small boat to cross the narrow channel between Albemarle (Isabela) and Narborough (Fernandina) Islands in the Galápagos Archipelago. They sailed all day in light winds and shifting currents in a gig, a large sailing skiff, to cross the four and a half miles of channel separating the two islands, and then an additional thirteen miles south along the volcanic coast. Having started that morning from Tagus Cove on Albemarle Island, they headed toward Mangrove Point on Narborough Island, where they arrived by 5:00 p.m. The three men were part of an expedition from California in search of giant tortoises, called *galapagos* in Spanish, the archipelago's namesake antediluvian reptiles much prized by natural history museums.

While crossing the channel, the men watched delicate Elliot's storm petrels hovering over choppy water, picking up tiny bits of food in their beaks while seldom alighting on the water. They proceeded due south in the channel between equator-straddling Albemarle Island and Narborough Island, just to the west. The volcanoes at the southern end of Albemarle loomed ahead, looking almost like a separate island through the haze. To their left, over the port side of the gig, rose the high volcanoes of northern Albemarle. Over the starboard rail, the single, massive, shieldlike volcanic peak of Narborough provided a reference point for dead reckoning. Black lava flows draped its slopes like secretions from an oozing volcanic pustule, leaving irregular green patches of older vegetation that had escaped fiery destruction by sequential lava flows. Several Galápagos penguins, incongruous birds on the equator now living far from their ancestral Antarctic home, bobbed and dove in the water among flocks of larger feeding seabirds. Brown noddy terns balanced on the heads of brown pelicans, waiting for an easy meal when small fish jumped out of the larger birds' bill pouches, for their bills hold more than their bellies can.

After beaching and securing the gig at Mangrove Point, two of the men, Rollo Beck and Joseph Hunter, set off on foot to hike overnight up the flank of the volcano, to which Narborough and all the islands in the archipelago owe their origin. The third man, a crew-member on the expedition, Frederick Nelson, stayed behind at the coast to watch over the boat and to await their return, perhaps in a day or two, depending on their luck.

Beck and Hunter hiked across rough volcanic terrain, going just about a mile inland in the short hour before sundown. In equatorial Galápagos, unlike temperate latitudes, darkness falls rapidly, with very little dusk, when the sun sets. Surrounded by the tropical darkness, Beck and Hunter tried to sleep on a patch of hard, smooth lava. Beck had only an oilcloth for a covering. A swarm of bloodthirsty ticks kept the men awake until midnight. Beck killed perhaps a hundred ticks, two at once sometimes. Just after dark, and again just before day-light, the men watched little Audubon's shearwaters flying and sporting around overhead.

At daylight Beck and Hunter set off together to continue their upward climb, but at noon they separated. Beck took two cans of brewed coffee and a canteen of water as he continued hiking to the top of the island, constantly looking for giant tortoises. Hunter headed down-hill to explore the lower elevations even as he collected bird specimens along the way, includ-ing finches and mockingbirds. By late afternoon, Hunter was back at the coast after a very hot walk across the black lava. In general, Hunter lacked the physical stamina of Beck, a man of indefatigable energy for field collecting. Hunter found Nelson collecting some flightless cormorants and lizards; they waited together for Beck to return. But Beck persisted upward, hoping to be rewarded with a significant, even fantastic, discovery.

Beck knew from years of experience that a sure method to locate giant tortoises was to look for their fresh droppings. On that day, he had found a few fresh lumps of tortoise excre-ment (containing, he noticed, undigested cactus and cactus spines), convincing him that he was on the right track. Hiking upward to the top of Narborough, he found himself in the geological context of "a narrow 'island' of lava of more ancient eruption than that over which the first stage of the journey was made." This lava "island" of black basalt rock, devastatingly hot in the tropical sun, had scattered cactus and a few bushes and vines, and tantalizing tor-toise clues. Beck wrote in his journal, "As I worked up through this strip of lava I saw a few old droppings of a tortoise. . . . Thinking that if a tortoise were down in this desolate patch, there would be many on top. . . . There were no signs of tortoises here, though iguanas were plentiful." At least they would not hear Beck approaching: from what Darwin had written about them, Beck knew the tortoises to be stone cold deaf.

Even more revealing than the fresh droppings was a rounded rock, roughly the size and shape of a small female tortoise. Here, male tortoises had mounted the rounded rock many times and left semen spread about, like Onan in Genesis casting his seed upon the ground. Beck quickly surmised from this seminal discovery that living female tortoises were likely extinct on the island.

Beck saw similar surrogate rock partners on other islands in the Galápagos, and these rocks always foretold that the male tortoises had become, well, desperate, as well as doomed to extinction. The female tortoises had been doomed long ago by their smaller size, not their inability to adapt to island life. They fell victim not to Darwinian natural selection, but to human artificial selection. Many years before Darwin first stepped foot on the Galápagos in 1835, whalers, buccaneers, and mariners in square-rigged tall ships and schooners stopped in the islands to get tortoises for food and to find fresh water. The smaller, lighter females were easier to transport from the highlands down to the coast. Sailors doing the collecting left behind the larger, much heavier males. Eventually, even the larger males would be taken from islands with dwindling populations until no individuals remained.

The story of mass slaughter of tortoises comes from other islands in the archipelago, particularly Albemarle Island, and not from Narborough. The tortoises of Narborough likely had been obliterated through volcanism rather than direct human exploitation. No clear scientific evidence exists that a robust and reproductively viable population of giant tortoises ever inhabited Narborough Island. Beck, the expedition leader and a legitimate expert on collecting Galápagos tortoises, knew this. He had been here several times before on other expeditions, specifically intending to collect tortoises from every island possible. Beck knew from the literature that no documented tortoise specimen had ever been collected from Narborough Island by him or anyone else. Beck stood there, looking down at the obvious biological clues of tortoise dung and splattered semen, oracles and omens of the fate of a species, trying to decide which branch of several old trails to pursue next. A tortoise had to be nearby. But where?

Any remaining male tortoises of Narborough Island might have been looking for love in all the wrong places, but Beck was confident he could track them down. On that April day in 1906, love was in the air and on the ground, and Beck determined to find at least one of the randy males. After a Galápagos tortoise-collecting trip in 1902, Beck had written, "Love affairs were in full progress during our stay, and the amorous exclamations of the males could be heard at a distance exceeding 300 yards, even in the thick forest." But if the Narborough males were feeling amorous in April 1906, they were keeping quiet about it.

Silence greeted Beck on the open lava fields and grassy forests of Narborough Island as he continued hiking toward the summit. Night fell a second time, and Beck slept alone, as did any male tortoise on the island. At first light the next day, April 4, 1906, Beck had a breakfast of hardtack, a can each of deviled ham and salmon, and a can of coffee. He made the short push to the top and was rewarded with a sweeping vista of an open, grassy plateau punctuated by tall cactus leading to the steep precipice into the volcanic caldera of Narborough. It was indeed excellent tortoise country.

Alone at the summit of the largest pristine island in the world, an island on which no known human had ever lived and from which no scientist had ever collected a single giant tortoise, Rollo Beck surveyed the vista for tortoises for several hours. He saw none. Big, brightly colored land iguanas were everywhere, but no tortoises. Being a shrewd tortoise collector, Beck descended the mountain to the area where he had previously found the tortoise dung. At that lower elevation, Beck quickly picked up the faint trail of a tortoise that had left tracks in the thin volcanic soil. Hot on its trail, Beck left his own footprints in the soil from his new pair of hobnail boots, although by now the jagged lava rock of Galápagos had completely worn down the hobnails.

A few more steps down the narrow trail, Beck finally found the large, old, and apparently frustrated male tortoise, and made biological history. Beck had reached the end of another long, hot day in the field. He sat down at sunset with the old tortoise grazing nearby. Unconcerned about Beck or anything else, the tortoise peacefully ate his last meal of brown Galápagos grass, taking big, purposeful, unhurried bites. Surveying the twilight vista extending all the way down to the coast and off to the horizon, Beck ate his own dinner of hardtack, canned salmon, and canned sardines, and drank one of the cans of brewed coffee, all opened with his field knife.

What Beck did next ironically validated naturalists' predictions that time was running out for giant tortoises. He collected himself for a few minutes before starting his work. He was calm and professional about what he had to do next, what he was paid to do next. An experienced collector of tortoises and birds of all sizes, Beck was the internal dynamo of the Academy expedition and particularly skilled at collecting giant tortoises for natural history museums. This tortoise would be just like all the others he collected, and yet different. But the old male tortoise remained oblivious to Beck. Later Beck wrote, "Going on some distance farther the old male was found slowly feeding on grass near the trail. Getting my pack, I ate supper and skinned the tortoise by moonlight" (see figure P.1).

That lone tortoise, having survived for years in forced bachelorhood, came to embody naturalists' prophecies of extinction. The old male tortoise probably made many futile conjugal visits to the rounded female-surrogate rock over the years, and Galápagos tortoises can easily live for 150 years. Those sad, unproductive couplings now symbolize human-caused changes to the archipelago's flora and fauna. Beck loomed over the tortoise holding a sizable knife, now worn dull from opening all the cans of food he had eaten over the past few days on Narborough and from use on other tortoises. His knife would transform the old male tortoise, who would never again fertilize the eggs of a female tortoise, from a living, breathing example of evolution into a one-of-a-kind museum specimen.

FIGURE P.1 Rollo Beck skinning a large specimen of Galápagos giant tortoise. While a somewhat gruesome image, this emphasizes the massive size of some specimens of Galápagos tortoise and the many hours of work needed to skin the specimen. A smaller individual walks in the lower right, awaiting its fate.

Beck skinned the tortoise to remove its massive innards, making it lighter for the arduous hike back to the coast, where Hunter and Nelson awaited. Skinning the tortoise turned it into a museum specimen. The quarter moon that shone down that night made Beck's five-hour task easier. The light of the moon reflecting off the tortoise would eventually mirror the light shone on the process of evolution in Darwin's islands. Much of what we now know about evolution depends on well-preserved museum specimens. And that is certainly true for our current evolutionary understanding of Galápagos tortoises. Even a single specimen can illuminate a dark corner on the stage of the evolutionary drama.

Their skinning by moonlight brought the tortoises one step closer to extinction, and Darwin's mechanism of island evolution one step closer to being understood. But Beck did not particularly care at that moment about Darwin, or natural selection, or evolution. Beck rarely mentioned the broader ideas of evolution in his field notes or in his scattered publications on natural history stemming from this and other expeditions. Those were ideas for other individuals at other times. Science came into the fieldwork of the eight young men on this expedition mostly through the sense that collecting was necessary for science—in some sense, collecting *was* science. Beck had been sent to the Galápagos to do a specific job with a specific goal. High on the volcano, Beck did exactly what his employer, the California Academy of Sciences in San Francisco, had sent him to the Galápagos to do: collect evolution (see figure P.2).

FIGURE P.2 Expedition leader Rollo Beck carrying a relatively small Galapagos tortoise and his shotgun. This tortoise is much smaller than the one and only specimen Beck collected on Narborough (Fernandina) Island in April 1906 and carried for many miles over impossible terrain back down to the coast.

Today, the tortoise Beck collected on Narborough sits motionless in a realistic taxidermist's pose in the tortoise room in the Herpetology Department at the California Academy of Sciences, the oldest museum west of the Rocky Mountains. Since Beck's moonlight excursion into extirpation high on the volcano, no tortoise has ever been seen alive on Narborough. Even the scientific name given to this single-specimen species is evocative of its fate—*Geochelone phantastica*. With stuffed limbs and neck, the specimen is both fantastic and a phantasm—an apparition of the imagination of the scientist who coined its Latin name in 1907, the Academy herpetologist, John Van Denburgh (1872–1924).

The specimen stands on a shelf, sentinel-like, representing an idea, a concept—that of a unique biological species that genetically evolved in the Galápagos from its ancestors. A small metal tag, perhaps half an inch by two inches, bearing its embossed California Academy museum number, 8101, is affixed to one leg with a length of wire. The male tortoise physically embodies Darwin's fundamental idea of evolution and natural selection. He was collected as an example of evolution, the idea that unites all of the Galápagos and that drives the thriving tourist industry today. The species on each island differ because of evolution. Paradoxically, this lone specimen's scientific value was proved time and again, by drawings and photographs of its shell, to fingernail clipping–sized pieces of dried flesh used for DNA amplification and retracing its island genealogy. Without it, we would be missing a chapter from the story of Galápagos evolution. Without it, Darwin would not be so thoroughly vindicated.

That moonlit night, Rollo Beck had indeed collected evolution, in the form of the namesake organism that epitomized evolution in the archipelago. More than a century later, other scientists continued to expand on Darwin's ideas using Rollo Beck's unique Narborough tortoise. Beck salvaged the last of a lineage, motivated by a sad prophecy. Science is actually better for it today, but it took a dull knife in April 1906 to advance the progress of science. Tied to a tree with a length of rope, one tortoise on a deserted volcanic island created a new branch in the evolutionary drama of the Galápagos. He was the last of his race, and the first.

NOTES

Page 1: *Early on the morning*: JSH, April 2–5, 1906; RHB, 101–104; LSA, 97–98; RWE, 57–58.

Page 1: *Crossing the channel to Narborough*: RHB, unpublished field notes in CAS Archives; LSA, 97–98; JSH, unpublished field notes in CAS Archives.

Page 2: *Hiking toward the top of Narborough*: Van Denburgh (1914, 301).

Page 2: *Stone cold deaf*: Darwin (1839, 465).

Page 3: *Until no individuals remained*: See http://darwinspigeons.com, accessed November 1, 2016.

Page 3: *No clear scientific evidence exists*: Van Denburgh (1914, 301).

Page 4: *Skinned the tortoise by moonlight*: RHB, April 4, 1906.

Page 5: *Eight young men*: Although there eleven men on board the schooner *Academy*, I have used the phrase "eight young men" coined by the museum director, L. M. Loomis, to describe the eight sailor-scientists apart from the other three men aboard: the navigator, the mate, and the cook. In this usage and in this sense, the other three men function as "invisible technicians," as described by Wylie (2009) and Morus (2016).

Page 6: *Use of DNA of Academy specimen 8101*: Poulakakis et al. (2015).

Page 6: *Tied to a tree*: Rollo Beck clearly knew he was collecting the very *first* tortoise known to science from Narborough Island, but he had no way of knowing it would also be the *last*. In the years since Beck, no one has ever seen, let alone collected, another tortoise from Narborough. Did Beck kill the very last individual of that island species? Probably not. In the mid-1960s, the entomologist David Cavagnaro visited Narborough and saw fresh tortoise dung on the island, but no tortoises. Karl Campbell gave the biologist Linda Cayot a photo of the scat (L. Cayot, personal communication, October 2014). The remaining tortoises could have died out naturally or been killed by a lava flow emanating from the island's caldera. And given the inaccessible nature of Narborough Island, there is still a very slight possibility that a few still roam the volcanic slopes.

1

Halfway to the Stars

THE FAME AND fortune of the Galápagos Islands as an evolutionary nerve center starts with a letter to Charles Darwin from his mentor and friend, the Reverend Professor John Stevens Henslow (1796–1861). As an undergraduate at Cambridge University from 1828 to 1831, Darwin took frequent constitutionals in the Cambridge countryside with Henslow, a senior professor of botany and geology. As a result, Darwin became known to students and dons as "the man who walks with Henslow." Their friendship was close and deep and lasted the remainder of Henslow's life. Henslow wrote the pivotal letter on August 24, 1831, and in it revealed that Darwin would be invited by Robert FitzRoy to participate as a naturalist in the upcoming voyage of HMS *Beagle* "to Terra del Fuego & home by the East Indies." Henslow wrote that he considered Darwin "to be the best qualified person I know of who is likely to undertake such a situation—I state this not on the supposition of yr. being a *finished* Naturalist, but as amply qualified for collecting, observing, & noting any thing worthy to be noted in Natural History." Henslow's supposition proved highly accurate, with dramatic implications for both Darwin and the Galápagos.

The Galápagos are certainly a beautiful physical, geographic place. But they also symbolize something, which makes them no mere islands. Darwin's voyage was only one of a series of expeditions by researchers in zoology, botany, and geology that thrust the Galápagos into the evolutionary spotlight. These researchers collectively made the Galápagos a showcase of evolution. The enchanted combination of intellectual factors is almost as elusive as the enchanted nature of the islands themselves. Geography is important, and so is chance.

The islands and their fauna lend themselves to hyperbole, perhaps because some of the organisms seem to be exaggerations of the truth. Swimming lizards, tropical penguins, cormorants that can't fly, fish-eating snakes, and tool-using and blood-eating vampire finches all inhabit the islands. These remarkable creatures, and many more, are what Darwin called "indigenous productions" and "aboriginal creations, found nowhere else."

For Darwin they were part of the "mystery of mysteries—the first appearance of new beings on this earth."

Today, if we're lucky and go to the right places, we can see as many of these organisms as remain alive today. The California Academy of Sciences expedition had the same goal.

The Academy's 1905–1906 expedition to the Galápagos achieved its success through careful planning and organization, and by the selection of competent fieldworkers as expedition members (figure 1.1). This pivotal expedition shaped the archipelago's history and the history of evolutionary thinking, particularly neo-Darwinism. The determined organizational efforts of the museum director and ornithologist, Leverett Mills Loomis (1857–1928), resulted in meticulous and methodical advanced planning for this expedition. Its clearly defined goals and objectives in the Galápagos Islands in the end met or exceeded Loomis's expectations.

Closer to home, the expedition guaranteed the future of an institution all but destroyed in the 1906 earthquake and fires that ravaged San Francisco. The narrative arc of the Academy expedition is little known, and the broad historical and scientific implications of this collecting expedition are little appreciated even by the majority of Galápagos aficionados. This book is an attempt to tell that story.

FIGURE 1.1 **Setting Sail for Darwin's Islands.** The crew of the newly renamed schooner *Academy* pose on deck in June 1905 in San Francisco (from left to right): Frederick T. Nelson (mate), Alban Stewart (botanist), Ernest S. King (field assistant), Rollo H. Beck (chief of the expedition), Joseph S. Hunter (mammalogist), (seated) J. J. Parker (navigator), Joseph R. Slevin (herpetologist), Edward W. Gifford (ornithologist), Washington Henry Ochsner (geologist and paleontologist), and Francis Xavier Williams (entomologist). Not pictured: James White (cook). This is the only known group photograph of the expedition crew. *Archives, California Academy of Sciences.*

The 1905–1906 expedition lasted seventeen months. The cadre of collectors spent a year and a day in the islands, making theirs the longest scientific expedition to the Galápagos in its history. The Academy expedition visited all of the thirteen major islands (several times in some cases) as well as many of the smaller islets. In comparison, Darwin had been in the islands for five weeks and visited just four islands. The Academy expedition collected some seventy-eight thousand biological specimens—more than any Galápagos expedition before or since. The collectors brought back 266 preserved specimens of giant tortoises, as well as numerous other reptiles, land birds and seabirds, mammals, insects, plants, land snails, and fossils. In effect, they brought the Galápagos back to San Francisco.

Charles Darwin immersed himself in scientific specimen collecting during the voyage of HMS *Beagle* from 1831 to 1836. For Darwin, the multi-year *Beagle* voyage proved to be the vehicle for his scientific insights, his newly amalgamated view of life. The voyage afforded him abundant opportunities to perceive connections and meaningful patterns between seemingly unrelated and random observations stemming from his methodical inspection and collection of flora and fauna. He gained some of these insights during the voyage, while other important parts of Darwin's personal synthesis occurred after HMS *Beagle* returned to England. The post-*Beagle* insights by Darwin's fellow naturalists, both botanists and zoologists, depended on access to his well-documented specimens. The acceptance and development of Darwin's theories were slow, gradual, and grounded as much by physical museum specimens as by ideas. Imagine how *little* we would know from the *Beagle* voyage if Darwin had *not* collected specimens.

The struggle that drives the narrative of Charles Darwin's mechanism of evolutionary change is not unlike the struggle that drives the narrative thread of great works of fiction. Struggle with the environment, which includes physical aspects such as food, climate, and space to live, creates conflict and opportunity. Struggle with other organisms in the environment can erupt into serious, life-and-death conflict. Sometimes the "other organism" can be your sibling or your nest mate. For Nazca booby nestlings, as we will see in chapter 9, the outcome of this life conflict is always deadly for the younger sibling. Scenes of great beauty in the Galápagos Islands can be locations of great conflict, and just as conflict drives the plots of good books and good movies, conflict also drives evolution. A life-altering, once-in-a-lifetime vacation can yield digital snapshots of evolution in action. Picture-postcard-perfect beaches and shorelines can reek of selective death, which is exactly what Charles Darwin had in mind.

For Charles Darwin, evolution formed the story that explained life. Evolution was a contest driven by competition to leave more genes in subsequent generations. He understood the mechanism of natural selection as a creative force that multiplied species and created new ones. His landmark book about the *origin of species* chronicled the creation of new species in an orderly ancestor-to-descendant chain of genetic events. Two of the simplest three-word definitions of evolution are "change over time" and "descent with modification," while any definition of evolution contains, at its core, the inevitable conflict between individuals or between species. When observing nature in the Galápagos, or elsewhere, look for conflict and you will see evolution in action.

The deserted volcanic archipelago that served as one of Darwin's many points of geographic inspiration during the *Beagle* voyage is now a permanent beehive of activity for local residents, numbering more than 25,000 and growing. The residents are joined by visiting scientists, resident scientists, and ecotourists from all over the world. As part of the enduring legacy of the 1905–1906 expedition, scientists continue to document patterns of evolution and have accumulated a sufficiently large number of specimens to elevate the islands to the status of living outdoor laboratory and museum of evolution.

Visiting and resident scientists, including zoologists, botanists, geologists, and conservationists, truly treat the Galápagos as a laboratory of evolution where they work on organisms of their specific interest. Scientists must obtain permission from the Ecuadorian governmental authorities to conduct research on specific organisms in clearly demarcated areas of the islands, which ecotourists are intentionally restricted from seeing as they navigate from island to island on vessels of various sizes. The scientists observe, measure, mark, release, and recapture individuals; some collect blood or tissue for genetic analysis. Rarely, if ever, do scientists today intentionally capture and preserve an animal for preservation in a museum—at least not the larger, emblematic animals such as birds and reptiles. Plants are another story, because a few leaves or flowers carefully picked off and pressed flat for drying on herbarium paper will not harm most plants. Insects are also commonly collected whole, because rarely can you sample just a part of a tiny insect for study. The islands' contemporary scientists and park wardens are intrinsically conservationists and good stewards of the land and the native menagerie.

How do we place into historical context the some seventy-eight thousand animal and plant specimens collected by the 1905–1906 expedition? Several authors over the years have grappled with the magnitude of the collection and have both misinterpreted the original intention of the Academy's expedition and the historical context of this collection and other museum collections worldwide. This Whiggish history, often written in a voice of moral outrage in the context of modern conservation, misses the target. No species were made extinct by the Academy expedition, with the arguable exception of the Narborough tortoise, which, even if this is true, had already been devastated by more than two hundred years of human exploitation. More birds are killed in a three-year period by the buses and pickup trucks zooming on the twenty-six-mile paved road between the town of Puerto Ayora and the Itabacca Channel, which separates Indefatigable (Santa Cruz) Island and South Seymour (Baltra) Island, than were killed by the entire 1905–1906 expedition. The published estimate is that, from 2004 to 2006, nine thousand birds were killed on the highway and that each vehicle on Indefatigable killed an average of seven birds per year. This killing of birds goes on day after day. Anyone riding in a vehicle from Baltra to Puerto Ayora has likely struck a bird. Just look out the windshield and watch. As conservationist Godfrey Merlen recently wrote, "It's ironic that Galápagos wildlife, the very magnet that brings the tourists, is being slaughtered by the taxi drivers who depend on the tourist dollars for their livelihood."

Yet, even if the work of the 1905–1906 expedition represents something that could never be, and should never be, repeated today, the collection remains worthy of celebration. One of the best sources of DNA samples for analysis by today's field biologists is, ironically, the museum specimens that researchers curated and cataloged from this and other past expeditions. And, hands down, the California Academy of Sciences' Galápagos expedition produced more specimens in use by today's investigators than any other trip. But at the time of the voyage, the Galápagos were not yet seen as the place that transformed our worldview of life on earth. They were not yet Darwin's islands. Darwin's finches were not even Darwin's finches in 1905; they were still just called Galápagos finches, and would be until 1935. For the Academy museum director, Leverett Mills Loomis, and the crew of the schooner *Academy*, the islands were not so much a place to study evolution as a place to collect evolution. These sailor-scientists were not there to gather data to support or refute evolution, but rather to collect the natural history products of evolution, the species found nowhere else on Earth.

When it comes to biological field collecting and museum collection building, to borrow a phrase, "quantity has a quality all its own." The notion that bigger natural history museum collections are better collections has direct implications for how scientific collectors operate in the field. When both the size of individual, world-record specimens and the size of an entire museum's collection matter, the effect on natural populations of organisms can be potentially devastating. Fortunately, in the case of the 1905–1906 expedition, we now know that, for the most part, the long-term effects on the islands' populations were minimal.

In their hearts and minds, and despite some varied scientific training, these young men viewed themselves as collectors and thought little about directly testing the veracity of evolution. Instead, they were driven by a shared sense of what great collectors do and a belief that they were serving the needs of science by building the museum collection. Their primary goal was to enhance the reputation of the California Academy of Sciences—their employer—by providing the museum with not just one specimen of everything from the Galápagos, or even ten specimens of everything, but as many specimens as possible. Their dedicated and competent efforts brought the Galápagos back to San Francisco.

To most people familiar with the city of San Francisco, the California Academy of Sciences is the big natural history museum in Golden Gate Park. But it was not always big, and it was not always in Golden Gate Park. Few people alive today would know the Academy as anything other than the large complex directly across from the de Young Museum and the Japanese Tea Garden, as it has been since 1916. Outside the Academy's front doors is the Music Concourse, with several fountains and rows of heavily pollarded elm trees, and the Spreckels Temple of Music, as the bandshell was called in 1899 when built. Visitors to the Academy on any given day include throngs of noisy schoolchildren on field trips, tourists from far and wide, and San Francisco and Bay Area residents. After viewing the Academy's exhibits, visitors can stroll outside along the Music Concourse among statues of Ulysses S. Grant, Ludwig von Beethoven, Giuseppe Verdi, Junipero Serra, Francis Scott Key (paid for by the Academy benefactor James Lick), and a naked Roman gladiator.

Since September 2008, visitors have wandered about the Academy's impressive new museum building designed by the Italian architect Renzo Piano. They enjoy the Steinhart Aquarium, the Morrison Planetarium, a rainforest exhibit, and numerous other permanent and rotating exhibits on natural history, including a substantial exhibit on the Galápagos in the *Islands of Evolution* exhibit. Yet visitors seldom learn that the Academy had the most modest of beginnings in downtown San Francisco well over 150 years ago. Few visitors know that the Academy's museum was originally located downtown on Market Street, and fewer still know much about the early years or trials and tribulations of this venerable San Francisco institution. Many visitors would be fascinated to know that the Academy's first president, the geologist Andrew Randall, was gunned down in 1856 on the streets of San Francisco over a $67,000 gambling debt. While less sensational, other aspects of the Academy's tumultuous history in the late nineteenth and early twentieth centuries are equally interesting.

Despite its humble and unpretentious beginnings as a small, underfunded institution, the California Academy of Natural Sciences, as it was first called, proved ultimately to be one of the dominant forces in San Francisco's cultural and intellectual history. The Academy, as an idea, was born in 1853. During those flush years of the California Gold Rush, when the Barbary Coast was a raucous and bawdy place, seven San Francisco gentlemen met under the dim glow of tallow candles in a poorly furnished room on Montgomery Street. They met in the heart of today's financial district to form a fledgling society for the study of natural history. Their stated intention on April 4, 1853, was to organize "an association for the development of the natural sciences." That they succeeded is a minor miracle.

Their nascent scientific body had every likelihood of failing, mainly due to a lack of funds. Several times in the early years the Academy's tenuous financial resources dwindled to nothing. There was no museum building, and the notion of the "Academy" as an institution could easily have ceased to exist. They survived the financial roller coasters of not one but two North American gold rushes in the nineteenth century: California in 1848–1855 and the Yukon in 1899. The Academy survived and continues to thrive today because of the ardor, dedication, and sacrifice of its founding fathers. The Galápagos Islands played an unexpected major role in that survival. The Galápagos benefited the reputation of the Academy, and the Academy benefited the reputation of the Galápagos.

In 1872, the Academy unanimously voted in Charles Darwin as an Honorary Member. In 1873, the Academy celebrated its twentieth year of existence. In that same year, the signature cable cars that universally symbolize the city of San Francisco began running up and down its hilly and foggy streets, halfway to the stars. Also in 1873, a major financial windfall propelled the Academy toward permanent success. The money came from Mr. James Lick (1796–1876), a wealthy land speculator with vast holdings dating from before the California Gold Rush. But Lick originally had very different plans for his vast wealth. Some might call it a "pyramid" scheme.

Lick accumulated untold anger toward and scorn for society throughout his life and planned a legacy that would make forgetting him difficult. Specifically, he planned to build a giant pyramid on land he owned at 4th and Market in downtown San Francisco as a monument commensurate in magnitude with his accomplishments and opinion of himself. But instead of building an actual stone pyramid larger than the Great Pyramid of Giza to honor and aggrandize himself after death, James Lick gave part of his vast fortune to the California Academy of Sciences, along with other institutions. The pyramid had been planned as a kind of rude gesture, akin to thumbing his nose at San Francisco, and California, and the world. Instead, Mr. Lick's focus changed from megalomania to philanthropy. With land and money from Mr. Lick, the nascent and perpetually insolvent California Academy of Sciences erected a seven-story professional office building and an adjoining natural history museum in 1891 on the site of the proposed pyramid on Market Street (figure 1.2).

The story of James Lick mirrors the story of the Academy, with a meager start and fortuitous growth. Mr. Lick was a carpenter and piano maker from Pennsylvania who over time amassed extensive landholdings in California, including all of Santa Catalina Island off the southern California coast. By enticing Domingo Ghirardelli to emigrate from Lima, Peru, in 1847, Mr. Lick became directly responsible for Ghirardelli chocolate being made in San Francisco. He had a general reputation as "an eccentric recluse, an avaricious skinflint, selfish, unlovable, unwashed, and unsocial." Others noted that, despite his eccentricities, he

FIGURE 1.2 **The New Academy.** Built with an endowment from the benefactor James Lick, the newly completed California Academy of Sciences building on Market Street in San Francisco in 1891, with rental offices facing the street and museum in the rear. This is the museum that Director Leverett Mills Loomis needed to fill with specimens to gain the respect of other long-established museums. *Special Collections, Stanford University.*

would be best remembered "for his strong will and determination, his ambition and drive, his honesty and generosity." After suffering a debilitating stroke in 1873, and at the urging of the Academy's president, George Davidson (1825–1911), Mr. Lick designated several San Francisco scientific and social institutions as financial beneficiaries of his great wealth.

Following Lick's stroke and the disposition of his fortune, he lingered for three years in his San Francisco hotel.

Lick's goal was to "promote the diffusion of Science and the prosperity and perpetuity of the Academy." Lick's strong atheist predilections led him to stipulate that the land be used strictly for scientific purposes, and never for political or religious purposes. The zoologist Edward S. Morse (a protégée of Louis Agassiz) clearly understood the significance of Lick's generous gift to the Academy, which elevated the institution to national and global significance. Morse stated in a June 1874 address at the Academy not only that it was the only endowed society on the Pacific coast, but that "the sum given to the Academy by Mr. Lick exceeded all the funds of all the natural history societies in the Atlantic states." The Lick bequest secured the Academy's future. The initial stumbling block for the Academy, which had only recently advanced its treasury into the black by some $1,500, was a requirement in the Lick land grant to come up with funds to complete the museum within two years. The original version of Lick's donation stipulated that the land on Market Street would revert to Lick if the building funds could not be raised.

Lick's bequest required the Academy to erect and maintain "a substantial and elegant brick edifice, three stories in height, with a substantial granite front faced with appropriate scientific emblems." The building ended up being seven stories high (figure 1.3). Lick wanted the building to be of "classic design" to distinguish it from buildings used for commercial or business purposes. But by the time it was ultimately finished, the building was, in fact, used to generate commercial income for the Academy. This wasn't exactly what Lick had in mind, but by the time the building was finished, Lick had been dead for fifteen years.

At the time of Lick's bequest, the Academy had crept along for twenty years, barely making ends meet with a membership that waxed and waned with the social, political, and financial tides of San Francisco. In its most extravagant projections, the Academy's trustees had not envisioned spending more than $25,000 to secure "proper accommodations for the institution." And now they were faced, in President Davidson's estimation, with raising $200,000 to build a suitable museum on Market Street land that was itself worth $150,000. The Academy had graduated from a shoestring operation into a mature and substantial institution, albeit one faced with raising unprecedented sums.

Throughout 1873, Academy members voiced concern that they would be unable to raise the necessary funds, and thus that their ticket to financial freedom would slip through their collective fingers. However, in October 1873, James Lick extended the deadline for completing the museum by ten years. In September 1875, he removed all restrictions from the deed, giving the Academy an open timetable for constructing the museum building.

But before the time restrictions were finally removed, and with additional square footage donated by Lick in July 1873, the amount needed for the museum building had grown steadily from $200,000 to $300,000. This caused even greater stress and uncertainty among Academy members and trustees. Now Judge S. Clinton Hastings, who had founded the Hastings College of Law in San Francisco in 1878, galloped to the rescue. Hastings proposed that he would join with "any twenty or any ten other gentlemen" to raise the money needed to build and maintain the "museum edifice." But with the deadline extended and Lick's financial restrictions removed, the Academy no longer needed Hastings's assistance. The Academy was now free to build what and when it wanted.

FIGURE 1.3 **Pre-Earthquake Academy Museum.** Interior view of the California Academy of Sciences on Market Street in San Francisco before the April 18, 1906, earthquake and fires. Note the crossed tusks on the second floor, and compare with Figure 10.4 after the earthquake and fire. *Archives, California Academy of Sciences.*

The new Academy museum went up between 4th and 5th Streets on Market Street in booming and bustling San Francisco, a location approximately equivalent—at least in the eyes of San Franciscans—to Paris's Avenue des Champs Élysées. In the town known as the "Paris of the West" and "Baghdad by the Bay," the Academy had a prime location within easy sight of the Ferry Building, still standing today on the waterfront at the foot of Market Street. Upon its completion in 1891, the seven-story building was divided into two parts. The Academy used the front portion to generate rental income, using the prime location to draw in retail and commercial tenants. The Academy's scientific interests were relegated to the back. On the first floor, a lecture hall featured free lectures by Academy members twice each month. The second, third, and fourth floors housed public exhibition halls, displaying the wonders of geology, zoology, and anthropology to curious visitors.

    With a brand new museum building, the Academy's board of directors and scientific cura-
tors were ready to expand their national and international reputations. For that, they would
need better collections.

The annual reports of museums, from the American Museum of Natural History in New
York, to the Field Museum of Natural History in Chicago, to the California Academy of
Sciences in San Francisco, commonly have a section on the extent of their scientific collections
(figure 1.4). Art museums do so as well, boasting about the size and diversity of their holdings.
No major natural history museum wants to have the fourteenth-largest bird collection in the
world. So size matters. In 2003, the California Academy of Sciences even titled its sesqui-
centennial annual report "18 Million Real Things" to emphasize how its scientific collections
validated and vindicated its role in the international scientific community. Even though the
Academy had been collecting since 1853, the current collection starts with the schooner full
of specimens from the Galápagos that returned to San Francisco in 1906, because the earlier
collections were destroyed by the earthquake. By 2014, the collection had grown to nearly 46
million specimens. Today, the Academy's curators and collections managers are not only keepers of the legacy of these irreplaceable biological collections but also champions of Galápagos
conservation and advocates for educational opportunities for Ecuadorian undergraduate and
postgraduate students pursuing research careers in the Galápagos. This 1905–1906 scientific
collection endures in its usefulness and relevance. All of the rest is lost to history.

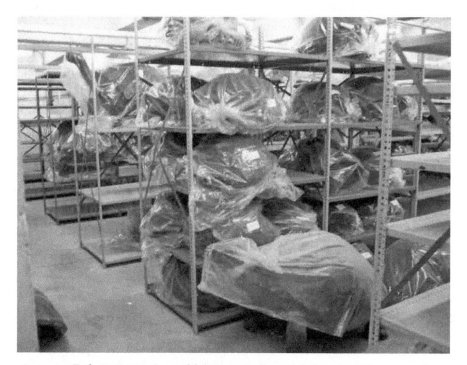

FIGURE 1.4 **Enduring Legacy.** Some of the 266 tortoise specimens from the 1905–1906 expedition,
still available for active research in the collections of the Herpetology Department at the California
Academy of Sciences in Golden Gate Park. *Archives, California Academy of Sciences.*

By the late nineteenth century, the Academy had grown into a mature institution from humble beginnings. It had acquired a building, research collections, and library and a competent scientific staff. The institution was poised to expand intellectually and geographically, and the Galápagos Islands were the ideal agent of expansion. In these exotic isles on the equator, made famous half a century earlier by Charles Darwin, the Academy's collectors could obtain abundant specimens of rare organisms to boost the prestige of their collections. But first, they needed a trial run.

Loomis and his colleagues sought status on the vanguard of natural history museums, perhaps rivaling that of the American Museum of Natural History in New York City. Loomis also coveted the fame and prestige bestowed on any museum possessing specimens from the exotic Galápagos Islands. Their agreed-upon and financed path to success involved conducting a series of extensive collecting expeditions, particularly to islands, to build up their biological and geological collections. These new and enlarged field collections would form the substance of detailed taxonomic monographs by the senior curatorial staff. The sheer size of the collections housed in drawers and cabinets in the museum on Market Street and the expected impact of the curators' scientific publications would provide the foundation for the museum's increased prestige and respect.

Loomis sought the largest seabird collection in the world and put the expedition doctrine right to work: specimen collecting would fill his new museum with research material. The first of the Academy's larger expeditions involving a sailing vessel sent to international waters was its 1903 expedition on the leased schooner *Mary Sachs* to a group of offshore Mexican islands known as the Revillagigedos. There, under the heat of the tropical sun, the Academy staked its academic claim to studying and collecting on remote eastern Pacific islands, and Rollo Beck honed his skills as an expedition leader (see figure P.2). This was the first international collaboration between Loomis as scientist turned consummate museum administrator and Beck as eminent field collector. The next several years would show how fruitful the Loomis–Beck collaboration would prove to be.

Loomis, an ornithologist, wanted to continue his personal research "so far as he could find time to do so," and he realized that Rollo Beck "would be a valuable aid in that work." Beck and his men would collect the specimens; Loomis and his colleagues would study them and publish scientific papers. It was a match made in heaven. Loomis took full advantage of Beck's skills, keeping one foot in the world of birds and the other in the world of museum administration.

Loomis envisioned the 1903 expedition as a training mission. He put the knowledge and skills of the Academy's curatorial staff to use in San Francisco and elsewhere in California while sending out young and somewhat inexperienced collectors after first training them in standard collecting and museum preparation techniques. Today, senior curators make up the core of most expeditions, with young, inexperienced staffers working as assistants. Scientists are often expected to participate in field expeditions to get that ineffable "boots on the ground" feeling, whether by collecting dinosaur remains in Mongolia or ants in Madagascar. It is, of course, infinitely easier to get to remote Pacific islands today than it was in 1905. A contemporary, well-funded, physically active curator might lose the respect of peers if she simply remained in her museum offices, publishing papers based on specimens others collected. But at the turn of the twentieth century, this sort of arrangement was entirely normal. Rollo Beck led a group of three or four students as his assistants, each paid "one dollar a day and expenses in the field." This was the going rate, equal to what the US Fish Commission paid its field collectors.

Loomis proposed a series of training sessions to prepare the young collectors to bring back high-quality specimens. The botanist Alice Eastwood (1859–1953) would instruct a student in how to collect and preserve plants; the herpetologist John Van Denburgh and the bird collector Rollo Beck would each teach a student how to collect reptiles and birds, respectively; and the anthropologist Alfred Kroeber (1876–1960) would instruct a fourth student in collecting anthropological specimens. Two years later, the Academy's Galápagos expedition would mirror this method of assembling a crew. Both expeditions relied on relatively young, but reliable men to collect specimens for a more senior group of curators who stayed back in San Francisco and subsequently published papers on the scientific results of the expedition. This plan worked smoothly, with two notable exceptions: at least two multi-year professional conflicts over specimens broke out, involving the botanist and paleontologist from the 1905–1906 Galápagos trip.

In January 1903, Loomis hired Rollo Beck as a field collector for $70 per month, plus expenses. This was Beck's first steady scientific job, although he had impeccable qualifications stemming from three previous collecting trips to the Galápagos Islands funded by Walter Rothschild, two of which even took place on the same ship, the *Mary Sachs*. Beck's personal characteristics (he was physically and mentally very tough), his previous field experience, and his familiarity with sailing vessels made him well qualified to head an expedition. Beck's formal education did not match his practical experience. Growing up in the small town of Berryessa, south of San Francisco near San Jose, Beck did not quite finish the eighth grade and left school at age twelve in 1883, the year after Charles Darwin died. For Beck, shipping out on a series of small schooners would prove to be, as whaling voyages likewise did for Herman Melville's narrator Ishmael, "my Yale College and my Harvard." Loomis placed his faith in Beck to accomplish the lofty expedition goals of the Academy.

The Revillagigedo Islands were remote and dangerous to visit. Therefore, the men thought it better to collect as much as possible before returning to the safety of San Francisco. "Several of the young men nearly lost their lives in attempting to make a landing" on the rock-bound coast of Clarion Island. This was one of the expedition's "most thrilling experiences" in these islands "over which the surf constantly breaks." Who would want to return to such a remote and dangerous place? Conservation, as we think of it today, was not on the minds of these young men as they collected for Loomis in this remote location.

The 1903 trip provides a sterling example of how biological conservation, as we came to know it in the late twentieth and early twenty-first centuries, was *terra incognito* for collectors of this era and generation. On San Benito, one of the islands in the Revillagigedo group, the young men secured, with great difficulty, a specimen of a little bird called the San Benito or McGregor's house finch. The significance of collecting this specimen was abundantly clear to the men. "There are only four or five birds of the kind left on the island," the *San Francisco Chronicle* reported in August 1903, "and they will probably be extinct by the time another expedition visits the place." The Academy visited again in 1925, and the species was last reported alive in 1938. It is now extinct. Speeding species to extinction was not the intention of these collectors, but it may have been a consequence. Not that it was easy to catch the birds. The San Benito house finch did not give up its rare feathers easily. "The bird brought back was chased all over the island and finally shot."

Although these bird collectors did not *intend* to cause an already rare species to become extinct, they wrote and acted as though they were powerless to prevent such an occurrence. Their thinking did not include the notion, so common today, of protecting species from extinction. They simply took extinction of rare species as inevitable. Consequently, they directed their actions toward collecting the last few for science, to be studied later. To them, the real tragedy would be if no preserved specimens were set aside in a museum cabinet to be handled, measured, and photographed at a later date.

The 1903 trip lasted four months on the schooner *Mary Sachs*, which left San Francisco at the end of April and returned in mid-August. Stopping first at Ensenada, the expedition proceeded to the San Benito Islands, San Martin, Natividad, and Guadalupe, and San Benedicto, Soccoro, and Clarion Islands in the Revillagigedo group. On this expedition the collectors secured about one thousand specimens of birds along with many valuable botanical specimens, and twenty gallons of alcohol were filled with snakes and lizards. Loomis was pleased and ready to run his longer Galápagos expedition, although its scheduled departure in November 1904 was delayed due to the Academy's failure to locate a suitable schooner.

For Loomis, the Academy's 1903 expedition to Mexico was just a practice run for a larger, longer, and more ambitious trip to the Galápagos. Both trips would have the same basic elements: a group of young men, a small schooner, a group of islands with exotic species, and sufficient time to collect thoroughly. The vigorous intellectual debate that followed publication of *The Origin of Species* only strengthened the allure of the archipelago. One could *read* the book and vicariously partake in the controversy, or one could *visit* the Galápagos and participate in the controversy firsthand. The Academy decided to both read and visit.

The big East Coast museums had already been to the Galápagos, and David Starr Jordan of nearby Stanford University, which had also sponsored a Galápagos expedition in 1898–1899, was a friendly rival to Loomis. This type of rivalry can be very productive, as scientists and institutions leapfrog over each other to acquire more numerous and better specimens from increasingly exotic locales.

Loomis oversaw the smallest details. He was an exacting planner and wanted to lay the groundwork to ensure the expedition's success. The Galápagos expedition would benefit his personal research on seabirds, and it would benefit the museum broadly. As they had in 1903, the senior curators at the California Academy of Sciences would send young collectors to the Galápagos while they themselves stayed home in San Francisco, with the ultimate goal of publishing papers on the results of the expedition. The Academy purchased a schooner, renamed it the *Academy*, and assembled a crew. Loomis insured the schooner *Academy* and, most important, her valuable scientific contents against loss and made plans for how the specimens would advance the expanding professional agenda of the California Academy of Sciences.

As Loomis explained to newspaper reporters who interviewed him prior to the ship's departure, this voyage to the islands would be longer than the 1903 expedition, and the collecting would be even more thorough. Clearly, this scientific voyage would bring the California Academy of Sciences fame and treasure, in the scientific sense.

On June 28, 1905, the eighty-nine-foot schooner *Academy* pushed off from Mission Street Pier 2, a modest dock adjacent to the Ferry Building. The schooner, with a crew of eleven—eight sailor-scientists and three crewmembers—sailed into the Pacific Ocean through the bridgeless Golden Gate and into the history of evolutionary biology.

For now, Loomis's work was done.

NOTES

Page 8: *Henslow wrote*: In the original letter, "your" is shortened to "yr."

Page 8: *Henslow's supposition*: Jenyns later published a monograph on the fishes collected during the 1831–1835 voyage of HMS *Beagle* (Jenyns 1842). Jenyns also published a memoir of Henslow (Jenyns 1862).

Page 9: *Indigenous productions*: VOB, 363; *Aboriginal productions*: VOB, 363; *Mystery of mysteries*: VOB, 363.

Page 9: *Determined organizational efforts*: Bishop (1929).

Page 10: *The cadre of collectors*: Wong (2003); James (2010, 2012).

Page 11: *Misses the target*: I greatly appreciate the historical sentiment expressed in the first paragraph of the preface to *The Bird Collectors* by Barbara Mearns and Richard Mearns, in which they express the sentiment of those of us who write about collectors and museum collecting.

Page 11: *Seven birds per year*: Jiménez-Uzcátegui and Bettancourt (2008).

Page 11: *Dollars for their livelihood*: Jiménez-Uzcátegui and Bettancourt (2008).

Page 12: *Few people alive today*: For two years in the mid-2000s the Academy had temporary quarters and moved back to Golden Gate Park in 2008.

Page 12: *Impressive new museum*: Wels et al. (2008).

Page 13: *Despite humble beginnings*: Hittell (1997, 11).

Page 13: *Charles Darwin as an Honorary Member*: Hittell (1997, 140).

Page 13: *The money came from*: Lick (1967); de Ford (1941).

Page 14: de Ford, 1941; and http://collections.ucolick.org/archives_on_line/James_Lick.html, accessed November 1, 2016.

Page 15: *Lingered in his San Francisco hotel*: Lick's hotel was called Lick House, located at the intersection of Montgomery and Sutter Streets in San Francisco. It was destroyed in the fire following the April 18, 1906, earthquake.

Page 15: *Lick's goal*: Hittell (1997, 172).

Page 15: *At the time of Lick's bequest*: Hittell (1997, 172).

Page 15: *Land value*: After the April 1906 earthquake and fires, the eighty-foot parcel on Market Street between 4th and 5th Streets was appraised at $500,000. Letter from L. M. Loomis to Andrew Carnegie, October 15, 1906, Council correspondence, California Academy of Sciences Archives.

Page 15: *But before the time restrictions*: Hittell, 1997, 153.

Page 17: *46 million specimens*: Annual Report, California Academy of Sciences, 2014, 2.

Page 18: *1903 expedition*: The schooner *Mary Sachs* had a substantial maritime record. See http://www.historymuseum.ca/cmc/exhibitions/hist/cae/nav73e.shtml, accessed November 1, 2016.

Page 18: *Offshore Mexican islands*: Hittell (1997, 414).

Page 18: *Loomis–Beck collaboration*: Dumbacher and West (2010).

Page 18: *A contemporary*: Even today, some well-respected curators are unable to get into the field much, if at all, for a number of reasons, including, above all, lack of funding.

Page 19: *Series of training sessions*: None of the eight young sailor-scientists on the Academy expedition were trained as anthropologists for the 1905–1906 trip, although E. W. Gifford subsequently became director of the Lowie Museum of Anthropology, now on the Berkeley campus of the University of California. (It was originally located in San Francisco, which is where Gifford served as the director; in 1991 it was renamed the Phoebe A. Hearst Museum of Anthropology).

Page 19: *Series of small schooners*: Melville (1851), last sentence of ch. 24, "For a whale-ship was my Yale College and my Harvard."

Page 19: *Surf constantly breaks: San Francisco Chronicle*, August 14, 1903.

Page 19: *McGregor's House Finch*: Also known by its Latin name, *Carpodacus mexicanus mcgregori* Anthony, 1897.

Page 19: *It is now extinct*: King (1981, 188); Fuller (1987).

Page 19: *Finally shot: San Francisco Chronicle*, August 14, 1903.

Page 20: *The big East Coast museums*: Larson (2001, 123).

Page 20: *Research on seabirds*: By 1905, Loomis had assembled the world's largest seabird collection in the Ornithology Department at the California Academy of Sciences by his own direct collecting and that of Rollo Beck, and through exchange with other field collectors and museums.

Page 20: *Assembled a crew*: In July 1952, the entomologist Edwin C. Van Dyke (who died later in 1952) spoke on tape with Francis Xavier Williams (a member of the 1905–1906 Galápagos expedition) and other curators at the California Academy of Sciences. Van Dyke stated that Loomis had approached him in 1905 and stated that Loomis "wanted an entomologist on that trip, he had to be a husky boy. I said there was only one that I knew that might be available and that was young F. X. Williams." Transcript of taped conversations between Edwin C. Van Dyke, Francis X. Williams, Edward S. Ross, Edward L. Kessel, and Hugh B. Leach, July 7, 1952, manuscript in California Academy of Sciences Archives.

2

Walking in Darwin's Footsteps

VISITORS TO THE Galápagos Islands walk, often literally, in Darwin's footsteps. They go ashore at some of the same sites Darwin did, and the boats they board navigate in the wake of HMS *Beagle*. Their experiences navigating between the islands, going ashore on an inflatable panga or Zodiac, and snorkeling in the shallows and hiking the visitor trails, engage all of their senses, making the Galápagos very real to them. A visit to the Galápagos involves travelers' bodies more than visits to most other vacation destinations. From trying to sleep on a rolling, pitching, and yawing vessel (or even walking the deck when under way in heavy seas), to kayaking, swimming, and birdwatching, to sampling new foods and trying out their rusty Spanish, ecotourists cannot escape the sense that they are distinctly *someplace else*. Galápagos visitors feel the islands are a special place, and they are correct.

Darwin experienced this same sense of dislocation during his voyage aboard the *Beagle*. His first stop was a rainforest in Bahia, Brazil, in 1832, some three years before reaching Galápagos:

> The day has passed delightfully. Delight itself, however, is a weak term to express the feelings of a naturalist who, for the first time, has wandered by himself in a Brazilian forest. The elegance of the grasses, the novelty of the parasitical plants, the beauty of the flowers, the glossy green of the foliage, but above all the general luxuriance of the vegetation, filled me with admiration. A most paradoxical mixture of sound and silence pervades the shady parts of the wood. The noise from the insects is so loud, that it may be heard even in a vessel anchored several hundred yards from the shore; yet within the recesses of the forest a universal silence appears to reign. To a person fond of natural history, such a day as this brings with it a deeper pleasure than he can ever hope to experience again.

Galápagos visitors snorkeling for the first time might have a similar reaction to the underwater world that Darwin had on first seeing the rainforest—or, for that matter, when they see any of the iconic Galápagos animals, such as tortoises and iguanas and finches, for the first time. In the Galápagos, Darwin is not arcane or obscure. He's just like us.

The Galápagos as a place and Charles Darwin as a person have become inextricably linked and inseparable. The geographic place and Darwin's ideas about species are permanently dovetailed in the minds of scientists and the general public. But one crucial, abstract idea ties Darwin and the Galápagos together: names. Organisms' scientific names are key to any discussion of evolution, and the species made famous by Darwin's voyage to the Galápagos have special names. Visitors traveling in the islands will soon hear the names lava lizard, lava gull, lava cactus, and lava heron. Before departing they will also hear of Darwin's finches, Darwin's cotton, and Darwin's sunflower. No trip is complete without mention of the Galápagos mockingbird, Galápagos storm petrel, and Galápagos hawk. A standing joke among Galápagos naturalist guides is that if anyone asks what a certain animal or plant is, and you don't know for sure, just add "lava" or "Darwin's" or "Galápagos" to the name of nearly any organism. Few people will detect your bluff.

These common, or vernacular, English names are one way to refer to plants and animals; the other way is to call them by their scientific or so-called Latin names. Most plants and animals in the Galápagos also have common or vernacular names in Spanish. The blue-footed booby, for instance, is locally known as the *piquero de patas azules*. But scientists, regardless of their native language, refer to these creatures by a universal scientific name: *Sula nebouxii* Milne-Edwards, 1882. Scientific names, usually based on Latin or Greek, are the lingua franca of scientific discourse.

The formal scientific naming and classification of new species is known as the science of taxonomy. The use of polysyllabic words of Latin and Greek origin adds to taxonomy's obscurity and inscrutability. But this tradition extends back at least into the mid-1700s. The agreed-upon convention, dating back to the publication of the Swedish botanist Carl Linnaeus's *Systema Naturae* in 1735, is to list two names, much as people have two names: a first name and a last name. This system, known as binomial nomenclature, combines a capitalized genus name (such as the dinosaur *Tyrannosaurus* or our own genus, *Homo*) with a lowercase species name (such as *rex* or *sapiens*). The plural of "genus" is "genera," but the singular and plural of species is the same. Today, and for more than one hundred years, the "official" scientific name also includes the name of the scientist who classified the species as such and the date of that classification. Written in complete form, our dinosaur would be *Tyrannosaurus rex* Osborn, 1905, recognizing that the paleontologist Henry Fairfield Osborn (1857–1935) described and named it in 1905. Our own scientific name is *Homo sapiens* Linnaeus, 1758. In practice, however, the scientist's name and date are usually dropped, and the generic name usually abbreviated, at least after the first mention. Thus, *Tyrannosaurus rex* Osborn, 1905 becomes the familiar *T. rex*.

Scientific names reflect the evolutionary uniqueness of any given species. Those with the same scientific name are viewed, ultimately, as biologically similar, while those with different scientific names are, by definition, biologically different. The mirror of taxonomy provides a reflection of evolution. Anyone who struggles with pronouncing scientific names in a guidebook struggles with understanding the very fabric of evolution.

Not all of the animals and plants Darwin observed and collected in the Galápagos had been classified and assigned scientific names at the time of Darwin's visit. Many of them had not yet been assigned a genus and species; some of them were, in fact, new to science. Charles Darwin visited the Galápagos in 1835, only to be followed seventy years later by the Californians on the schooner *Academy*. In the more than one hundred years since their return to San Francisco, Darwin's name has become even more closely associated with the islands, largely because of taxonomic research on the species he first wrote about in *The Origin of Species*. In 1935, the British ornithologist Percy Lowe (1870–1948) gave the name "Darwin's Finches" to the birds that, in Darwin's day, were simply known as Galápagos finches (figure 2.10). A little more than a decade later, the Englishman David Lack (1910–1973) published a book, *Darwin's Finches*, that cemented the connection between Darwin and these birds of "first-class biological importance." They have since become textbook icons of evolution.

Darwin's name continues to be associated with the islands through the names of plants and animals. To give only a few examples:

Darwin's cotton—*Gossypium darwinii*, an endemic species of cotton
Darwin's daisy —*Darwinothamnus tenuifolius*, an endemic daisy
Scalesia—a genus known as the "Darwin's finches of the plant world"
Carpenter bee—*Xylocopa darwinii*, an endemic carpenter bee

The late Stephen Jay Gould opened an essay in *Natural History* magazine about Charles Darwin and the *Beagle* voyage with a deceptively simple question, "Who was the naturalist on board HMS *Beagle*?" Most readers would confidently answer, "Charles Darwin," but Charles Darwin was not, as it turns out, the official naturalist aboard the *Beagle*. That job

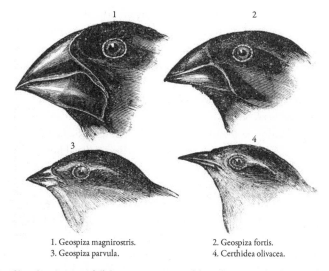

1. Geospiza magnirostris.   2. Geospiza fortis.
3. Geospiza parvula.   4. Certhidea olivacea.

FIGURE 2.1 **Profiles of Evolution.** Of all the iconic organisms of the Galápagos Archipelago, the finches that bear Darwin's name owe much of their academic fame to material collected by the 1905–1906 expedition. Source: Public domain. There are three places that list this illustration as part of the public domain: https://en.wikipedia. org/wiki/Darwin's_finches#/media/File:Darwin%27s_finches_by_Gould.jpg; http://darwin-online.org.uk/ converted/published/1845_Beagle_F14/1845_Beagle_F14_fig07.jpg; http://www.biodiversitylibrary.org/ page/2010582#page/393/mode/1up.

title rested with the ship's surgeon, while Darwin sailed as a "supernumerary" passenger to accompany Captain Robert FitzRoy during the five-year voyage. FitzRoy chose Darwin as a gentlemanly companion to stave off the isolation, loneliness, and ennui that afflicted his predecessor on the *Beagle*, Captain Pringle Stokes.

Riffing on Gould (who was in turn riffing on an old Groucho Marx bit about Grant's Tomb), I'll pose a related question: Why did the *Beagle* sail around the world? The immediate reasons for this voyage, which lasted from 1831 to 1836, are likely either "natural history collecting" or "coastal surveying in preparation for British expansionism." The *Beagle* certainly accomplished both of these missions, but for the real motivations one need look no further than suicide, kidnapping, and pedophilia. All this and more can be garnered from Peter Nichols's insightful biography of Robert FitzRoy, *Evolution's Captain* (2003).

A startling trajectory of events led up to Darwin's participation in the *Beagle* voyage and arguably the ultimate genesis of his revolutionary ideas about biological evolution by natural selection. The tale begins with the suicide of Captain Pringle Stokes in the cold and gloomy latitudes of Tierra del Fuego in 1828 on an earlier *Beagle* voyage. Stokes's suicide catapulted twenty-four-year old Lieutenant Robert FitzRoy into command of the *Beagle*. Before returning to England, the new captain kidnapped four Fuegian Indians in retaliation for a stolen longboat. FitzRoy transformed their presence on the *Beagle* into a divinely inspired living experiment in theology and anthropology. He aimed to transform the Fuegians into civilized and educated gentlefolk, proving the power of Christianity over the "savage beast." Of the four Fuegians, one twenty-year-old man (Boat Memory) died of smallpox on arriving in England. The other three were housed at the Walthamstow Infants School, just north of London. Two of the survivors were male: York Minster, about twenty-six years old, and Jemmy (or Jeremy) Button, about fourteen years old. The third was a young girl about nine years old named Fuegia Basket. Word soon got out that FitzRoy's human experiment had not gone as planned. Eight months after their return to England, those close to the situation at Walthamstow realized that "the sweet child [Fuegia Basket] adored by so many was having sex with a hulking, unreformable savage [York Minster]." One can only imagine the physical and emotional pain suffered by Fuegia Basket as the result of FitzRoy's failed experiment.

With his experiment proving a failure, FitzRoy desperately needed to return the Fuegians to their homeland lest his reputation with the Queen be compromised and his naval career run aground. Thus, FitzRoy set in motion the 1831 departure of the *Beagle* under the guise of more hydrographic surveying for the Royal Navy. FitzRoy realized how the earlier trip had affected Pringle Stokes and knew that suicide lurked in his own family's past. FitzRoy kept a vigilant eye on his own internal demons.

The voyage transcended its ignominious origins to became a resounding scientific and personal success for both FitzRoy and Darwin. The Fuegians survived the voyage back to Tierra del Fuego and returned to their somewhat complicated lives in that region. The hydrographic surveying, exploration, and mapping accomplished during this second *Beagle* voyage ensured FitzRoy's future success in the Royal Navy and his good favor with the Queen. Mission accomplished. Darwin and FitzRoy collected specimens and made observations that later proved crucial for Darwin's work as the best-known naturalist to sail on the *Beagle*.

The Galápagos Islands, located hard on the equator due west of Ecuador in the Pacific Ocean, became the most famous and celebrated destination of the 1831–1835 *Beagle* voyage, largely through the publication of Darwin's account of the trip, published in 1839. The islands got their own chapter, wedged chronologically between South America and Tahiti, in Darwin's travelogue. Today, that book is often simply and commonly called *The Voyage of the Beagle*, a title it acquired coincidentally in 1905. It had a longer name when it first appeared in 1839: *Journal of Researches into the Geology and Natural History of the Various Countries Visited by H.M.S. "Beagle."* In the several editions of *Voyage of the Beagle*, Darwin described what he saw on the four islands he visited and speculated obliquely and tangentially about several evolutionary topics that would later come to have greater importance. But while *Voyage* brought Darwin fame, the crowning achievement in Darwin's life was the 1859 publication of *On the Origin of Species*. This book presented Darwin's mechanism of biological evolution by natural selection and inextricably forever linked Darwin and the Galápagos Islands.

Although not as well known as Darwin's finches, Galápagos mockingbirds were very important to Darwin's thinking. Everywhere he went, he recorded what he saw and where. Darwin accurately kept track of which island each type of mockingbird was from, and that made all the difference to his later ideas about evolution by natural selection. These mockingbirds were arguably more important to Darwin's development of his theory of evolution than the finches because they stimulated his thinking about differences in beaks and feathers between birds on different islands.

Today there are four known species of mockingbirds in the Galápagos, but Darwin knew of only three. During the voyage, Darwin collected specimens in much the same way that Beck and his crewmates would do in the 1905–1906 expedition. Back in England, the British ornithologist John Gould used the specimens to identify three new species of mockingbirds, all of them previously undescribed in the scientific literature. Darwin turned to Gould, one of the most famous ornithologists of his day, for insight into his bird specimens, much as he turned to other experts in zoology and botany for insights into animals and plants. From the time that the mockingbirds landed by chance on these islands millennia ago, down through the many generations as they slowly and sequentially diverged from each other, the mockingbirds had no scientific names and therefore no separate identity. When Gould said in 1837 that the birds were "new to science," he clearly meant, and Darwin clearly understood, that these species of birds were being documented for the first time. On the basis of his encyclopedic knowledge of bird taxonomy and distribution, Gould knew these birds were found nowhere else on Earth. They were unique to the Galápagos, and that meant they had evolved in the islands (see figure 2.2).

Gould's naming decision, the crucial and functional step of scientific taxonomy, carried with it at least two broad and inescapable implications: the mockingbirds were previously undiscovered by any scientist or naturalist, and they occurred only in the Galápagos. Darwin carried this logic further. If the birds were *endemic* to the Galápagos, they must have *evolved* there from earlier continental ancestors. Gould's naming decision provided the crucial push for Darwin to accept what he called the *transmutation of species*. By this nineteenth-century phrase, Darwin meant what we now call "evolution."

The mockingbird species had originated in the Galápagos Islands. The Galápagos were the birthplace of many new species. They were a hotbed of evolution. But what is a species,

FIGURE 2.2 **Three Mockingbirds.** Following the *Beagle* voyage, the three named species of Galápagos mockingbirds (a fourth has since been named) were of seminal importance to Darwin in formulating his evolutionary ideas. Source: © The Natural History Museum / The Trustees of the Natural History Museum, London, used with permission.

exactly? And what were Gould's criteria for recognizing new species? And what, in turn, were Darwin's criteria for species that might have originated by his natural mechanism of selection? Even today, when we hear about endangered or threatened species, we rarely stop to ask what it means to refer to something as a "species." But it was certainly important to Gould and to Darwin for the Galápagos specimens.

So what *is* a species? Taxonomists easily get caught up in hand-waving about what species *names* can tell us after they are named. Even today there is no standard percentage in genetic makeup to tell one species from another. The gold standard remains the ability, or inability, to produce fertile offspring. Of course, we cannot tell this from preserved specimens, which takes us back to the realm of scientific judgment. For museum specimens named in the traditional way without DNA analysis, a new species is what a competent taxonomist says a new species is. So donkeys and horses are separate species, because, together, they produce only sterile mules. To get more mules, you have to start over again with male donkeys and female horses. The perennial problem for nondomestic animals, like living Galápagos finches, mockingbirds, and tortoises, lies in demonstrating whether very different-*looking* animals, or animals that live in very different geographic settings, can in fact *reproduce* or not. Working with preserved museum specimens does not make this task any easier. Darwin himself addressed the notion of "species" to set the stage in *On the Origin of Species* for his discussion of how new species arise. "No one definition [of the term 'species'] has satisfied all naturalists; yet every naturalist knows vaguely what he means when he speaks of a species."

Initially designated as new to science by Gould, the iconic finches had far less impact on Darwin than is commonly believed, eclipsed as they were in their influence on Darwin by the archipelago's three species of mockingbirds. Even the iconic Galápagos giant land tortoises have much greater influence today on our thinking about evolution and conservation than they ever did for Darwin in his day. The tortoises owe their rise to the status of evolutionary icons to the eight young men of the 1905–1906 expedition and to the detailed anatomical work done long after Darwin died in 1882. But all of this work on finches and mockingbirds and tortoises figures prominently in our thinking today because someone sat down with an array of museum specimens and patiently and meticulously described them as new to science. In other words, someone engaged in the tedious activity of taxonomy based on museum specimens. Our enjoyment of the Galápagos today, and our understanding of the flora and fauna, are due to these taxonomic specialists.

The scientific names of Galápagos species tell evolutionary stories. These organisms are visibly a part of the Galápagos story that we hear about from naturalist guides or that we read about in species identification guidebooks. The naming of species resides in the domain of zoological or botanical taxonomy. This naming, as a science, should reflect the actual evolutionary history of a group of animals or plants. For example, Darwin's colleague John Gould could have decided, as a bird taxonomist, that the various beak shapes of finches and mockingbirds spread out before him on the table represented merely a wide range of varieties, not actual species. Domestic dogs, for instance, vary tremendously, but Chihuahuas, Pomeranians, Yorkshire Terriers, Great Danes, Irish Wolfhounds, and Rottweilers are all dogs: *Canis familiaris*. Using dog variation as a benchmark of speciation, Gould could have easily lumped together the most widespread members of Darwin's mockingbirds, the Galápagos mockingbird, *Mimus parvulus*, with the type from Charles (Floreana) Island, *Mimus trifasciatus*, and the one from Chatham or (San Cristóbal) Island, *Mimus melanotis*, all of which he named in 1837. If it worked for dogs, why not for mockingbirds or finches? Of course, in the case of dogs, Gould could have reassured himself that all of these varied dogs can, in theory, interbreed and produce viable offspring. He could not do that experiment with Darwin's dead mockingbirds. But they ended up being three dead birds and one big idea.

In the absence of living, breeding animals, experienced taxonomists rely on their judgment to resolve the conflict between "lumping" together similar forms into one species or "splitting" up variation into multiple species. If the obvious beak variation that Gould saw in Darwin's Galápagos mockingbirds was really nothing more than the kind of variation one sees in domestic dogs, then all the mockingbirds could be "lumped" together into a single, highly variable species. In Darwin's terms, this would have suggested that the amount of evolution that took place over time in the archipelago, the descent with modification that Darwin himself wrote about, was much *less* remarkable. The existence of a single unique mockingbird species would have suggested a simple scenario of modest evolution, but the presence of numerous species shows a complicated story of substantial evolution and divergence from a common ancestor. In the Darwinian understanding of taxonomy, species equals evolution.

Gould could also have used cruciferous vegetables as his species measuring stick, citing the great range of variation within a single species seen from broccoli to cauliflower, from brussels sprouts to kale, kohlrabi, and collard greens. But both dogs and brassicas share another characteristic besides wide variation: both have been developed through artificial, rather than natural, selection. Several years prior to publishing *The Origin of Species*, Darwin had noticed how domestic breeders of pigeons, cows, and dogs used selective breeding practices to emphasize certain desired characteristics. In *The Origin of Species*, Darwin used artificial selection as a direct analogy to natural selection. We can observe the results of *artificial* selection every time we visit the grocery store to look over the abundant fruit and vegetable offerings or settle in to watch a recap of the Westminster Dog Show. In the Galápagos, we can observe the results of *natural* selection.

The 1905–1906 Galápagos expedition had as a fundamental and explicit goal the collection of specimens to be studied by taxonomists back at the museum in San Francisco. Those taxonomists would, in the years following the expedition, pass their individual and personal judgments on Galápagos birds, tortoises, lizards, snakes, plants, mammals, insects, and mollusks. In each case, the taxonomic decisions translated into evolutionary stories, where species equal evolution. Every taxonomist faces the same decision Gould did: whether, on the basis of years of experience, to lump obvious physical variations into a single species or split off groups of specimens with "significant" variation into separate species. If a group contained many species found nowhere else on Earth, then considerable evolution must have preceded the field collecting. The evolutionary scenarios would depend on the measured judgment of taxonomic specialists for years to come, many of whom were the resident curators at the California Academy of Sciences. The majority of birds unique to the Galápagos Islands were given their scientific names—and thus their evolutionary history—by zoologists who never visited the islands.

By the time Charles Darwin died on April 19, 1882, he had done more for the fame and modern reputation of the Galápagos Islands than a thousand advertising agencies or tourist bureaus. He had, in fact, made this obscure volcanic archipelago, with its odd assortment of plants and animals, the centerpiece of one of the most revolutionary ideas in the history of human thought: biological evolution by means of natural selection.

Darwin made the Galápagos Islands more than just a static museum of evolution—more than a mere *showcase* of evolution, as they are often described today. He made them the *proof* of evolution. Galileo's believers relied on images in telescopes and sketches of stars and planets in the heavens. Einstein's believers would rely on complex mathematical and computational proofs on sheets of paper or computer screens. But Darwin's believers would rely on finches, tortoises, iguanas, and mockingbirds that you could walk among. In the Galápagos, you could reach out and touch evolution.

Although it's always difficult to evaluate the significance of any one visitor's voyage, it seems reasonably safe to say that Darwin is the single most important person to have visited the Galápagos. Unlike most figures in the history of science, we know what Darwin was doing on nearly every day of his life because he left behind a thorough paper trail of notebooks and correspondence. Historians of science have analyzed Darwin's formal and informal writing to document when ideas occurred to him and how he arrived at those ideas. And from these sources of documentation, historians know that the case of the mockingbirds weighed heavily on Darwin's mind.

Charles Darwin clearly knew what he was up against when he sat at his writing chair in the ground floor study of his stately country home, Down House, located sixteen miles south of London, penning the first draft of *On the Origin of Species*. He wrote nearly daily for many months during 1858 and into 1859, with a big window to his left opening on sweeping views of an extensive lawn, his brick-and-glass greenhouse for plant experiments, and the Sand Walk, where he retreated daily for pensive, therapeutic strolls. Walking on that comfortable, level path, keeping track of the number of loops he walked by kicking aside black chert cobbles, Darwin knew he was fighting an uphill battle for the hearts and minds of his fellow Victorian naturalists.

The slow, Darwinian pace of ecotourism travel in the Galápagos is nicely mirrored in Darwin's writing about the islands in *The Origin of Species*. His sentences are unhurried, if a little ponderous and convoluted in the style of the day, but they reveal why Darwin considered the islands central to his ideas about evolution. Darwin knew that distinct species in a single genus tend to live some distance apart, but he was having difficulty deciding how far was far enough. If obviously related birds lived far apart in Britain and Norway, how different did the birds need to be in feathers or beak or song before ornithologists would call them separate species? Darwin noted this "dilemma of difference" with Galápagos birds when he wrote, "Many years ago, when comparing, and seeing others compare, the birds from the separate islands of the Galápagos Archipelago, both one with another, and with those from the [South] American mainland, I was much struck how entirely vague and arbitrary is the distinction between species and varieties." Is a bird on the South American mainland a separate species from its obvious relative in the Galápagos, or are the two merely varieties of a single species? For Darwin's ideas to hold true, the birds of the Galápagos had to be separate and unique species. They were different in the Galápagos because they *evolved* in the Galápagos.

Ten chapters later in *The Origin of Species*, Darwin returned to the topic of Galápagos birds and island evolution. He was taken with the observation that small oceanic islands had high proportions of endemic species compared with much larger mainland areas. Although islands have fewer total species than the mainland, they have a disproportionate number of unique species. Why? Darwin postulated that new arrivals to islands would be "eminently liable to modification" when competing with long-established island residents. Darwin's notion of competition as a driving force in evolution would turn out to be one of the sharpest points of disagreement among ornithologists studying the evolution of his mockingbirds and, later, his finches.

Darwin also noted that the amount of evolutionary divergence of land birds was markedly greater than that of seabirds. He came very close to summarizing current evolutionary thought related to the higher rates of gene flow in ocean-dwelling birds than in land birds when he wrote, "Thus in the Galápagos Islands nearly every land-bird, but only two out of the eleven marine birds, are peculiar; and it is obvious that marine birds could arrive at these islands more easily than land-birds." The frequent flyers carried all the genetic baggage of their relatives in distant mainland or island populations, and this constant influx of genetic material, or gene flow, is now known to impede the process of evolution. Isolated species evolve faster than well-traveled species, and Darwin saw this pattern in 1859.

By the time *The Origin of Species* appeared in print on November 24, 1859, Darwin had resigned himself to his inability to change the minds of what he called "experienced naturalists" concerning the mutability of species, especially when those naturalists held long-standing views contrary to his own. For those unconvinced Victorian naturalists, often of an older generation, species were as immutable as the continents were fixed in place. Darwin knew the doubters would quickly fall back on expressions such as the "plan of creation" and "unity of design" and that they would elevate unexplained and marginal difficulties in his novel mechanism of natural selection to support their predictable rejection of his theory. For Darwin, species were mobile in time and space, just as he was mobile on his salutary circuits around the Sand Walk.

Darwin's hope for vindication of his radical new idea rested with what he called "young and rising naturalists," the kind who had open minds about the transmutation of species and who could impartially look at both sides of the evolution argument before reaching conclusions. The eight young naturalists from the California Academy of Sciences of the 1905–1906 collecting expedition were just what Darwin had in mind. They were not going on a collecting expedition to just any group of volcanic islands. They were headed for classic ground: the Galápagos Islands. The Academy collectors played a pivotal role in cementing the name of Darwin to the Galápagos Islands, and the specimens they collected would be the intellectual glue. The islands would become Darwin's islands, just as biological evolution by natural selection would become Darwin's great intellectual innovation.

NOTES

Page 23: *Day passed delightfully*: Darwin (1959, 20).
Page 25: *Namesake birds*: Lowe (1936); Steinheimer (2004).
Page 25: *First-class biological importance*: Lowe (1936, 310).
Page 25: *Ever Since Darwin*: Gould (1981, 28).

Page 25: *Naturalist aboard the* Beagle: A contrary interpretation of Darwin's role on HMS *Beagle* holds that he really was the naturalist. See van Wyhe (2013).

Page 26: *Darwin was chosen by FitzRoy:* Pringle Stokes shot himself in the head on August 1, 1828, and died on August 12, 1828. FitzRoy's experiment with the Fuegians went well, for a while. Although one of the four died, the remaining three proved a great success and were soon put on display in London as a kind of living anthropology diorama. It was as though the pages of *National Geographic* magazine came alive in Victorian London.

Darwin played an important functional role on board the *Beagle*. FitzRoy was also aware that suicide ran in his family. His uncle, Lord Castlereagh, had committed suicide by slitting his throat in 1822. Knowing what happened to Pringle Stokes as a result of isolation from the other men and the pressure of command on board the *Beagle*, combined with his family history, FitzRoy decided to take along a "supernumerary" passenger. This educated man of suitable class who was not in the Royal Navy would be someone FitzRoy could dine with and confide in during the long voyage. Thus, Charles Darwin became a kind of mental health committee for FitzRoy to ward off his inner demons, a plan that was only temporarily successful. FitzRoy's inner demons could be restrained but not vanquished.

Page 26: *Fuegia Basket*: Nichols (2003).

Page 27: *Gould as famous ornithologist*: Tree (1991).

Page 27: Tree (1991).

Page 29: *No one definition*: OOS, ch. 2, "Variation under Nature."

Page 31: *Gave flight to his ideas*: Durham (2012).

Page 31: *Entirely vague and arbitrary*: OOS, ch. 2, "Variation under Nature."

3

Setting Sail from San Francisco

LIKE DARWIN ON the *Beagle* voyage in the nineteenth century, at the start of the twentieth century the California Academy of Sciences needed a vessel for *its* expedition. The vessel of choice was a schooner. In comparison with square-rigged vessels, schooners have several distinct advantages for the type of coastal work and navigation between multiple islands that was required to meet the goal of the scientific exploratory interests of the Academy. A schooner's fore and aft rig made it highly maneuverable and relatively easy to sail, both with and against the wind, giving it the ability to navigate into and out of confined harbors and bays that larger square riggers cannot negotiate safely. A clear advantage of a schooner's fore and aft rig was its ease of operation, which meant the vessel could be sailed by a smaller crew than would be needed for a comparably sized square rigger. Almost all the sail handling could be done from the safety and convenience of the deck. Seldom did crewmembers have to climb aloft in difficult weather situations. The schooner *Academy* also had a massive lead centerboard that could be raised up for sailing into shallow bays, an invaluable feature for working in Galápagos waters. These were the very characteristics built into the schooner when the William E. Woodall shipyard in Baltimore constructed her as the *Earnest* for the US Coast and Geodetic Survey in 1875. The Coast Survey was the maritime equivalent of the Lewis and Clark Expedition, surveying the three coasts of the United States after its establishment by President Thomas Jefferson and Congress in 1807. During the schooner *Earnest*'s service for the Coast Survey, the crews determined longitude, took tens of thousands of depth soundings and surveying angles, constructed coastal surveying signals, and set up tidal stations. This was much the same as the work of the bark HMS *Beagle*, which was roughly the same length (ninety feet) and built for the same purpose as the schooner *Earnest*. But the *Beagle* was a square-rigged vessel, and the *Academy* was fore and aft rigged.

The *Earnest*'s crews spent twenty-three years surveying portions of both the east and west coasts of the United States during the crucial years following the Civil War when the complex interactions of politics and science were played out on the national scene. This schooner, and her civilian scientists working under the direction of US Navy officers, mirrored the long rivalry between the civilian Coast Survey and the US Naval Hydrographic Office. The *Earnest* was active during the Age of Sail and was instrumental in the growth of marine resource science at the end of the nineteenth and beginning of the twentieth century. When the Coast Survey changed its name in 1879 to the Coast and Geodetic Survey, the *Earnest* was at the peak of her hydrographic career (see figure 3.1).

While rounding the Horn in 1878 en route to San Francisco, the crew of the *Earnest* gathered equatorial sea surface temperature data suggesting a previously unreported El Niño event in the eastern Pacific Ocean, undoubtedly having biological ramifications in Ecuador's Galápagos Islands.

Following those twenty-three years of coastal surveying in Florida, Maine, Washington, and Alaska, the schooner *Earnest* was transferred in 1898 to the navy for use in Puget Sound in Washington State and in San Francisco Bay. The naval service of the *Earnest* as a coastal tender and training vessel was unremarkable, save for her going hard aground in San Francisco Bay in 1904. In his capacity as US president, Theodore Roosevelt approved the public sale of the heavily damaged vessel—which was docked for repairs at the Mare Island Naval Shipyard, in Vallejo, California—to the highest public bidder. The dimensions of the

FIGURE 3.1 **A New Schooner for Science.** The schooner *Earnest* after its construction in Baltimore for the US Coast Survey in 1875. This schooner would be purchased in 1904 by the California Academy of Sciences and renamed the *Academy* after the museum. Ultimately, Academy Bay on Indefatigable (Santa Cruz) Island was named by the crew of the 1905–1906 expedition for the schooner. The name "Academy" in Galápagos refers at times to the boat, the bay, and the museum. *NOAA Central Library, Silver Spring, Maryland.*

Academy, as the schooner was renamed in 1905 for the Galápagos expedition, were modest: overall length of eighty-nine feet, beam of twenty-three feet, draught of seven feet, three inches, and a displacement of eighty tons (figure 3.2).

Within a year of the 1904 grounding in San Francisco Bay, the California Academy of Sciences purchased the schooner for $1,000 on May 12, 1905, in a closed-bid auction, changed her name to *Academy*, and dispatched her on the seventeen-month voyage of scientific collecting and exploration to the equatorial Galápagos Islands (figures 3.2, 3.3, and 9.2).

FIGURE 3.2 **Maritime History Personified.** The eighty-nine-foot schooner *Academy* just before departure for the Galápagos Islands in June 1905, taken at the Mission Street pier in San Francisco. The original is a glass negative in the J. Porter Shaw Library at the San Francisco Maritime Museum. The *Academy* was originally built as the schooner *Earnest* in 1875 in Baltimore, in the William E. Woodall boatyard for the US Coast Survey. The schooner sank off Isle au Haute, Maine, in 1876, was repaired, and sailed around the Horn nonstop to San Francisco in 1878, passing two hundred miles west of the Galápagos Islands through one of the largest El Niño events on record. Until 1901, the *Earnest* was in service with the Coast Survey (which became the Coast and Geodetic Survey), with service in the San Juan Islands, Washington, and into Alaskan waters. While in service as a US Navy training vessel in San Francisco Bay in 1904, the schooner was tossed ashore on Yerba Buena Island, then sold to the California Academy of Sciences for use as a scientific research vessel. She spent seventeen months on the 1905–1906 Galápagos expedition of the California Academy of Sciences, and ultimately disappeared from maritime records on December 31, 1915, in Balboa, Panama, when sold after tragic events unfolded during an ill-fated gold-hunting expedition to Tierra del Fuego at the southern tip of South America. Note: On the wharf behind the aft mast of the schooner is the Harbor Emergency Hospital (partially obscured by the mast). *Glass negative at the J. Porter Shaw Library, San Francisco Maritime Museum.*

Contained within this apparent maritime tragedy after running aground, and within sight of the San Francisco waterfront, was the nucleus of a new life for Woodall's now twenty-nine-year-old *Earnest*. During this great expedition of 1905–1906 in the waters of the volcanic archipelago, her crew of "eight young men" served as sailor-scientists and functioned as species-seekers, assembling an unrivaled collection of scientific specimens.

THE GALÁPAGOS JOURNEY BEGINS

Early on the foggy morning of June 28, 1905, Leverett Mills Loomis stood on the dock at Mission Street Pier 2 in San Francisco and oversaw the final preparations for departure of the eighty-nine-foot schooner *Academy*. Behind him, and just a few blocks up Market Street from the waterfront and the Ferry Building, was the six-story museum of the California Academy of Sciences, a testament to the fifty years of financial and intellectual upheavals the Academy had survived as a struggling institution. Loomis had acquired expedition funding, purchased and outfitted a suitable schooner, and assembled a competent field party. There was nothing left for him to do but stand on the dock and wave goodbye (see figure 3.3).

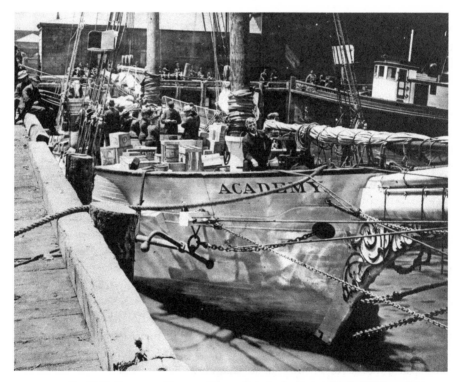

FIGURE 3.3 **Ready for Departure.** The expedition entomologist, Francis Xavier Williams, in the bow of the schooner *Academy* just before departure from San Francisco in late June 1905. Family members of the crew and dignitaries are gathered on board for the rechristening of the schooner as the *Academy*. Food and equipment are scattered on deck, to be stowed below before the vessel sets sail. *Archives, California Academy of Sciences.*

At 10:00 that June morning, the tugboat *Relief* made fast alongside the schooner, and thirty minutes later they hauled out from the Mission Street bulkhead. The tug and the schooner moved steadily across the calm, early-morning water of San Francisco Bay and headed straight through the Golden Gate and into the wide Pacific Ocean. Aboard the schooner, the group of eight sailor-scientists and three crewmembers were about to have their wooden vessel tested at sea for the first time since she had been badly damaged in a violent March 1904 storm in San Francisco Bay. The personal mettle of the men would also be put to the test.

The men of the museum expedition were poised, like Jason and the Argonauts, to set off on an epic journey. Their quest was not the mythical Golden Fleece, but the very real biological bounty of the Galápagos Islands, particularly the gigantic land tortoises. Just as Jason had assembled a group of experts and specialists to sail with him on the *Argo*, so had the museum director, L. M. Loomis, and the expedition leader, Rollo Beck, assembled biological specialists to sail on the schooner *Academy*. By the time these ocean travelers returned home, Jason to Thessaly and Beck to San Francisco, they would have many stories of adventures and misadventures to tell.

THE QUEST BEGINS

When the *Academy* headed out to sea, bound for the Galápagos Islands, the Golden Gate presented a very different sight from what we see today. In 1905, the Golden Gate was bridgeless, and another thirty years would go by before the landmark Golden Gate Bridge would be built. With its 1849 Gold Rush fame long eclipsed, the Golden Gate was merely the name given to the entrance to San Francisco Bay between the pair of rocky headlands flanked by the Civil War–era Fort Point on the San Francisco side and the open rocky coast of Marin County on the north. The last military cannons were removed from Fort Point in 1900, and the red brick fort stood then, as it does today, as a silent sentinel over the wind- and wave-swept entrance to San Francisco Bay.

The weather was slightly hazy that day in 1905 as the coastal fog burned off. A moderate westerly breeze greeted the men and the schooner after their rough passage through the "potato patch," a short stretch of choppy, swirling water that stands, like Charybdis, as an obstacle to any vessel entering or leaving San Francisco Bay. The potato patch exists in the narrow finger of ocean that lies just west of where the Golden Gate Bridge is today. The intersection and refraction of ocean waves and swells in that narrow finger of water between the dangerous, rocky headlands of San Francisco and Marin County creates a chaotic mix of steep, choppy, and unpredictable seas. Hence, the potato patch is this deceptively short stretch of water between the relative calm of San Francisco Bay and the open ocean beyond Point Bonita and Point Lobos. Many an ocean voyage to and from San Francisco has started, or ended, on a raucous maritime note because of the frequently violent wave action in the potato patch. The start of the voyage of the *Academy* was no exception.

Near the end of the potato patch and almost into the open ocean, the schooner passed the Point Bonita Lighthouse to starboard with its second-order Fresnel lens shining brightly and warning mariners as it had done since 1877. The foghorn in a small building next to the lighthouse built in 1903 was bleating out its low-frequency, fog-penetrating warning as the

Academy went past that morning. Farther out, a lightship anchored offshore marked the entrance to the Golden Gate. The schooner passed it two miles distant over the starboard beam. In the first hour of the expedition, the schooner had emerged from two dangers, the headlands and the potato patch, before she could head south on the open sea, just as the *Argo* had passed between Scylla and Charybdis before continuing its journey.

Once the *Academy* was beyond the lightship, the open ocean welcomed the men with a fabled good omen of mariners: several black-footed albatrosses flew by, although they did not pay the men any particular attention. Two hours and fifteen minutes after leaving the dock at the foot of Mission Street, the *Academy's* crew raised all of her lower sails and let go the tug. The adventure was under way.

As its sails filled with wind for the first time in more than a year, the *Academy's* white pine mast bent slightly under the strain. The Baltimore-built schooner had a new lease on life in her natural and preferred environment, the open sea.

"at first you're afraid you're going to die"

For most of the eight young men, their inexperience at sea became painfully obvious in those first few hours sailing south from San Francisco. The effect of the rolling, pitching, and yawing of the schooner in the potato patch can be summed up in one word—seasickness—and within those first few hours, it overtook most of the party. They undoubtedly learned the meaning of the old phrase "At first you're afraid you're going to die; then you're afraid you're going to live." Their initiation had begun.

Richard Henry Dana wrote about seasickness during another celebrated sea voyage in his classic *Two Years Before the Mast*. Ordered to climb the rigging to tie down some sails, Dana lamented, "I had not got my 'sea legs' on." He also was dreadfully seasick, with hardly strength enough to hold onto anything in the pitch dark. Although the crew of the schooner *Academy* did not have to climb aloft to adjust the sails (one of the advantages of a schooner rig over a square-rigged vessel), they still suffered. Dana persevered in his nighttime sail-handling duties as he held on with all his strength. "I could not have been of much service," Dana recalls with vivid accuracy, "for I remember having been sick several times before I left the topsail yard, making wild vomits into the black night, to leeward." Dana's shipmates appreciated his thoughtfulness, and aim, under such adverse conditions. No one wants to clean up after a seasick sailor.

Dana also wrote about the feeling of superiority that healthy crewmembers felt over others less fortunate, particularly passengers on board his vessel, the brig *Pilgrim*. "I will own," Dana wrote smugly, "that there was a pleasant feeling of superiority in being able to walk the deck, and eat, and go aloft, and compare one's self with two poor, miserable, pale creatures, staggering and shuffling about decks, or holding on and looking up with giddy heads, to see us climbing to the mast-heads, or sitting quietly at work on the ends of the lofty yards." Dana summarized the personal dynamics between those sadly afflicted and those more fortunate crewmembers whose equilibrium had already been reconciled with the chaotic movement of the ocean. "A well man at sea," Dana astutely observed, "has little sympathy with one who is sea-sick; he is apt to be too conscious of a comparison which seems favorable to his own manhood."

Without the modern convenience of Dramamine, Bonine, or Marazine pills to swallow or scopolamine patches to discretely paste behind their ears, several of the young men on the *Academy* quickly fell victim to the rocking of the schooner in the open ocean. Only three of the sailor-scientists, Beck, Slevin, and Gifford, continued to function while the others suffered badly. As the youngest member of the expedition, Gifford did not mince words in his private diary when he wrote, "Was not seasick." Gifford's manhood was not to be challenged. For the unfortunate others, the terrestrial equilibrium of inner ears was quickly disrupted and stomach contents were emptied on deck, or in the water if they had Dana-like accuracy. Charles Darwin suffered tremendously from seasickness on the *Beagle* voyage, writing in his diary in July 1832, "The weather has been most provoking; light variable breezes, a long swell, & I very sick & miserable.—This second attack of sea-sickness has not brought quite so much wretchedness as the former one. But yet what it wants in degree is made up by the indignation which is felt at finding all one's efforts to do anything paralysed." Loomis's hand-selected "husky young men" were off to a shaky start, barfing in the wake of Charles Darwin en route to the renowned Galápagos Islands. The eastern Pacific Ocean churned, and so did their stomachs.

One black-footed albatross continued to follow the schooner as she headed south toward Mexico. Yet a good omen can also make a good museum specimen. After ten days of following the *Academy* this particular black-footed albatross became the men's first specimen. Albatrosses are thought to embody the souls of departed sailors. Despite the bad luck, ill winds, and freezing temperatures brought on by shooting an albatross in Samuel Taylor Coleridge's 1798 poem *The Rime of the Ancient Mariner*, this bird was just the first of many albatrosses collected by the men en route to, from, and in the Galápagos. For Beck and company, the fabled mariner's companion embodied the beginning of a newer and better research collection for Dr. Loomis back in San Francisco. They were immune to the ill will of the tutelary spirits that might exist in the eastern Pacific Ocean (see figure 3.4).

The *Academy's* first stop was the Mexican town of Ensenada, just south of San Diego on the west coast of the Baja California peninsula. To reach Ensenada, the schooner first sailed due west out of the Golden Gate until she reached the vicinity of the Farallon Islands some two dozen miles offshore. Then she simply "turned left." This expedition had two broad objectives, formulated by Loomis and carried out by the young collectors: first, to collect biological specimens from a handful of islands off the Pacific coast of México, and second, after stopping at Cocos Island, to proceed to the main objective: the biological mother lode of the Galápagos Islands. Restraint in collecting was not part of the group's vocabulary or part of their plan. Their mindset is difficult to fathom today, especially for anyone with a predilection for conservation. In the absence of any other preservation option, the men on the schooner *Academy* were determined to bring the Galápagos back to San Francisco.

On their first full day out from San Francisco, a moderate westerly breeze pushed the schooner along at a comfortable five knots. At noon, the *Academy* approached Point Conception near Santa Barbara, sailing comfortably about 120 miles out to sea. During the second "dog watch," between 6:00 and 8:00 p.m., the main peak-halyard block broke, so the mainsail was lowered and made fast for the night. The schooner's aging rigging and equipment were bound to break periodically, and this was the first section of her running rigging to give way.

FIGURE 3.4 **Collecting at Sea.** Rollo Beck on an earlier expedition catching an albatross. Beck perfected the technique of rowing out a distance from the becalmed schooner and spreading chopped bait on the surface, then rowing back and shooting seabirds drawn to the bait. This way he was able to collect birds that were otherwise difficult to spot. *Archives, California Academy of Sciences.*

With increasing wind the next day, the men sailed the schooner in a maneuver called "wing and wing," holding out the foresail to starboard and the mainsail out on the port side. Wing and wing is an efficient and common way to sail rapidly downwind because it maximizes the sail area exposed to a following wind. The seasick members of the party pulled themselves out of their bunks below deck to lend a hand at the halyards. An even stronger wind after sundown made lowering the mainsail prudent. The aging schooner, now in her thirtieth year, proved to be a handful. Water leaked considerably from the main deck and made conditions rather uncomfortable below decks. Two of the men, Slevin and Nelson, the mate, spent time below covering up their stores to prevent water damage. They also judiciously plugged leaks between deck planks with empty sacks.

By the fourth day out from San Francisco, on July 1, members of the expedition felt better due to calmer seas and reduced wind, and Williams, the entomologist, although still seasick, appeared at the table for the first time. Spending time below deck was probably an unpleasant experience for Williams and his fellow sufferers. Nor was the area below deck a haven for Richard Henry Dana during his two years before the mast. He considered the intense odors lingering there to be as bad as the movement of the ship itself. "That inexpressibly sickening smell, caused by the shaking up of bilge water in the hold," Dana explained of his own stomach-wrenching experience, "made the steerage but an indifferent refuge from the cold wet decks."

If the smells below deck on the *Academy* resembled those on similar vessels, such as Dana's ship, the *Pilgrim*, then "the horrible state of things below" certainly worsened already weakened stomachs. Dana graphically described using the nauseating stench well-ing up from below deck to actually induce vomiting, an unpleasant task at best. "I remem-ber very well," Dana painfully recalled, "going to the hatchway and putting my head down, when I was oppressed by *nausea*, and always being relieved immediately. It was an effectual emetic."

By afternoon on July 1, the *Academy* was almost becalmed and the men wasted no time commencing with scientific collecting for the museum. Slevin and Beck lowered the skiff into the water and rowed out some distance to entice seabirds to approach with table scraps from the galley. Beck, the consummate bird collector with so many years of experience on similar trips with other schooners, knew this baiting or chum-ming technique would be effective. He shot about eleven petrels and saw many shear-waters and black-footed albatrosses flying past the schooner. This early success would have delighted Loomis back at his desk in the museum in San Francisco, because these types of pelagic tube-nosed seabirds, the "tubinares" or tubenoses, as they are called, were his ornithological specialty. With his extensive field collecting days behind him, this forty-seven-year-old ornithologist sent off young, able-bodied men to do the col-lecting while he stayed behind to manage the increasingly complicated scientific aspects as well as the business affairs of his growing enterprise. Through his office window at the museum, Loomis saw horse-drawn streetcars, pigeons, and crowds on Market Street. Looking through ocean-sprayed portholes, the men on the schooner saw barkentines, albatrosses, and whales.

In the 1890s, Loomis spent extended periods documenting the remarkable annual migra-tion of seabirds off Monterey, California, some 150 miles south of San Francisco. He wrote a series of five papers on these pelagic seabirds, which remain among his enduring con-tributions to ornithology. But now, the rigors of fieldwork were left to the younger men, whose combined efforts would, if everything worked as planned, enrich Loomis's growing museum.

Staying well offshore as they sailed south to Mexico, the schooner steered clear of major hazards to navigation, such as the California Channel Islands and the treacherous Cortes Bank seamount that rises from the ocean floor to within half a fathom of the surface some fifty-five miles southwest of San Clemente Island. Later in the voyage, other hazards became more difficult to avoid.

The first landfall of the expedition was made at 2:00 a.m. on July 2 when the men approached Todos Santos Bay. Standing offshore until daybreak, they shaped course for Ensenada, their first destination. Yet by 6:00 a.m. they were again almost becalmed, sur-rounded by schools of bonito or skipjack tuna and pods of whales. The calm water was good for the seasick members, but bad for their southern progress. With a freshened breeze, they made the port of Ensenada by early afternoon and dropped anchor in four fathoms off the wharf. Four days and three hours out of San Francisco, they were pleased with the first leg of what they knew would be a long trip.

The captain of the port came aboard, and they were "given practique," the clearance for a ship to proceed into port after complying with health regulations or quarantine. The com-bination of the calm water of the port and the warm weather put all the members in a good mood, even those who had been so wretchedly stricken with *mal de mer*.

Unlike Captain Ahab in Melville's *Moby-Dick*, who had to nail a gold coin to the mast of the *Pequod* in order to entice his men to hunt for the white whale, Rollo Beck needed no incentive to entice his eager young collectors. They willingly set upon their prey as soon as they got ashore.

After leaving Ensenada, as the schooner headed south for San Martin Island off the coast of Baja California, the fabled albatross became much more than a mariner's companion. Thirty miles west of Mexico's San Martin Island, black-footed albatrosses were very common. On July 10, 1905, nineteen birds were seen at once and, of those, "eleven were killed inside of five minutes." The souls of long-dead sailors were again brought aboard the schooner to join the expedition, one for every living member of the crew.

APPENDIX

BIOGRAPHICAL SKETCHES OF THE EIGHT YOUNG MEN ON THE SCHOONER *ACADEMY* DURING THE 1905–1906 GALÁPAGOS SCIENTIFIC COLLECTING EXPEDITION OF THE CALIFORNIA ACADEMY OF SCIENCES

While Darwin was in the Galápagos, he functioned in the same way as the eight young men on the 1905–1906 expedition. Darwin was twenty-six years old, about the same age as three of the men on board the schooner *Academy*: the mammalogist, J. S. Hunter, at twenty-five; the herpetologist, J. R. Slevin, at twenty-four; and the schooner's mate, F. T. Nelson, at twenty-eight. Darwin was still years away from being the great theorist we know today. He was a young man intent on collecting plants, animals, and rocks without firmly understanding the meaning of what he was collecting until much later. For each of the eight sailor-scientists, I read everything they ever wrote (including personal letters, field notes, and unpublished manuscripts), sent away for their death certificates, visited their graves, and met with or spoke on the phone with any of their living relatives I could locate.

Rollo Howard Beck: Chief of the Expedition and Master of the Vessel
1870–1950

Despite his background as a San Jose, California, farm boy with an eighth-grade education, thirty-four-year-old Rollo Beck proved to be the right man to lead the 1905–1906 Galápagos expedition of the California Academy of Sciences. Beck had been a birder and collector since boyhood, had an established professional relationship with the Academy, and, most significantly, had already participated in three collecting expeditions to the Galápagos, starting in 1897. On two of these, he had been the leader. Thus, Beck knew the collector's trade, and he knew the Galápagos and its fauna arguably better than any other American, from his boots-on-the-ground experience. He also knew how to lead a group of young men on an extended expedition. In the field, Beck was indefatigable, able to thrive in conditions most would find physically debilitating and intimidating.

After his return in 1906, Beck went on to prove himself one of the greatest collectors of his time. His intrepid wife, Ida Menzies Beck (1883–1970), traveled with him on two years-long expeditions, the Brewster–Sanford Expedition around South America (1912–1917) and the Whitney South Seas Expedition (1920–1929). Beck summarized his internal motivation as a specimen collector when he stated, "Looking back at the various expeditions which I have headed, it appears remarkably strange to me that every one of the principal directors for whom I have labored had the magnetic qualities that kept me eternally striving with utmost endeavor to get the best material at whatever locality I happened to be working" (Beck, 1936, 6). One wonders if the magnetism he perceived in others was in fact his own.

Beck's remarkably productive life of specimen collecting ended on November 22, 1950, in Merced, California. He was survived by his wife, Ida Menzies Beck; they had no children.

Sources

Anonymous, 1950. "Rollo H. Beck, Famed Naturalist, Dies Here at 80." *Merced Sun-Star*, November 24, 1950.

Beck, R. H. 1936. [Autobiographical sketch]. In R. C. Murphy, *Oceanic Birds of South America*. New York: Macmillan, 1: 3–7.

Mearns, B. and Mearns, R. 1998. *The Bird Collectors*. New York: Academic Press, 472 pp. (Rollo Beck, 337–344).

Pitelka, F. A. 1986. "Rollo Beck: Old-School Collector, Member of an Endangered Species." *American Birds* 40(3): 385–387.

Rothschild, M. 1983. *Dear Lord Rothschild: Birds, Butterflies and History*. Philadelphia: Balaban, 398 pp.

Edward Winslow Gifford: Ornithologist
1887–1959

At seventeen, Edward Gifford was the youngest member of the 1905–1906 Galápagos expedition and served as the expedition's ornithologist, although everyone collected birds when possible. Gifford was born on August 14, 1887, in Oakland, California. Without formal education beyond high school, Gifford became a full professor at the University of California at Berkeley and the director of its Museum of Anthropology at the University of California located in San Francisco (originally the Lowie Museum of Anthropology, now the Phoebe A. Hearst Museum of Anthropology on the Berkeley campus). Later in life, a lack of formal education not being an impediment, he became an archaeologist and anthropologist. In these roles Gifford made important contributions in kinship, ethnography, mythology, folklore, ethnology, material culture, and physical anthropology. He is closely associated with the discovery of Lapita pottery documenting the peopling of Pacific islands in Micronesia, Polynesia, and coastal Melanesia. He led or participated in expeditions to Mexico, Tonga, Fiji, New Caledonia, and Yap. He also extensively studied California Native American tribes, being present at the autopsy of Ishi, the last member of the Yahi tribe of the Yana people.

After a long and productive academic career, Gifford died on May 16, 1959, in Paradise, Butte County, California, beside his wife and field companion, Delila Sara (Giffen) Gifford.

Sources

Anonymous, 1959. "Edward W Gifford, Anthropologist, 71." *New York Times*, May 15, 1959.

Foster, G. M. 1960. "Edward Winslow Gifford, 1887–1959." *American Anthropologist* 62(2): 327–329.

Foster, G. M. and McCown, T. D. 1956. University of California Museum of Anthropology, Report to Chancellor Clark Kerr, for the year ending June 30, 1956. University of California, Berkeley, 15–22.

Heizer, R. F. 1959. "Edward Winslow Gifford: 1887–1959." *American Antiquity* 25(2): 257–259.

Joseph Slayton Hunter: Mammalogist
1879–1972

Joseph Slayton Hunter was born in Lincoln [?], Nebraska, on August 9, 1879. His father, Joseph Hunter, was originally from New York State and his mother, Mary Crocker, was originally from Vermont.

Beginning just five months after the return of the 1905–1906 Galápagos expedition, the aptly named Hunter spent the next forty-two years working for the California Fish and Game Commission, becoming the head of the California State Bureau of Game Conservation when it was reorganized in 1935. His proudest accomplishment in the area of conservation was his major role in the establishment of the California game refuge system in 1927. Hunter also taught ornithology at the University of Nebraska and worked as the entomologist for San Mateo County, California, handling pest control.

Joseph Hunter died at the age of ninety-two on January 20, 1972, in San Mateo.

Source

Neal, E. 1949. "Joe Hunter, State Game Chief, to Retire after 42 Years' Service." *San Jose Daily News*, August 11, 1949.

Ernest Samuel King: Herpetology Assistant
1886–1948

Ernest Samuel King was born December 16, 1886, in San Jose, California, a member of a pioneer San Jose family. His father was John S. King, for whom King Road in San Jose was named. His mother was Bessie (Dill) King, and he had a brother named Orlo King.

A simple and poignant summary of Earnest King's life appeared anonymously in the September 1948 issue of the *Academy Newsletter*, which stated, "In the spirit of adventure and with a love and experience of nature, Ernest S. King—fresh from high school graduation—shipped for the Galápagos Islands as assistant herpetologist on the Schooner *Academy*, June, 1905. During this 17-months expedition of the California Academy of Sciences, the boy King proved a capable man in the field and became the lifelong friend of his shipmates and of the Academy, even though he did not afterwards follow a career in science." King was eighteen years old when the schooner *Academy* departed from San Francisco.

King was office manager and salesman for J&H Handcraft Metals, as well as a past vice president of the Santa Clara County Humane Society. He was a Freemason and member of San Jose Lodge No. 10 of the Free and Accepted Masons. King died at his home in San Jose on July 1, 1948, at sixty-one years of age. He was survived by his wife Anna E. (Berthold) King (1887–1962) and preceded in death by his daughter, Vivian King.

Sources

Anonymous. 1948a. Death notice. *San Jose Mercury News*, July 3, 1948.

Anonymous. 1948b. "Ernest S. King." *Academy Newsletter*, no. 105, September 1948, 4.

Washington Henry Ochsner: Geologist and Conchologist
1879–1927

Washington Henry Ochsner was born in Prairie du Sac, Wisconsin, on July 4, 1882, according to his death certificate issued by the Oregon State Health Division, but his date of birth is listed as July 4, 1879, in his own hand on his application to attend Leland Stanford Junior University dated August 27, 1904, less than a year before the departure of the schooner *Academy* from San Francisco. The reason for this discrepancy is not known. Perhaps the death certificate informant, his widow, Hilda Ochsner, was not aware of his actual birth year. Ochsner was the first to die of the "eight young men" who served as sailor-scientists on the 1905–1906 expedition. Prior to attending Stanford and going on the expedition, he had attended the University of Kentucky and the University of Wisconsin.

Following completion of the 1905–1906 expedition, Ochsner re-enrolled in Stanford University and submitted a petition to receive undergraduate geology course credit for work he did on the expedition. He wrote, "Ask that I be granted ten hours university credit for field work under Geology VIII—original work in paleontology. The work consisted of collecting and exploration done on the recent scientific expedition to the Galápagos Island from June 1905–December [*sic*] 1906. My work was carried out under the direction and supervision of Dr. J. P. Smith [James Perrin Smith, 1864–1931]." Ochsner graduated in May 1909 with an AB degree in geology and mining from Stanford, in the same graduating class as the malacologist S. Stillman Berry (1887–1984). At the time of his death, he was a Freemason and member of Berkeley Lodge No. 363 of the Free and Accepted Masons.

Just a month before the expedition, on Sunday, May 28, 1905, Ochsner was described in the *San Francisco Chronicle* as "another Stanford postgraduate [he had not yet graduated], [who] will represent the sciences of geology, paleontology and conchology, especially with regard to the origin of the islands and their rock-written history, which at present is entirely unknown, particularly in the detail of the comparison of the fossil and living shells." After the expedition returned to San Francisco, Ochsner became embroiled in a multi-year intellectual battle with William Healey Dall (1845–1927) over who had the right to publish on Galápagos fossils that Ochsner had collected. This controversy did not end until both men died in the same year and were united posthumously as coauthors of several molluscan species (see Dall and Ochsner, 1928a,b). Persistent geological fieldwork by Ochsner resulted in the discovery of oil (after his death) in the Kettleman Hills in central California. Ochsner died on April 11, 1927, in Portland, Oregon, of complications from pneumonia.

Sources

Anonymous, 1909. "Stanford Seniors Bid Adieu to Alma Mater." *San Francisco Call*, May 18, 1909.

Anonymous, 1927. Death notice and obituary. *San Francisco Chronicle*, April 15, 1927.

Anonymous, 1937. "Business: Kettleman Kitty." *Time* magazine, January, 11, 1937. http://www.time.com/time/magazine/article/0,9171,757286,00.html, accessed July 18, 2011.

Dall, W. H. and Ochsner, W. H. 1928a. "Tertiary and Pleistocene Mollusca from the Galápagos Islands." *Proceedings of the California Academy of Sciences* 17: 89–140.

Dall, W. H. and Ochsner, W. H. 1928b. "Land Shells of the Galápagos Islands." *Proceedings of the California Academy of Sciences* 17: 141–185.

Joseph Richard Slevin: Assistant Herpetologist

1881–1957

Joseph Richard Slevin was born on September 13, 1881, in San Francisco, California. His father, Thomas E. Slevin, was an avid amateur ornithologist, a member of the California Academy of Sciences, and a close friend of George Davidson (1825–1911), a past president of the Academy from 1871 to 1887. Thus, Slevin had early and strong connections to the California Academy of Sciences that lasted a lifetime. Slevin attended school at St. Ignatius High School in San Francisco and at St. Mary's College in Kansas, but did not attend graduate school. Starting in 1904 in a training role at the Academy under John Van Denburgh (1872–1924), he rose to the position of curator in 1928 following Van Denburgh's death. Slevin worked at the Academy for fifty-three years and was renowned for his extraordinary character and devotion to the welfare of the Academy. Slevin served in the US Navy as a submarine officer in World War I and also made multiple voyages with the Oceanic Steamship Company. Unable to enlist in the US Navy during World War II, Slevin worked tirelessly in the Academy's instrument shop on contract work for the navy.

Slevin became an expert in the scientific and exploration history of the Galápagos Islands. Slevin's interest in herpetology took him on field expeditions to Baja California, Australia, Central America, and throughout California, and he published extensively on specimens he collected, which are held in the Academy and other museums.

Slevin died unexpectedly on February 15, 1957, at home in San Francisco. He was survived by his son, Joseph Gregory Slevin. His wife, Imelda Slevin, preceded him in death.

Sources

Anonymous, 1957. "Obituaries: Joseph R. Slevin; S.F. Curator." *San Francisco Chronicle*, February 16, 1957.

Anonymous, 1957. "Joseph R. Slevin, 1881–1957." *Academy News Letter*, no. 207, March 1957, 2–3.

Miller, R. C. 1959. "Joseph Richard Slevin, 1881–1957." *Copeia* 1: 84–85.

Slevin, J. R. 1959. *The Galápagos Islands: A History of Their Exploration*. Occasional Papers of the California Academy of Sciences, 150 pp.

Slevin, J. R. and Leviton, A. E. 1956. "Holotype Specimens of Reptiles and Amphibians in the Collection of the California Academy of Sciences." *Proceedings of the California Academy of Sciences*, ser. 4, 28: 529–560.

Alban Stewart: Botanist
1875–1940

Alban Stewart was born in Wellington, Missouri, on January 14, 1875. His father, also named Alban Stewart, was originally from Virginia, and his mother, Laura Jones, was a native of Missouri.

Alban Stewart received his BA degree from the University of Kansas in 1897 and an MA degree from the same university in 1899. Stewart worked as an assistant paleontologist at the University of Kansas in 1897–1899 and as an assistant at the US National Museum from 1900 to 1904.

After the Galápagos expedition, Stewart received his PhD from Harvard University in 1911, working under the direction of Professor B. L. Robinson. Shortly after the return of the schooner *Academy* from the Galápagos Islands, Stewart became embroiled in a bitter dispute lasting many years with the director of the Academy at the time, Leverett Mills Loomis (1857–1928), and continuing with Barton Warren Evermann (1853–1932) over whether he would be paid by the Academy to work up the botanical material collected during the expedition.

Stewart died on May 31, 1940, in Tallahassee, Florida. At the time of his death he was head of the Department of Bacteriology at the Florida State College for Women, which would become Florida State University. He was survived by his wife, Ethel C. Stewart.

Sources

Anonymous. 1938. "Stewart, Prof. Alban." In *American Men of Science*, edited by J. M. and J. Cattell, 1360. New York, Science Press.

Francis Xavier Williams: Entomologist
1882–1967

Francis Xavier Williams was born on August 6, 1882, in Martinez, California. His family moved to San Francisco in 1887, and he quickly became immersed in natural history, particularly in insects. He became acquainted with research at the California Academy of Sciences early in life through the entomologist and physician Edwin C. Van Dyke (1869–1952). Williams was an undergraduate at Stanford University during the 1905–1906 Galápagos expedition, graduating in 1908. He worked as a plant inspector at the Ferry Building in San Francisco from 1908 to 1910, then went to the University of Kansas as an assistant curator of entomology, obtaining an MS degree there in 1913. He then moved to Harvard University to accept an assistantship under Professor William Morton Wheeler (1865–1937) and received a DSc degree under Wheeler in 1915.

In May 1916, Williams moved to Honolulu, Hawaii, as assistant entomologist at the Hawaiian Sugar Planters' Experiment Station. He traveled widely in the United States, conducting research in entomology, and in Guatemala, Australia, Philippines, Ecuador, Trinidad, British Guiana, Barbados, Brazil, New Caledonia, and East Africa.

Williams is the only member of the 1905–1906 expedition for whom a tape recording of his voice exists in the Archives of the California Academy of Sciences. Toward the end of his life, he was closely associated with the San Diego Museum of Natural History. Williams died, following a stroke, on December 16, 1967, in Chula Vista, California. He was preceded in death by his wife, Louisa Lewis (Clark) Williams, who died in 1965.

Sources

Arnaud, P. H., Jr. 1970. *Lists of the Scientific Publications and Insect Taxa Described by Francis Xavier Williams (1882–1967)*. Occasional Papers of the California Academy of Sciences, no. 80, 33 pp.

Grammer, A. R. 1947. "A History of the Experiment Station of Hawaiian Sugar Planters' Association." *Hawaiian Planters' Record* 60(2–3): 177–228.

Zimmerman, E. C. 1969. "Francis Xavier Williams, 1882–1967." *Pan-Pacific Entomologist* 45(2): 135–146.

NOTES

Page 34: *Schooner "Earnest" constructed for US Coast Survey*: The Woodall shipyard was located on the Baltimore waterfront at Tide Point, adjacent to the Domino Sugar factory, between Key Highway East and the water of the Patapsco River, southeast of the Inner Harbor.

Page 34: *Maritime equivalent*: Manning (1988).

Page 35: *Biological ramifications*: James (2003).

Page 38: *The wide Pacific Ocean*: "In the Council, thanks were tendered to John D. Spreckels for the use of a steamer in towing the vessel of the Academy's scientific expedition to the Galapagos Islands out to sea" (Hittell 1997, 448, July 17, 1905) The sugar magnate John D. Spreckels had donated the services of his powerful tugboat—one last bit of corporate support for this impressive scientific venture.

Page 39: Dana (1840, 8, 72).

Page 40: *Schooner in the open ocean*: The movement of a vessel on the water can be described as occurring in three directions, although these movements can, by virtue of the interplay between vessel and swells, occur simultaneously: (1) yaw, which is rocking in the bow-to-stern direction; (2) roll, which is side-to-side movement; and (3) pitch, which is overall up-and-down movement. When yawing, rolling, and pitching occur simultaneously, seasickness frequently ensues.

Page 40: *Was not seasick*: Gifford diary, June 29, 1905. The propensities of Parker, the navigator, Nelson, the mate, and White, the cook, to seasickness is unclear. They were the professional old salts on this trip, and presumably their sea experience made them immune.

Page 40: *Darwin suffered tremendously*: http://beagleproject.wordpress.com/2012/07/09/darwins-raisins/, accessed August 15, 2012.

Page 40: *The main peak-halyard block*: This is a pulley (the block) with a line running through it (the halyard) used by crewmembers to raise the mainsail on a schooner.

Page 41: Dana (1840, 8).

Page 42: *Dana graphically described*: Dana (1840, 8); emphasis in the original.

Page 42: *Almost becalmed*: Rollo Beck originated and perfected the field collecting technique of shooting seabirds from a small boat that was rowed out a distance from a larger ship. The technique was adopted by Robert Cushman Murphy, for whom Beck later collected off the coast of South America. Murphy described the grisly business of shooting large numbers of birds in another part of the world, 36° south latitude, in the open ocean roughly east of Buenos Aires. On an early November day in 1912, Murphy and the cooper Jose Correia (who would in the 1920s work as Rollo Beck's field assistant in the South Pacific) went out in a dory from the whaling vessel *Daisy* with their shotguns. "I began the slaughter when the first petrel flew within range," recounted Murphy, "and soon Correia and I were banging away together, endeavoring to collect the greatest variety of specimens from among the bewildering thousands of waterfowl" (Murphy 1947, 129).

Page 42: Tubinares have been renamed Procellariiformes; they include the albatrosses, petrels and shearwaters, storm petrels, and diving petrels.

Page 43: *Eleven were killed*: Gifford bird notes.

4

Before It Is Too Late

THE NOTION THAT time was running out clearly motivated Loomis and the men on the Academy's Galápagos expedition. From their perspective, either they would collect as many of the encountered specimens as possible, or it would be a "damnable shame" that species went extinct before scientific documentation could occur. In the conservation logic of their time, animals were better off dead and preserved in a museum than left to the whims of reckless humans and introduced animals. The 1905–1906 expedition would combine museum preservation in the face of imminent extinction with detailed anatomical study for the sake of pure science.

Both of the administrative brains behind the expedition—the ornithologist Loomis and the ever-practical, boots-on-the-ground expedition leader, Beck—were committed to doing anything they could to document the fauna, even before biodiversity conservation became trendy, popular, and de rigeur in wild places around the world. While certainly not recognizable as "conservationists" in the current sense of the term, they faced some of the same problems as contemporary conservationists. As pragmatic realists, both Loomis and Beck faced difficult decisions familiar in today's biodiversity conservation work. Which species, out of many candidates, should be preserved? How should limited financial resources be allocated to save species? What techniques were appropriate to save species from extinction? Should the focus be on the species or their habitats?

Loomis and Beck answered these questions differently than we would today. Instead of protecting species for a survival that they could not guarantee, they collected specimens on behalf of science and for the future. They documented biodiversity by creating a library of specimens, an archive of specimens. Their approach, which was common to zoological and botanical expeditions conducted by museums in the late nineteenth and early twentieth centuries, has been vindicated many times over in the century since the *Academy's* triumphant return to San Francisco. Loomis and Beck showed that salvage collecting was the best scientific prospect for an otherwise pessimistic situation.

Contemporary newspaper stories giving some context for the 1905–1906 expedition identified two agents of extinction as feral dogs and cats that had been carried to the remote Galápagos on ships. Other alien mammals created their own components of environmental chaos: rats, pigs, cows, goats, and donkeys made an ass of the delicate ecological balance that evolved in isolation. Once the Academy men were in the islands and finding few of the expected marine iguanas at Finger Point, known as Cerro Brujo today, on Chatham Island, where Darwin went ashore in September 1835, Joseph Slevin wrote, "It is probably the dogs have them well cleaned out." Direct human activities, such as mass killing of giant tortoises for oil or food, were also implicated. The prediction was that when the numbers of feral animals dramatically increased, they would soon turn to eating the native plants and animals. "The result has been," the *San Francisco Chronicle* stated in August 1903, "that priceless specimens of birds have been destroyed wholesale. The world of science has awakened to the state of affairs, and strenuous efforts are being made in various parts of the world to secure specimens *before it is too late.*" Loomis and his field collectors were on a scientific mission inspired by a desire to short-circuit oblivion (figure 4.1).

The sense of urgency associated with the imminent extinction of Galápagos organisms can be traced back at least to Georg Baur's (1859–1898) writings in 1891. In that year Baur, a paleontologist at Yale, traveled to the Galápagos as part of an expedition that he sponsored. Unlike Beck and his crewmates, Baur was more interested in the islands' geography and the implications of geology for understanding Darwinian evolution than he was in collecting per se.

Baur's observations reflect the fevered pitch at which he lived his life and which likely brought on his premature death at age thirty-nine. Even before his departure for the islands, he argued for more intensive collecting efforts on the Galápagos, writing, "Such work ought to be done *before it is too late.* I repeat, before it is too late!" The consequences of ignoring his plea would be that "the natural history of the Galapagos will be lost, as it has unfortunately been lost in so many islands." Wherever this biological damage was done, it was done "irreparably!" The species would be "lost forever." In effect, Baur was stating the guiding philosophy of salvage collecting.

Rollo Beck, too, understood the dire situation of Galápagos wildlife, particularly that of the tortoises, firsthand. By the time that Loomis hired him to lead the Academy's expedition in 1905, he had already been to the islands three times, funded by the private industrialist Walter Rothschild. Beck's devastating photographs of tortoise killing fields, taken in 1897 on his first journey, inspired subsequent journeys. Each time, Beck saw evidence of environmental destruction all around him.

Walter Rothschild (1868–1937) played an important role in Galápagos history, not by visiting the islands, but by funding collecting expeditions. He readily took to natural history as a child, and that passion eventually led to the establishment of a substantial private natural history museum at his home in Tring, England. He was a scion of the Rothschild banking family and, after the death of his father, became the second Lord Rothschild, in 1915. Rothschild had an odd penchant for giant tortoises. Largely through Beck's field efforts, he amassed a sizable collection of living and preserved giant tortoise specimens at his museum and home. In world history, what became known as the Balfour Declaration of 1917 expressing support for a Jewish homeland was a letter to Rothschild from the United Kingdom's foreign secretary, Arthur James Balfour (1848–1930). Rothschild's museum, some forty

FIGURE 4.1 **A Damnable Shame.** Two photos, both likely taken by Rollo Beck, emphasizing the human depredation on Galápagos tortoises. The call to action was raised by Georg Baur and acted upon by the California Academy of Sciences. (a) Tortoises slaughtered for their oil. This photo from the 1897–1898 Webster–Harris expedition was the basis for Lord Rothschild's sad prophecy of "a damnable shame" if species were "fast disappearing," and thus this photo and information about the wanton destruction of Galápagos organisms served as the main motivation for the 1905–1906 expedition from San Francisco. (b) Man with a machete with tortoises in a lava rock enclosure. *Archives, California Academy of Sciences.*

miles northwest of London, started as the Walter Rothschild Zoological Museum and was renamed in 2007 as the Natural History Museum in Tring, part of London's Natural History Museum. It was the largest zoological collection ever amassed by a private individual, with impressive numbers of bird skins, birds' eggs, butterflies, beetles, mammals, reptiles, and fishes. He needed collectors like Rollo Beck to populate his museum, opened in 1892, a year after the California Academy of Sciences opened in 1891. Beck played the same role early in his field collecting career for Rothschild as he did for Loomis in 1903 and 1905–1906, and later for other museums such as the American Museum of Natural History in New York. In a shifting of zoological fortunes, Rothschild was forced to sell the majority of his bird collection in 1932, some 225,000 specimens, when blackmailed by an aristocratic former mistress.

Beck noted the effects of human visitation throughout the islands. About Albemarle (Isabela) Island, Beck wrote, "It is only within the last two years that the home of these very large tortoises has been invaded by man, but the rapidity with which they are being killed, and the reason for their destruction, leaves us little hope that they will survive any longer than did the American bison after the hide hunters began their work of extermination." Beck documented, firsthand, the effect of wild dogs on tortoises: "We saw a number of skeletons of tortoises that the dogs had killed, and noticed that, as a rule, the females being the smallest were the ones to suffer. However, we saw a couple of males over three feet long, showing that when hunger is keen enough even the large ones are killed." Humans and the predatory alien species they introduced to the islands were clearly responsible for the ongoing decimation of tortoise populations.

Beck continued his observations on wild dogs, writing, "After seeing on this mountain [Sierra Negra on Albemarle Island] dozens of tortoises of good size, one wonders where all the small ones are; but after spending a few days a-foot and seeing the many wild dogs in that region—descendants of those left years ago by sailing vessels—we can only wonder that so many of the large ones remain. From the time that the egg is laid until the tortoise is a foot long, the wild dogs are a constant menace, and it is doubtful if more than one out of 10,000 escapes. We certainly saw none, and the natives told us that the dogs ate them as fast as they were hatched."

Beck witnessed human depredation as well and how that made the wild dog problem worse. "While at the ranch [above the town of Vilamil on Albemarle] where nearly 50 men were at work, we were amazed at the reckless and heartless manner in which some of the natives destroyed the tortoises. The proprietor [of the ranch] informed us that only the males were killed, but we noticed that the working people made little distinction in the sexes when killing for food. Some evenings, two or three men coming in from different directions would each carry in his hand a small piece of tortoise meat, and a pound or so of fat with which to cook it. Of each tortoise killed not over five pounds of meat would be taken, the remainder being left for the wild dogs that swarmed about." Beck further indicts the local Ecuadorians for waste and destruction, observing, "One Saturday evening I had occasion to go down the trail a mile or so, after some of the natives had departed for the shore settlement [Puerto Vilamil], where all the women and children lived. I found a large tortoise, three feet six inches long and hundreds of years old, which had been cut open with a machete, but apparently not more than three pounds of meat had been taken from it. A little farther on lay a dead female, from which nothing had been taken save a string of eggs and very little meat."

Beck documented, in words and photographs, the damage done by oil hunters—not speculators in petroleum, but local Ecuadorians who hunted the tortoises for their oil. Like whale oil, tortoise oil was a valuable commodity in the maritime economy. "I took two photographs at the water-hole, where lay the largest number of tortoise skeletons. There were about 150 skeletons at this pool, and a half-mile away, in another depression, were about 100 more. While there were more skeletons at these two places than we saw elsewhere, frequently 10 or 15 were observed in other basins where the tortoises had gone for water." He explained how the oil collectors worked. "After making camp near a water-hole, and killing the tortoises there, the collector brings up a burro, throws a couple of sacks over the pack-saddle, and starts out to look for more tortoises, killing them wherever found. A few strokes of the machete separate the plastron [the interlocking bones on the belly side] from the body, and 10 minutes' work will clear the fat from the sides. The fat is then thrown into the sack, and the outfit moves on." Beck's concern and disdain are palpable in these descriptions.

Why were the Ecuadorians killing tortoises at such a furious pace? Money. Beck noted, "When the burro is well laden, man and beast travel back to camp, where the oil is tried out. The small ones are seldom killed, because they have but little fat. By daily visits to the few water-holes during the driest season, in the course of a month the hunters get practically all the tortoises that live in the upper part of the mountain" (figure 4.2). Beck was already calculating in his head the countdown to extinction. He was frustrated and wanted to be the scientific savior of the tortoises, not their destroyer.

Taking his observations of tortoise slaughter to the obvious conclusion of the tragic trajectory he witnessed all around, Beck further observed and predicted, "When we first stepped

FIGURE 4.2 **Tortoise Collecting.** Instead of carrying the heavy tortoises to the coast for transport out to the waiting schooner, the men sometimes used local horses, donkeys, or mules as beasts of burden. *Archives, California Academy of Sciences.*

ashore at the settlement [Vilamil], we saw a number of casks lying on the beach, and learned on inquiry that they contained 800 gallons of tortoise oil. In a large boat, under a nearby shed, were 400 gallons more. While we were there the boat sailing between the island and Guayaquil left for that port with those casks and a cargo of hides. The value of the oil in Guayaquil was about $9.00 (American) per 100 pounds. While the tortoises are as plentiful as we saw them, this price yields a fair profit to the hunters, but two more raids such as that shown in the photograph will clear that mountain of all the fair-sized tortoises upon it, and then the oil business is ended." Then the tortoises would be ended as well. Beck knew tortoise extinction was coming if the combined killing by oil hunters, killing by humans for meat, and killing by wild dogs was not ended soon.

Beck also witnessed tortoise destruction farther north on Albemarle Island at Tagus Cove, a common visitor destination today. Shifting his observations to the northern portion of Albemarle, Beck noted that the land iguanas had been exterminated, although they were faster and better able to escape than were the tortoises. Near Tagus Cove there were many tortoise trails, indicating to Beck that hundreds of tortoises used to live there. But on his visit in 1897, he could barely find one, demonstrating that the species on that part of the island had been extensively used as food for whalers. So here was another serious source of depredation, in this case from occasional visitors to the islands seeking whales to harpoon.

On Isla Pinzón (Duncan Island in Beck's narrative) he noted a further threat to tortoises. After locating thirty live tortoises that would be sent back to Europe for Rothschild's collection, Beck hit on a conundrum. He noted his collecting party had located no very small specimens. This time, the culprits were large rats, "of recent introduction, and now common everywhere on the island," which "eat the young as soon as they are hatched." Humans, dogs, rats—all posed serious threats to tortoise populations in the late nineteenth century, and all were documented by Rollo Beck.

Beck's devastating photographs of the killing fields of tortoise carcasses so disturbed Walter Rothschild back in England that he was compelled to act. As only an occasional visitor to the islands, Beck was helpless to end the senseless killings that threatened the Galápagos tortoises. All he could do was collect. For Beck, scientific collecting was a last-ditch effort amid a mass-death scene. Scientific collecting was a form of scientific triage. Although Beck and his men could not preserve the species in the wild, they would preserve museum specimens for posterity. He knew this *before* the 1905–1906 expedition, and his observations and photographs were a primary motivating factor for the Academy museum director, Loomis, to undertake the expedition in the schooner *Academy*.

In the world of late-nineteenth-century natural historians and museum curators, collecting was the way they conducted their science. Consider Elliott Coues (1842–1899) (last name pronounced "cows"), the most prominent and respected ornithologist at the end of the nineteenth century. After Coues died, an obituary described his 1872 book, *Key to North American Birds*, as having "probably done as much to advance the interests of ornithology in this country as any other work." Because Coues's book was the standard guidebook and reference work on ornithology of the day, Rollo Beck, Joseph Hunter, and Edward Gifford, the field collectors whose primary emphasis was ornithology, learned bird-collecting techniques from it. For bird men, it was their bible. Coues was a founder of the American Ornithologists Union in 1883 and an editor of its journal, the *Auk*.

In *Key to North American Birds*, Coues gave basic directions to amateur bird collectors and professional ornithologists about how to conduct fieldwork, including how to build a bird skin collection for either personal or professional study. Coues opened the first section of the book, on collecting and preserving birds, by announcing, "The Double-barreled Shot Gun is your main reliance." Few birders today would even know how to use a shotgun, let alone want to use one. Times and standards change. Even a well-known portrait of the namesake of the National Audubon Society, John James Audubon (1785–1851), shows him with his shotgun, standard equipment for birders of his day.

Today, both amateur birders and professional ornithologists might consider a good pair of binoculars or a spotting scope their "main reliance." Contemporary birders frequently compare notes on these essential tools when they are out in the field with like-minded colleagues; they read reviews and compare binocular and spotting scope models in magazines and online. As with shotguns, considerable money can be involved. Birders discuss the relative merits of the best magnification, lens coating, objective lens diameter, exit pupil, field of view, weight, and water resistance. Coues's recommendation on how much to spend on a shotgun echoes what birders today consider when buying binoculars or a spotting scope: "Get the best one you can afford to buy; go the full length of your purse in the matters of material and workmanship." Anyone who has purchased cheap binos and then been disappointed might find herself agreeing with Coues, at least on the value of high-quality equipment. Coues was not an anticonservationist; he was simply a product of his time.

Coues advised birders to collect "all you can get," for example, "fifty or one hundred of any but the most abundant and widely diffused species." Since widely diffused or common species could be collected at a later date, collectors should focus on relatively uncommon or even rare birds. Coues explained, "With a few possible exceptions . . . enough birds of all kinds exist to overstock every public and private collection in the world, without sensible diminution of their numbers." This set the stage for field collectors like Beck and for museums like the California Academy of Sciences. Both Rothschild (who never set foot in the Galápagos) and Baur (who spent two and a half months in the islands in 1891) implored museums to collect as many Galápagos tortoises as possible before they became extinct. Coues's approach to bird collecting gave the green light to this kind of approach.

Viewing nature as an inexhaustible source of specimens eliminates restraint from the collector's mindset. For Coues and collectors of his day, collecting did not—could not—permanently damage a species or the environment. We now know, of course, that this assumption has not always been correct, whether with birds or any other kinds of threatened species.

Coues proclaimed that his method of collecting birds improved the mental and physical health of the bird collector. He explained, "It is Unnecessary to speak of the Healthfulness of a pursuit that, like the collector's occupation, demands regular bodily exercise, and at the same time stimulates the mind by supplying an object, thus calling the whole system into exhilarating action." A hard day of shooting specimens would rejuvenate the bird collector's mind, body, and spirit. Coues depicted the practice of field ornithology as tantamount to an overall health regimen. Perhaps a contemporary researcher afflicted with malaise might consume a bowl of the latest high-fiber cereal from Mr. Kellogg in Battle Creek, then spend the day field collecting in high spirits. Ornithology was good for what ailed you, like taking the waters at a health spa.

Loomis was well aware that scientific collecting in the Galápagos Islands was crucial before it was too late. Reports were arriving, partly through Rollo Beck himself, that populations of Galápagos tortoise had suffered tremendous depredations from historical and ongoing wholesale slaughter. Culprits included the likes of pirates, buccaneers, privateers, whalers, and other mariners, as well as local Ecuadorian residents of the islands, wild dogs, and introduced rats. Loomis applied the same rational to the 1903 expedition the Academy had sent to the offshore Mexican islands. That expedition sought "to secure for science some of the valuable and fast-disappearing specimen[s] of animal and plant life from the practically unexplored islands in the Pacific." A motivation for this, the Academy's first international schooner-based expedition, was that "island life all over the world is fast disappearing."

Following the collecting trip to the Mexican islands, the Academy originally planned to send additional collecting expeditions to Tiburon Island in the Gulf of California and then to a series of islands off the coast of Chile. These two expeditions never materialized. Instead, the Academy mounted its expedition to the Galápagos. Loomis instructed his crew to collect as many specimens as possible, and collect they did. To contemporary eyes, the sheer number of specimens collected is by far the most striking thing about all of these expeditions, but especially the 1905–1906 journey to the Galápagos. All told, Beck and his crew returned to San Francisco with at least 78,000 specimens, including 266 giant tortoises. It comes as no surprise that this has elicited some criticism.

In mid-2014, a controversy broke out over the significance and collecting of museum specimens. An editorial letter in the journal *Science* advocated that field scientists "revisit and reconsider field collecting practices and policies" due to the danger of extinction of certain species with small and often isolated populations. Although the letter did not specifically mention the Academy's specimens, its collections epitomized the kinds of practices the letter writers had in mind. Curators at the California Academy of Sciences and more than sixty other international research institutions spanning six continents vehemently disagreed with this characterization of their work. The original letter argued unconvincingly that the collection of scientific specimens played a significant role in species extinctions, pointing to examples of now-extinct birds, frogs, and plants. In response, the curators called attention to the many uses of museum specimens, not the least of which documented the arrival of diseases (historical specimens of Galápagos mockingbirds aided this), the thinning of bird eggs by DDT, and population changes due to climate change.

Other critics have more specifically targeted the 1905–1906 expedition, claiming that scientific collecting had effects on Galápagos wildlife that were equal to or worse than human depredations. John Hickman pursued this faulty and biased logic in his 1985 book, *The Enchanted Islands: The Galapagos Discovered*. Hickman contrasted the behavior of the local inhabitants, who depended on the tortoise slaughter for "fresh meat" and a steady income, with the scientists who added "to the general depredation." He wrote, "Many [local inhabitants] had good reason for taking advantage of an obvious source of food in a hard world where even bare necessities were scarce. There is less excuse for the scientists who, in ignorance or stupidity, took many hundreds of tortoises for their collections, and continued doing so until as recently as the nineteen-thirties."

Hickman's critique echoed one made by Ian Thornton in his 1971 book, *Darwin's Islands: A Natural History of the Galapagos*. "Time and time again," he wrote, "scientists, after collecting on an island, declared the tortoises of that island to be extinct, only for some later expedition to discover survivors, which were promptly skinned and carried away as precious specimens of a 'dying' race. These 'last survivors' were collected from Duncan [Pinzón] by four different expeditions in 1897, 1898, 1900, and 1901, and yet the Academy Expedition of 1905–06 discovered eighty-six tortoises on that island, which they killed and removed for study; over sixty of these were females." He continued, "In comparison with the thousands removed as provisions in the previous two or three centuries, these numbers admittedly are small; but it must be remembered that the populations from which they were taken were already drastically reduced and struggling for their very survival. Indeed, the paradox is that this was the very reason given for their removal!" This was, of course, exactly the point.

What these critiques fail to recognize is that, in the absence of governmental and NGO infrastructure, conservation, preservation, and restoration are simply not possible. This is especially true in the remote, distended Galápagos archipelago. Today, through the gallant efforts of the Galápagos National Park Service and the Charles Darwin Foundation, the islands serve as a global model of biological and ecological conservation. In the context of 1905, however, the Academy's field collectors were neither thoughtless nor overenthusiastic. They were actively engaged in preventing the "damnable shame" that seemed inevitable based on the known trajectory of tortoise exploitation in the archipelago.

The entomologist Howard Evans has attempted to put the expedition's emphasis on collecting in a more positive light, albeit with the benefit of hindsight: "This seems a considerable slaughter, but it should be remembered that at this time tortoises and other native animals were being devastated by visiting whalers and by the settlers, and goats and other introduced animals were destroying habitats rapidly. The work of the expedition provided basic knowledge on the fauna that ultimately led to successful conservation efforts." It would take another generation and a new perspective on conservation before concrete efforts would be made to preserve the biota of the Galápagos.

Darwin himself noted the devastation of the tortoise populations, and his writing also inspired Loomis to conceive of the expedition. Darwin summarized the cynical view of islanders and sailing vessels: "The inhabitants, although complaining of poverty, obtain without much trouble, the means of subsistence. In the woods there are many wild pigs and goats; but the staple article of animal food is supplied by the tortoises. Their numbers have of course been greatly reduced in this island [Charles], but the people yet count on two days' hunting giving them food for the rest of the week. It is said that formerly single vessels have taken away as many as seven hundred, and that the ship's company of a frigate some years since brought down in one day two hundred tortoises to the beach."

And yet Darwin himself had also contributed to the tortoises' demise. Eating and enjoying tortoises was an experience that the crews of the *Academy* and the *Beagle* shared. While staying in the upper region of James (Santiago) Island with Sims Covington, Darwin wrote, "We lived entirely upon tortoise-meat." No complaints were lodged about the food, and Darwin fell back on his recent South American experience when he wrote, "The breast-plate roasted (as the Gauchos do *carne con cuero*) with the flesh on it is very good." Darwin admitted to further impact on the tortoise population when he described how "the young tortoises make excellent soup; but otherwise the meat to my taste is indifferent."

Biological conservation, as we know it today, was not yet a viable option for the collectors from San Francisco, either conceptually or institutionally. The giant tortoises of Narborough (Fernandina), and of all the islands in the archipelago, made the islands fertile ground for studying and collecting evolution. The sizes and shapes of their shells were the very first hint Darwin had been given in 1835 about subtle differences in organisms between islands. The tortoises had spilled their seed and their blood on this "little world within itself," this "satellite attached to America," as Darwin himself called the islands. For now, and until conservation infrastructure was put in place, "preserving" tortoises meant pickling them in an arsenic-laced embalming fluid, not preserving them alive in the wild for future generations.

Loomis and Beck interpreted the situation as a choice between two deaths: a nameless, faceless death relegating each individual to oblivion by thoughtless island residents or wild beasts, or a documented, scientifically relevant death that would benefit the corpus of scientific knowledge for all time. For Loomis and Beck, this was an easy choice, historically necessitated by the lack of any protective measures for the organisms.

Beck described a possible route to saving the Galápagos tortoises well before the terms "conservation," "preservation," and "restoration" became the lingua franca of the islands. Based on his experiences in the late nineteenth century, he stated, "A very few years will probably see the extinction of two or three of the present living species, and while a few specimens of the others may linger for a much longer time, they, too, are bound to disappear under the attacks of their enemies." Similarly, he advocated the relocation of tortoises to zoos, where the public might watch "the ponderous movements of a 500-pound tortoise, hundreds of years old." In each case, the emphasis was on obtaining specimens before it was altogether too late.

Today, as it has for more than forty years, the "Fausto Llerena" Giant Tortoise Breeding and Rearing Center located at the Charles Darwin Research Station and the Galápagos National Park Service successfully accomplishes this much-needed work. Dozens of tortoise eggs are incubated and hatched annually; then the tortoises are allowed to grow big enough in captivity to resist predators. Entire islands have been cleared of feral predators and competitors, opening the path for the successful reintroduction of the archipelago's namesake. Rollo Beck would approve.

But in 1905, scientific collecting *was* the equivalent action of today's conservation and preservation. Rollo Beck documented the problems and challenges facing wildlife in the Galápagos with his writings and photographs. Rollo Beck, Walter Rothschild, and Georg Baur issued clarion calls to save Galápagos tortoises from inevitable destruction by publishing and publicizing the facts of destruction. To preserve the Galápagos tortoises, they preserved the Galápagos tortoises.

In an archipelago famed for its giant tortoises, Lonesome George is probably the most famous tortoise in the world. People travel from all over the world to see the gigantic Galápagos tortoises. But the tortoise whose story is the most compelling is Lonesome George, who was nearly shot by conservation workers when he was discovered living on Abingdon (Pinta) Island, not quite half a degree north of the equator in the Galápagos. His discoverers, Manuel Cruz of the Darwin Station and Francisco Castañada, a National Park warden, at first mistakenly took him for a goat moving about under a palo santo tree. During those brief moments when Cruz and Castañada trained their hunting rifles squarely on Lonesome George, history nearly repeated the events of April 5, 1906, when Rollo Beck killed the last tortoise on Narborough Island. By calmly holding their fire, Cruz and Castañada spared the life of a solitary individual and avoided the oblivion of an entire species. A sigh of relief must have followed their realization that the mistaken object of eradication was actually a significant scientific discovery.

But an escape from the rifle would not be enough to save the tortoise population on Abingdon Island. For the rest of his life, park wardens searched for a suitable mate for Lonesome George. The matchmaking effort was intended to ensure that his genetic material would be passed to future generations, but a mate could not be found. When Lonesome George died in 2012, he died single. Keeping the tortoise alive was easier than finding him a mate.

Shortly after his discovery in 1972, the Ecuadorian government posted a reward of $10,000 for anyone who could find a female Pinta tortoise, either on Abingdon Island itself or in a zoo somewhere in the world. The reward was never collected.

In an ironic twist, the race of tortoise most similar in shell shape to Lonesome George is named for Rollo Beck; today it lives on Wolf Volcano on Albemarle Island. Rothschild named it on the basis of specimens collected by Beck in 1901. This race of tortoise bears the ponderous official scientific name *Geochelone elephantopus becki* (Rothschild, 1901). Rothschild's name and the date are in parentheses to indicate, through a sort of taxonomic bookkeeping, that this geographic race was previously placed in a different genus when it was first described in the scientific literature. For many years all the forms of Galápagos tortoises were classified in the genus *Testudo*. Eventually they were transferred from that genus to the genus *Geochelone*.

In another ironic twist, Rollo Beck and Joseph Slevin were able to find and collect only three tortoises from Abingdon Island in 1906 that were "very fat, and showed the effects of good living." Beck and his collecting party considered tortoises to be very rare on Abingdon and did not attempt to collect additional specimens. Lonesome George was quite likely alive when Beck visited, but if so, he and his team couldn't find him. Beck's group was unable to penetrate the dense and thorny coastal vegetation of Abingdon Island.

Local Ecuadorian fishermen, when working in the waters around Abingdon Island as recently as the 1950s, landed on the island and slaughtered tortoises for food. Clearly, there were a number of tortoises living there during and after the 1905–1906 expedition. Lonesome George dodged the bullet, or more likely the machete, more than once in his turbulent lifetime.

Articles appearing in the *San Francisco Chronicle* in the weeks and months before the 1905 voyage began made sure that readers understood the urgency of the situation. The headline of a well-illustrated article by Edward Berwick that appeared exactly a month before the *Academy* left San Francisco proclaimed, "Expedition Which Goes After Wonderful Specimens, Fast Becoming Extinct on Galapagos Islands, Is the Most Important Ever Sent Out From the Pacific Coast." Berwick explained this urgency and the motivation for the expedition. "Before their entire extermination is effected," the article stated about Galápagos organisms, "the Academy of Sciences desires to obtain such a series of specimens of all these doomed species as shall adequately represent them in its museum for the benefit of students of all nations, for all time to come." Part of the justification for heading south was that specimens of "effete and exterminated races[,] being practically priceless," were in danger. The gun had sounded and the race was on.

Edward Berwick's prescience in his newspaper feature reflected Loomis's grand ambition. In the time between his arrival at the Academy in 1894 and the 1905–1906 Galápagos expedition, Loomis had managed to build a thirty-thousand-specimen bird collection "recognized as the best in the United States." This recognition stemmed from both its size and the diversity of species it contained. Scientists traveled long distances to use the collections. Loomis stated, by way of the *Chronicle*, "The present expedition is expected to add to it largely." The Academy's bird collection was "unsurpassed" in terms of numbers of specimens of local land birds and waterbirds and in terms of "completeness of series, in beauty and finish of preparation and in exactness of arrangement." Loomis had plenty to crow about to the press and his professional colleagues. The Galápagos expedition would only make the collection stronger.

When the 1905–1906 expedition members returned to San Francisco, the wooden world that had served as their home for the past seventeen months contained another world— that of the Galápagos and its organisms. The expedition party did not just collect two of everything, or ten of everything, or even one hundred of everything. They collected as many of everything as they could during the available time. At first glance, this exuberant approach to collecting seems extreme or environmentally irresponsible, but that interpretation lacks an understanding of the crew's objectives and Loomis's desires. The eleven young men worked shoulder to shoulder on classic ground to fill as many of the schooner's cubits as possible with the treasures of the Galápagos.

Once back in San Francisco, the specimens entered a new world, the world of scientific investigation. Museum specimens concretize the otherwise abstract concept of species. Spread out on a table, specimens can be carefully measured, photographed, and illustrated. Researchers can sample DNA from fragments of remaining tissue adhering to the inside of a bird skin, a tortoise shell, or a pressed plant leaf. Modern laboratory researchers have amplified the field efforts of the 1905–1906 expedition with cutting-edge DNA techniques, which are constantly changing. From an analysis perspective, museum specimens yield more, and vitally different, information than living specimens observed from a distance, or only momentarily when captured for a blood sample before being released. Preserved museum specimens yield their secrets for many years, to many generations of researchers, on behalf of their living descendants. And while the crewmembers set out to collect rather than to investigate Darwinian theory, their work would play a central role in increasing our understanding of evolution in the years to come.

NOTES

Page 52: RWE, 47.

Page 52: *The result has been*: San Francisco Chronicle, Friday, August 14, 1903; emphasis added.

Page 52: *Sense of urgency*: Baur was then at Clark University in Worcester, Massachusetts, but later at the University of Chicago until his death.

Page 52: *Fevered pitch*: Baur (1891); emphasis in original.

Page 54: *Beck noted the effects*: Beck (1902).

Page 54: *Beck continued his observations*: Beck (1902).

Page 54: *Beck witnessed*: Beck (1902).

Page 55: *Beck noted*: Beck (1902).

Page 55: *Where the oil is tried out*: This refers to flesh being boiled out in cast-iron try-pots in a tryworks, as on a whaling ship.

Page 56: *Tortoise slaughter*: Beck (1902).

Page 56: *Humans, dogs, rats*: Beck (1902).

Page 56: *Mindset of field collectors*: For information about Elliott Coues see http://en.wikipedia .org/wiki/Elliott_Coues, accessed June 30, 2010.

Page 56: *After Coues died*: For obituaries of Elliot Coues, see Smith (1900), Victor (1900), and Marble (1900). The complete title of Coues's 1872 book is reminiscent of the long title of Darwin's landmark 1859 book *On the Origin of Species by Means of Natural Selection, or the Preservation of Favored Races in the Struggle of Existence*. The title of Coues's 4th edition (1894) is *Key to North American Birds Containing a Concise Account of Every Species of Living and Fossil Bird at Present Known from the Continent North of the Mexican and United States Boundary, Inclusive of Greenland and Lower California*. The twelve-page "Historical Preface" to the 1894 edition is a mini-course in ornithological history.

Page 56: *Reference work on ornithology*: Beck cites Coues in his autobiographical sketch published in Murphy (1936).

Page 57: *Proclaiming in functional terms*: Coues (1894, 1).

Page 57: Brinkley (2009).

Page 57: *Let alone want to use one*: Weidensaul (2008); Cutright and Brodhead (1981).

Page 57: *Standard equipment for birders*: Callahan (2014); see 74–75 on the 1887 Winchester shotgun and birders' use of the shotgun.

Page 57: *Best one you can afford*: Coues (1894, 1).

Page 57: *Coues advised birders*: Coues (1894, 13).

Page 57: *Collectors of his day*: Barrow (2000).

Page 57: *Good mental health*: Coues (1894, 19).

Page 58: *Unexplored islands: San Francisco Chronicle*, August 14, 1903.

Page 58: *Mounted their expedition*: The Academy planned to visit Tiburon Island in February 1904. Here they would be able to acquire "the richest of specimens" on an island inhabited by people "nearest to the aboriginal type of any that have lived in the United States." Despite tales that these people were "cannibals, bloodthirsty and savage," the proposed expedition would collect natural history specimens and "these Indians and their lore will be closely studied." As mentioned in the text, this expedition did not materialize. *San Francisco Chronicle*, Friday, August 14, 1903.

Page 58: *Significance of museum specimens*: For the original editorial letter in the "Perspectives" section of *Science* see http://www.sciencemagazinedigital.org/sciencemagazine/18_april_2014?pg=4#pg36, accessed April 19, 2014.

For the response: http://www.sciencecodex.com/collecting_biological_specimens_essential_ to_science_and_conservation-134261, accessed May 1, 2014.

For another response: http://www.sciencecodex.com/scientific_collections_play_vital_role_in_ conservation_biology-134266, accessed May 1, 2014.

For yet another response: http://ns.umich.edu/new/releases/22196-collecting-biological-specimens- essential-to-science-and-conservation, accessed. For the press release from the California Academy of Sciences see http://www.calacademy.org/newsroom/releases/2014/science_response.php, accessed May 1, 2014.

Page 58: *Faulty and biased logic*: Hickman (1985).

Page 59: *Time and time again*: Thornton (1971, 137, 141).

Page 59: *Collecting in a more positive light*: Evans (1985, 207–208).

Page 59: *Preserve the biota of the Galápagos*: In 1934, during the presidency of José Maria Velasco Ibarra, an Ecuadorian law was passed declaring the Galápagos Islands a National Preserve. Velasco Ibarra was president of Ecuador four times, in 1934, 1944, 1952, and 1968. http://www.biography.com/people/jos%C3%A9-mar%C3%ADa-velasco-ibarra-37034, accessed November 3, 2016.

Page 59: *Two hundred tortoises to the beach*: VOB, 361.

Page 59: *Young tortoises make excellent soup*: VOB, 362.

Page 60: *Attacks of their enemies*: Beck (1902).

Page 60: *Abingdon Island*: The Latin taxonomic name of Lonesome George and any of his fellow species, which derives from the older name of his home island, is *Geochelone abingdonii* (the species name is pronounced abb-ing-DOHN-ee-eye, with both of the letters at the end pronounced separately).

Page 61: *Female Pinta tortoise*: See http://www.theguardian.com/environment/2012/jun/25/lonesome-george, accessed November 3, 2016. See also Nichols (2006).

Page 61: *ponderous official scientific name*: Now changed according to the rules of taxonomy to *Chelinodis becki* (Rothschild, 1901).

Page 61: RWE, 83.

Page 61: *Slaughtered tortoises for food*: Reynolds and Marlow (1983).

Page 61: *Newspaper report*: See *Los Angeles Herald*, June 24, 1905; and Berwick, "Expedition which goes after wonderful specimens, fast becoming extinct on Galapagos Islands, is the most important ever sent out from the Pacific Coast." *San Francisco Chronicle*, May 28, 1905.

5

Collecting Evolution

EVOLUTION WAS IN the air, along with love, pollen, and the ubiquitous Bay Area fog, in the summer of 1905. In the years and months leading up to the 1905–1906 Galápagos expedition, and continuing to this very day, biologists and museum curators increasingly based their work on biological evolution outlined by Charles Darwin. But not all of them were doing so in 1905. In learning centers throughout California, the California Academy of Sciences, the University of California at Berkeley, Stanford University in the South Bay, and up north in Santa Rosa, evolution was being written about and discussed and experimented with. The topic of biological evolution dominated the intellectual climate of the greater San Francisco Bay Area, as it did the rest of the country and much of the world.

Evolution depends on inheritance. In the 1910s and the 1920s, researchers at Thomas Hunt Morgan's fly lab at Columbia University solidified the foundations of genetics, turning the rediscovery of Gregor Mendel's work on plant genetics in the spring of 1900 and others into an entire discipline based on fruit flies. In the 1930s and 1940s, a group of scientists in the fields of population and ecological genetics, paleontology, evolutionary biology, ethology, and botany connected the new genetics with Darwinian evolution in an intellectual movement that came to be known as the evolutionary synthesis.

But in 1905, both what we now refer to as "classical genetics" and "the evolutionary synthesis" were far in the future. Most biologists accepted Darwin's theory of evolution even while they disagreed over its precise mechanisms, but not about evolution itself. The study of natural populations was as important to them as laboratory studies of fruit flies or plants, and scientific collecting expeditions documented the raw material of evolution: variation between individuals.

During the summer of 1904, on the campus of the University of California at Berkeley, a ferry ride across the San Francisco Bay from the California Academy of Sciences, the Dutch botanist and geneticist Hugo de Vries—one of three scientists credited with recognizing the importance of Mendel's work—delivered a series of twenty-eight lectures explaining his evolutionary work in terms of the factors of variability, inheritance, selection, and mutation. The year before, while Loomis planned the Academy Galápagos expedition, the English biologist William Bateson won the Darwin Medal for his work on the theory of organic evolution, which was based on his examination of variation and heredity. In the year before that, in 1903, the American biologist Thomas Hunt Morgan published his landmark treatise, *Evolution and Adaptation*. But Darwin's notion of natural selection was not universally accepted; in fact, the evolutionists de Vries, Bateson, and Morgan explicitly rejected it. They preferred a mutation mechanism of evolutionary change. As an alternative to the natural selection of Darwinism, they favored genetic mutations that could cause morphological jumps, or saltation. The mutationist view allowed for change to take place in discrete jumps, while the Darwinist view advocated slow, gradual change. Darwin even had a Latin phrase to describe his opposition to saltation: *Natura non facit saltum*, or "Nature does not make jumps." The gradual and minute change from generation to generation favored by natural selection was the essence of Darwin's views. The mutationists leapfrogged right over Darwin when de Vries discovered and named genetic mutations between 1901 and 1903. De Vries wrote in the preface to his UC Berkeley lectures, "The current belief [Darwinian theory] assumes that species are slowly changed into new types. In contradiction to this conception the theory of mutation assumes that new species and varieties are produced from existing forms by sudden leaps." The debate was eventually resolved in Darwin's favor, but as the schooner *Academy* left San Francisco, even Director Loomis was leaning toward the mutationist camp over the Darwinian school of thought. The eight young men inherited this controversy as they sailed south.

Anyone in the natural sciences who was paying attention to the literature in the field knew that Darwin was being examined and tested, even if he was not yet vindicated and completely accepted. This new wave of publications made abundantly clear that biologists could not understand evolution, particularly the emergence of new species, without understanding variation between individuals. Did new species sprout up suddenly, as a result of genetic mutation, as the new Mendelians claimed? Or did they emerge gradually over time, as environmental conditions slowly magnified the effects of small variations, as proposed by Darwin's original theory of natural selection? The best hope of finding out came from collecting expeditions like the Academy's 1905–1906 Galápagos trip.

In San Francisco, at the helm of the California Academy of Sciences, Loomis was motivated twofold by Darwin, but not in the usual ways. On the one hand, Loomis was acting on the well-founded fear that species in the Galápagos, especially the islands' namesake giant tortoises, were fast disappearing due to human depredation. Loomis knew that previous collecting expeditions had been too short, lacking the necessary time to collect a complete and representative suite of museum specimens. Loomis intended to change that with *his* expedition. He was going to do it right. But his men needed to get to the islands before it was too late, before many of the unique Galápagos flora and fauna went extinct. For Loomis and every naturalist of his day, the archipelago was made famous "by Charles Darwin, whose account of the voyage of H.M.S. *Beagle* has become a classic amongst students of nature." It was time to leave San Francisco and sail south.

On the other hand, Loomis was likely opposed to Darwinist natural selection, and perhaps to evolution itself; his scientific writings (mainly on the tube-nosed seabirds such as albatrosses, petrels, and shearwaters) reveal little about his views on evolution and natural selection.

When Loomis did write about evolution, he limited himself to variation; his choice of words and examples suggests a deep skepticism about Darwinian evolution. In a publication about waved albatross plumage, for example, Loomis argued that the variation exhibited in birds might best be understood as dichromatism (two colors), with darker phases and lighter phases linked by intermediates. Each color form could be subjected to different types of natural selection and lead to new, different species. Loomis rejected the idea that plumage variation could be the result of natural selection and that light and dark populations might represent "incipient species," caught in the act of diverging and becoming new species, a lighter species and a darker species.

Loomis held strong religious beliefs stemming from his New England *Mayflower* ancestry and Puritan stock and from his early years in Ohio, where his father was a Presbyterian minister. This alone, however, does not explain his evolutionary views, as many leading American Protestants embraced evolution around 1900. He was a bit of a stickler, though. A lengthy obituary in the *Auk* in 1929 stated, "From this ancestry he derived the unswerving devotion to what he held right and an inability to compromise that marked his later life." Loomis's mother "held such pronounced views on the observance of Sunday" that in deference to her, members of expeditions under his direction, including the 1905–1906 Galápagos expedition, were forbidden to collect on Sunday. Many religious people accepted evolution, but Loomis was apparently not one of them.

The young scientific crew of the *Academy* may have been more open-minded about evolution. Nothing in any of their writings suggests a firm commitment to either side in the unresolved debate over the mechanisms of natural selection in the early twentieth century. Their open-mindedness was tempered by the pragmatic and immediate concern of collecting specimens that would be worked up after the expedition by the naturalists back in San Francisco, much as Darwin had worried about the "experienced naturalists" who would take issue with his writing as he dipped his pen in the inkwell and daily strolled the Sand Walk at Down House.

That the Academy collectors maintained an ambivalent stance on evolutionary theory, while simultaneously collecting specimens that could either prove or disprove evolutionary mechanisms, can be attributed at least partially to their youth and inexperience: none of the crewmembers were established researchers who had concerned themselves with matters of evolutionary theory in the forty-seven years since *The Origin of Species* had been published. Nor did they hold authoritative decision-making positions in the academic and museum hierarchy. They were not the testers of theories; they were the collectors of data. They were not the architects, but the brick carriers.

Rollo Beck, the expedition's leader and the preeminent natural history collector of his day, held only an eighth-grade education. Beck was not concerned with supporting or rejecting what Darwin had fretted about at Down House. Instead, he and his fellow expedition members would collect the crucial specimens that populated Darwin's classic grounds. They would seek information on species occurrences on particular islands and in particular habitats. But whether or not they intended it, the cache of specimens that Beck and his capable group of young naturalists brought back to San Francisco would vindicate Darwin many times over in future scientific analyses.

Today, some 180 years after Darwin's visit in 1835, the Galápagos Islands are widely known as Darwin's living outdoor laboratory of evolution. By 2004, a record-breaking one hundred thousand people had visited the Galápagos on ecotourism vacations, many of them drawn to the islands made famous by the published work and well-deserved reputation of Charles Darwin. That number has now surpassed 215,000. Travelers today want to see what Darwin saw, they want to experience what Darwin experienced, to stand in the volcanic birthplace of a revolutionary and tumultuous idea.

In addition to the cachet and charisma of Charles Darwin, advertising agencies and tourist bureaus had another promotional dream come true. Most of the spectacular wildlife of the Galápagos is perfectly fearless and doesn't move an inch, even if you approach. The Galápagos was custom-made for tourism. Walt Disney could not have done better had he tried. But the islands and organisms, and all the drama of predators and prey, volcanoes and ocean currents, blistering tropical sun and torrential El Niño rains, are real. This is not Frontierland or Tomorrowland or Adventureland. This is where animals and plants live and die in a harsh, unforgiving environment. This is the vibrant crucible of new species.

As remote and obscure as the Galápagos Islands were, and remain, Darwin permanently gave these islands what every aspiring real estate agent prizes: location, location, location. Because of the intellectual connection to Darwin, who is sometimes described as responsible for giving biologists something to do, the islands have become a frequent vacation spot for biologists longing to see the place he made famous. Ecologically minded tourists from around the world visit these fabled islands, but not as they would one of the Hawaiian Islands at an all-inclusive mega-resort with golf courses, tennis courts, swimming pools, and world-class restaurants. Visitors to Galápagos visit as Darwin himself did, aboard a small ship with a slow-paced, island-hopping itinerary. Almost all visitors to the Galápagos re-create the voyage of the *Beagle*, whether they know it or not. Almost all visitors become Darwin, whether they know it or not. Species are transformed and people are transformed. It is the nature of the place.

But the technical problem is this: just like the followers of Galileo and Einstein, the followers of Darwin are seeking something elusive, intangible, and largely invisible. Today's Darwinians are seeking to observe a *biological process* that takes years and years to play out. The drama they are drawn to might occur high up in a tree, deep down in a burrow, or even under water. So most visitors to the islands must make do, when they visit for a week or ten days, with seeing the *products* of evolution by natural selection—the finches, mockingbirds, cacti, tortoises, iguanas, and flightless birds that populate the small cluster of islands. Even the most careful observers can rarely catch evolution or natural selection in action. Their challenge is to recognize evolution in action in the field. One can see populations change over time, but is this change evolutionarily significant? Will the change lead to the production of new species? Most of us have to settle for the results, the box scores, if you will, after the game has finished and winners and losers can be clearly identified.

Today, Darwin's finches perch prominently in the evolutionary narrative and illustrations of most high school and college introductory biology textbooks.

Around the time of the 1905–1906 expedition, scientists investigating evolution or the origin of species were unlikely to mention natural selection directly or possibly even to refer to Charles Darwin. Instead, they almost certainly observed and commented on *variation* within a species. Variation was the nexus of field and laboratory investigation. Museum specimens are ideal for documenting variation, because numerous specimens can be spread out on a table, meticulously measured at a leisurely pace, far from the rigors of heat, wind, and rain. For naturalists of the time, as now, variation is any aspect of an organism that could be measured and observed, from bone length to feather color to the size and shape of any body part. A prominent example in the literature of the time was the size and length of the pinchers on the backsides of pincher bugs or earwigs, a common insect in gardens. Variation was then, and is now, viewed as the raw material that over time could lead to new species. "Variation" was code for "evolution."

In 1894, William Bateson (1861–1926) wrote, "To collect and codify the facts of Variation is, I submit, the first duty of the naturalist." And he went on to explain how variation is an important component of the origin of species. He continued, "Whatever be our views of Descent, Variation is the common basis for them all." He hoped that naturalists would "follow the serious study of Variation, and so make sure a base for the attack on the problems of Evolution."

Looking back on the journey, the expedition's entomologist, Francis Xavier Williams, wrote in 1911 that "all the islands and many of the 'mere rocks' of the group were visited at least once, and a number, several times, and from different points and during various seasons." Williams contrasted this thoroughness with prior expeditions: "Thus the Expedition, equipped for the special purpose of studying and collecting specimens of natural history, was able to bring together a far larger and more varied assemblage of specimens than was collected perhaps by the sum total of all the previous expeditions to these islands." In other words, it wasn't just that the *Academy* brought back *more* samples; the *Academy* also returned with *more varied* samples—an essential distinction.

But Williams himself might not have adhered to Darwin's mechanism of evolution. When addressing the notion of how the grasshoppers of the Galápagos came to be differentiated as species and races on various islands, Williams wrote, "Natural selection I don't think plays an important part here." Perhaps time would shed more light on Williams's opinion, as he himself suggested. "There still remains," he wrote, "an immense field for further investigation there, and the only manner in which a satisfying knowledge of the natural history of these interesting islands could be obtained, would be by residing in the Archipelago for several years, and studying the fauna in all its relations in a most thorough and systematic manner."

Williams could not possibly have imagined the ways the Galápagos have been studied since. Museum specimens have been used, for example, in conservation genetics to study the history of insecticide resistance. Museum specimens are also commonly used in studies of morphology (the bumps and hooks and knobs of anatomy), biogeography (where organisms occur north, south, east, and west), and ecology (who does what to whom when species interact). But perhaps most important here, museum specimens are used to provide evidence for gene frequency change owing to natural selection. This type of study is the modern embodiment of the legacy of Charles Darwin. Gene frequency change *is* evolution.

All this was ahead of Williams, his fellow crew members, and their ship full of specimens. First, they had to get to the Galápagos.

NOTES

Page 65: *In the years after 1910*: Bowler (1984).

Page 66: *The current belief*: de Vries (1905).

Page 66: *Made famous by Charles Darwin*: LSA, 5.

Page 67: *Protestants embraced evolution*: Roberts (1988).

Page 67: *Forbidden to collect on Sunday*: Larson (2001) considered Rollo Beck's connection to the Methodist Church as the source of the Sunday hiatuses of specimen collecting, but the evidence favors Loomis as the source of this ban on activity.

Page 69: *Serious study of variation*: Bateson (1894, ix).

Page 69: *Studying and collecting*: Williams (1911).

Page 69: *Grasshoppers of the Galapagos*: Williams 1, 6. Unpublished notes, Archives of the California Academy of Sciences.

Page 69: *There still remains*: Williams (1911).

6

Galápagos at Last

FINDING THE GALÁPAGOS proved a little trickier for the men aboard the schooner *Academy*. On the morning of September 17, 1905, when the crewmembers licked their wounds from the previous day's roughhousing shenanigans during their crossing-the-equator ceremony, the schooner drew unusually close to the lee shore of the coast of Ecuador. That morning they ran into heavy weather and shipped green water over the bow, flooding the cabin. The *Academy* sailed only a few miles from the mainland coast, and Slevin noted that the coastline appeared "precipitous and covered with forests." Late in the afternoon they steered close to Manta Bay, bringing into view a coastal village of white adobe houses with their corrugated iron roofs clearly visible. They saw various sailing vessels, including a three-masted ship lying at anchor in the bay. It was the closest the schooner got to the South American mainland during the entire trip.

Why Parker, the navigator, and Beck, the vessel master, chose to navigate this close to mainland Ecuador is unexplained, but not inexplicable. Normally a sailing master does not allow his vessel to get so close to what is called a "lee shore," where a change in wind direction or a drop in wind speed can bring a sailing vessel dangerously close to running aground. Without specific business to conduct in any coastal town of Ecuador and without an auxiliary engine, Parker and Beck had no reason to sail the schooner so far to the southeast after leaving Cocos Island. Likely, Parker and Beck were having trouble determining their longitude and needed to make visual contact with the mainland to establish their exact location at 81° west before sailing 8.5° due west for the Galápagos themselves. Perhaps they initially planned to go ashore at Manta, or elsewhere, and changed their minds, although that is unlikely. No one ever stated the exact reason in the ship's log or in any of the field notes. At least the men got a glimpse of the mainland before heading west for their actual island destination.

On September 19, 1905, the schooner quietly sailed past Cape San Lorenzo and "turned right" in order to head due west for the Galápagos. This maneuver set the *Academy* off on a port tack, with the prevailing wind coming up from the south. The schooner made a steady five to six knots for the next four days, no doubt due to Twain's "down-hill" sailing advantage. Before long the crew would be enthusiastically walking in Darwin's footsteps and fervently collecting specimens for Loomis's museum (figure 6.1).

After more than two months at sea, the *Academy* was on its final leg from Cocos Island to the Galápagos, after skirting the mainland of Ecuador for navigational precision. With mild winds, the schooner made good progress, and a noticeable coolness was in the air. The South American continent slowly disappeared behind the crew. For five days, from September 18 to 23, the schooner maintained a steady westward course just south of the equator as she approached Galápagos. Several flying fish landed aboard one night and were turned over to the cook. "We found them excellent eating," recalled Slevin, and no doubt they made a welcome addition to an otherwise limited diet.

The *Academy* arrived in Darwin's Islands on September 24, a week later than Darwin had arrived seventy years before (figure 6.2). The *Beagle* had approached the islands on September 15, and after anchoring in what is now called Wreck Bay on Chatham (San Cristóbal) Island, Darwin went ashore for the first time on September 16. The *Academy* nearly duplicated Darwin's first island landfall, but the fickle equatorial winds changed the crew's plans. Instead they were forced to divert her course to Hood (Española) Island.

FIGURE 6.1 **A Bird in the Hand.** The exuberance of the expedition leader, Rollo Beck, is readily visible as he collects burrowing seabirds. His shotgun is propped up on the left. *Archives, California Academy of Sciences.*

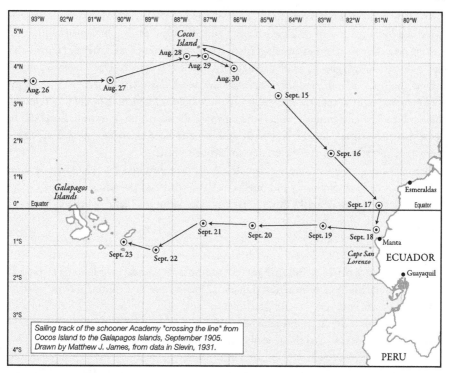

FIGURE 6.2 **Sailing Track.** The crew of the schooner *Academy* had to resort to finding the islands using dead reckoning, due to a faulty chronometer, by sailing from Cocos Island to the coast of Ecuador, where they picked up the equator near Cape San Lorenzo and sailed due west to reach the archipelago. Illustration by M. James. Source: Matthew J. James, used with permission.

Despite Charles Darwin's close association today with all things Galápagos, his initial impressions were quite negative when he first went ashore on Hood Island on September 17, 1835. He wrote in what would become known as the *Voyage of the Beagle*: "The dry and parched surface, being heated by the noonday sun, gave to the air a close and sultry feeling, like that from a stove: we fancied even that the bushes smelt unpleasantly." Modern ecotourism operators and promoters assiduously steer clear of Darwin's succinct one-line summary: "Nothing could be less inviting than the first appearance." Darwin seemed to physically recoil from the land and all the organisms there. Fortunately, even for Darwin, first impressions are not always lasting impressions.

Herman Melville recorded parallel impressions in his 1846 *The Encantadas*, where he wrote, "Take five-and-twenty heaps of cinders dumped here and there in an outside city lot; imagine some of them magnified into mountains, and the vacant lot the sea; and you will have a fit idea of the general aspect of the Encantadas, or Enchanted Isles." Even for Melville the geology was obvious: "A group rather of extinct volcanoes than of isles; looking much as the world at large might, after a penal conflagration."

Even the lava landscape that captivates geologists today was to Darwin "a broken field of black basaltic lava, thrown into the most rugged waves, and crossed by great fissures, [it] is everywhere covered by stunted, sun-burnt brushwood, which shows little signs of life."

To Darwin, the plants he encountered and collected on the rocky and rugged islands were "wretched-looking little weeds," the tortoises "antediluvian," the marine iguanas "hideous-looking creature[s] . . . stupid and sluggish in [their] movements," and the land iguanas "an ugly yellowish-brown species" with a "singularly stupid appearance." Darwin's initial observations about the islands, and their unusual flora and fauna, make it seem as though the Victorian naturalist's experience was a bit like the vacation from hell. Nor did the "few dull-colored birds" he encountered while walking about the islands impress him. Little did he know that these simple-looking finches would come to bear his name and later appear in every biology textbook as a classic example of island evolution.

Generations of scientists and visitors alike have demonstrated that the Galápagos are indeed inviting, both physically and intellectually. By the 1845 edition of the *Voyage of the Beagle*, Darwin was aware of the evolutionary significance of the islands when he wrote:

The natural history of these islands is eminently curious, and well deserves attention. Most of the organic productions are aboriginal creations, found nowhere else; there is even a difference between the inhabitants of the different islands; yet all show a marked relationship with those of America, though separated from that continent by an open space of ocean, between 500 and 600 miles in width. The archipelago is a little world within itself, or rather a satellite attached to America, whence it has derived a few stray colonists, and has received the general character of its indigenous productions.

Years later, Darwin reflected on the entire voyage, including the Galápagos: "The voyage of the *Beagle* has been by far the most important event in my life and has determined my whole career. I have always felt that I owe to the voyage the first real training or education of my mind." For Darwin, the vacation from hell turned out to be superior to Ishmael's "Yale college and [his] Harvard," and constituted a veritable graduate research program that he relied on for the remainder of his very productive life. Further exploration and study of the Galápagos revealed the untruth of the islands' stereotyped reputation as "inhospitable, deserted stone blisters in a broad ocean." Darwin revised history, in more ways than one.

Meanwhile, the group of eight young sailor-scientists and three crewmembers on the *Academy* who had set sail from San Francisco eighty-eight days earlier had a very different Galápagos experience than Darwin. Only Beck had visited the islands previously, and all of them eagerly commenced collecting specimens ashore on Hood Island, as Slevin wrote, after they "enjoyed their first close up view of the Galapagos, a region that never ceased to be of interest during an entire year that followed." They dropped anchor for the first time in Galápagos waters in Gardner Bay at about 9:00 a.m. on September 24, 1905. Again, no gold coin needed to be nailed to the mast (as Captain Ahab did to encourage his men to hunt for Moby Dick) to entice the collectors to begin filling the schooner's coffers with rare and exotic specimens for the California Academy of Sciences (figure 6.3). Darwin had already made the islands famous, and the Academy men were overjoyed to be basking in the oppressive heat that so repelled Darwin.

The first footprints the men left on a Galápagos beach were at Gardner Bay on Hood Island. The wide bay impressed the men with its "dazzling white sand of the coral beach" and its numerous sleeping sea lions, which they approached "without any fear on their part and

FIGURE 6.3 **Documenting the Expedition.** Rollo Beck set aside his shotgun for his camera here at Punta Suárez on Hood (Española) Island during their visit to the first island in the archipelago in September 1905, photographing blue-footed boobies. *Archives, California Academy of Sciences.*

even slapped with one's hands." The beach at Gardner Bay remains the same today and is a popular visitor site, although the Galápagos National Park and its naturalist guides would severely admonish any tourist with the temerity to slap a sea lion.

At Gardner Bay, while the collectors busied themselves ashore, Nelson overhauled the mainsail and kept the schooner shipshape. Throughout the seventeen-month Galápagos expedition, the aging schooner *Academy* required almost daily maintenance. Despite her distinguished pedigree, she was, after all, a thirty-year-old vessel that had seen hard service with the US Coast and Geodetic Survey in the waters of Alaska and Washington from the late 1870s through the 1890s. The wear and tear started to show, and most of the repairs fell on Nelson in his role as mate.

NOTES

Page 71: *Shipping green water*: On any vessel when the bow plunges into an oncoming swell in heavy weather and the thick flow of water over the bow takes on a green color due to the light from the storm.

Page 71: *Covered with forests*: LSA, 33.

Page 72: *Darwin went ashore*: Even today, September 16 is celebrated as "Galápagos Day" by the UK-based Galápagos Conservation Trust, and a similar celebration is observed by Ecuadorians and others living in the islands.

Page 73: *Darwin's initial impressions*: VOB, 359.

Page 73: *Parallel impressions*: Melville (1856).

Page 74: *Singularly stupid appearance*: VOB, 357–386.

Page 74: *Indigenous productions*: VOB, 363.

Page 74: *Determined my whole career*: Darwin (1958, 76).

Page 74: *Vacation from hell*: Lopez (1998, 51).

Page 74: *Eagerly commenced collecting*: LSA, 34.

Page 75: *Dazzling white sand*: LSA, 35.

7

Confluence

WHERE GEOGRAPHY AND IDEAS MEET

TODAY, WHEN ONE approaches Galápagos in a jet plane, each brown island in the archipelago resembles half a walnut shell on a dinner plate of deep blue ocean. The islands are isolated on several levels, and isolation is what makes the Galápagos interesting. Organisms are isolated from their continental relatives, and that promotes the genetic processes of evolution espoused by Darwin and biologists today. Scientists who visit the islands are isolated from family and friends, and that promotes introspection. The isolation of Galápagos fosters and stimulates both new species and new ideas. Genetically, organisms experience reduced contact, or gene flow, with compatriots from the mainland. This reduced gene flow promotes genetic change over time, which in turn promotes evolution. At a fundamental level, evolution is largely the result of reduced gene flow, of island isolation. This view of evolution is newer than Darwin's original ideas, because he did not know about genes or about gene flow. But new research and new ideas, the neo-Darwinian synthesis, build upon Darwin and give us our view of Galápagos today.

We come to Galápagos to see firsthand those biological changes resulting from isolation. We end up somewhere out in the islands in a cabin or stateroom on a ship at an anchorage in a bay off the coast of an island in an archipelago far removed from anywhere familiar. It is the perfect setting for observing nature and taking notes and photographs, just like the members of a scientific expedition. We are connected to the past and immersed in the present.

The Galápagos Islands are easy to locate on a map but quite difficult to reach in reality. If you hunt for the equator on a world map, allow your eye to fall on the prominent bulge on the west coast of South America. Mentally draw a horizontal line, left and right around the world, and you'll have the equator. It gets the number 0° because it is, by definition, neither north nor south. It is the dividing line between north and south, and has a mythical reputation among mariners. Just a short distance to the left of that bulge in South America sit the Galápagos Islands, tiny specks of land straddling the equator.

Another method also works. If you know the geography of the United States, you can run an imaginary line due south from Chicago through Memphis and New Orleans. It doesn't have to be a very exact line, but keep going south and you'll reliably hit the Galápagos out in the Pacific Ocean.

The imaginary north-to-south line from Chicago to the Galápagos and the better-known east-to-west band around the equator cross paths at a geographic confluence, a symbolic intersection or crossroads. The confluence is like an imaginary Hollywood and Vine in Los Angeles, or Times Square at Broadway and 7th Avenue in New York. Major street intersections in large cities are famous because something of note is located there. With the Galápagos confluence, there is nothing there except ocean, a bit more like Oakland in Gertrude Stein's characterization. Chicago, Memphis, and New Orleans straddle an imaginary line down the middle of the United States, and the southward extension of that line eventually hits the Galápagos, just as sure as if the train called *The City of New Orleans* continued due south over the Gulf of Mexico, across the Yucatán Peninsula, through Guatemala, and into the eastern Pacific Ocean. The last stop on this imaginary express train would be not only at the intersection of invisible latitude and longitude lines but also at the intersection of intangible human ideas. To be more precise, Chicago isn't exactly 90° west of Greenwich—it's only about 87° west—but it's still easy to locate at the southern tip of Lake Michigan. So the north-to-south line made by the actual tracks of *The City of Chicago* train is a little bit off-kilter. Even the Galápagos are not exactly on the 90° line; at least there is no solid ground or island right where the 90° line crosses the 0° equator. The exact "X marks the spot" falls unobserved and unmarked far out in the open ocean in the northeast portion of the archipelago.

The four compass directions of north, south, east, and west converge in this rocky archipelago (as they do everywhere on Earth), but here they are accompanied by the persona of Charles Darwin, lumbering giant tortoises, the origins and concepts of evolution, and little finches with multi-shaped beaks.

The Galápagos sit 90° west of a crucial starting point for Earth measurement: the prime meridian, which extends from the North Pole to the South Pole through the town of Greenwich, England. This method of measuring the Earth puts the Galápagos exactly one-fourth of the way around the world from Greenwich and the mother of all meridians. Head the same distance in the opposite direction and you'll hit Bangladesh, Bhutan, China, Mongolia, and Russia north of the equator, and a lot of open ocean in the Indian Ocean to the south, until you hit Antarctica. The prime meridian through Greenwich is the conventional starting line for going around the world clockwise or counterclockwise, east to west, or west to east. It also defines the time zones of the world. Heading west from Greenwich past 90° at the Galápagos, and then 180° at the International Date Line, and continuing on to 270° and back to Greenwich brings the traveler full circle and a full twenty-four-hour day to completion.

Measuring the world east and west with reference to Greenwich, as we do today, is a fairly recent convention. As a standard reference point and invisible north-to-south line, it was formally decided on only some twenty years before the 1905–1906 expedition. A committee decision was reached by nearly unanimous agreement at the 1884 International Meridian Conference held in Washington, DC (France, Brazil, and the Dominican Republic hopped on the bandwagon a few years later). Before 1884, several prime meridians existed in various countries for various reasons, including one that ran right through Washington, DC, itself. The 1884 conference sought to standardize Earth measurement and is counted as an important accomplishment during the presidency of Chester A. Arthur (1829–1886), the twenty-first and one of the lesser-known presidents.

Why Greenwich, England? The decision was partly pragmatic: most of the world's shipping used British-made navigation charts and *The Nautical Almanac* for reference, and mariners determined their longitude with chronometers set to Greenwich Mean Time. The other factor was historical: the Royal Navy's Captain James Cook (1728–1779) and other eighteenth-century pioneering navigators made the first accurate world charts, drawn with reference to a zero meridian at Greenwich.

All of the imaginary north-to-south lines running from the North Pole to the South Pole on a schoolroom globe or map are called lines of longitude. Due to the curved surface of the Earth, all the longitude lines get closer and closer together as they approach the poles, and eventually come to a point at the poles. One of these imaginary thin black lines crosses your current location, no matter where you are right now, whether sitting in an orthodontist's office in Chicago or bobbing on an ecotourism boat in the Galápagos. If you happen to move even a small distance east or west, a slightly different line of longitude will cross your location. Getting away from longitude is not possible, no matter how fast you run. Like a shadow on a sunny day, longitude and latitude lines form a sort of geographic shadow: wherever you go, a line is there.

Each time one of the whole-number longitude lines crosses a whole-number line of latitude, the intersection is called a confluence. These would correspond to some of the thin black lines on a globe or map. There is even a web page devoted to travelers who post pictures of the terrain at each confluence, verified and documented with a snapshot of a hand-held Global Positioning System (GPS) receiver showing the whole numbers of the particular confluence. A total of 64,442 geographic confluences exist on Earth, making a crosshatched pattern over the entire Earth, like a blanket made of graph paper. With so much of the Earth's surface covered by water, 70 percent or so, it is no surprise that only 21,541 of the confluences are on land, and the other 38,411 crisscross on open water. If you're doing the math, the remaining 4,490 are on the Arctic ice cap.

Visiting these confluences on land has become a bit of sport, similar to geocaching, where participants use Global Positioning System coordinates to find containers, or geocaches, at specific locations all over the world. When I recently checked the 90° west and 0° equator confluence on the web page confluence.org, no intrepid traveler had yet "bagged" that one near the Galápagos. With nothing out there but wide-open ocean and the horizon in the distance, you'd need the time, money, and inclination (and a well-fueled boat) to chug all the way out there from one of the small boat harbors in the Galápagos. Once reaching the invisible confluence, you'd have to be satisfied simply floating around for a while, taking snapshots in the north, south, east, and west directions, in the traditional method of documenting when one "bags" a particular confluence. All your photos would show the same thing: ocean. A photograph of your GPS unit displaying 0° and 90° west would be your final proof. The problem is there's nothing much to see out there, except perhaps some migratory seabirds, which would make some people connected to the 1905–1906 expedition quite happy. Two men who had an intense interest in migratory seabirds and would have been thrilled to visit a confluence as obscure as the one in the Galápagos were the organizers of the 1905–1906 expedition, the Academy's director and ornithologist, Leverett Mills Loomis, and his hand-picked chief of the expedition, Rollo Howard Beck.

The Galápagos Islands lie not only at the intersection of invisible latitude and longitude lines, but also at the confluence of important ideas. A common aphorism states that Darwin gave biologists something to do. And with the Galápagos, he gave them a place to do it. Beyond the geography of graph paper, the Galápagos are positioned at the intersection of major ocean currents in the eastern Pacific Ocean. These currents provide contradictions and unexpected contrasts. The unexpected geographic contradiction of sunbaked volcanic islands and cool Antarctic ocean currents is contrasted with several biological contradictions: iguanas that feed and swim comfortably in the ocean, penguins and pelicans perched side by side on the volcanic rocks, giant tortoises that floated or drifted over from South America on vegetation rafts easily described by visitors as "antediluvian" apparitions, as they seemed to Darwin on his first encounter. Flowing up from Antarctica is the Peru or Humboldt Current, appropriately named for the traveler and adventurer who inspired Darwin. It is really the cold water of the Humboldt Current that makes the Galápagos Islands so unusual. Ocean currents flowing south from Panama and west in the South Equatorial Current are more mundane, as warm ocean currents go. Even the east-flowing subsurface Equatorial or Cromwell Countercurrent is not unlike an oceanographic lava tube transporting cold water to the islands from the mid-Pacific. The confluence of cold water and warm water and the upwelling of nutrients make the islands productive and foggy for parts of each year, not unlike the San Francisco Bay Area from which the 1905–1906 expedition departed.

The coolness of the air and water had also struck Darwin as odd, considering the proximity of the islands to the equator. Darwin knew that "the singularly low temperature of the surrounding water, brought here by the great southern Polar current," distinguished this region from other equatorial regions. Darwin and the Academy group, once they arrived in the islands, would have welcomed any lowering of the temperature.

NOTES

Page 77: *This view of evolution*: Gene flow is not universally important in species formation (Ehrlich and Raven 1969).

Page 79: *Factor was historical*: Sobel and Andrewes (1998); Johnston et al. (2015, 6–35).

Page 80: *Singularly low temperature*: VOB, 358.

8

On Death, and Life, in Galápagos

⌒

WHILE STUDYING THE log books and published results of the 1905–1906 expedition, I have been repeatedly struck by how many parallels can be drawn between the experiences of the eight young sailor-scientists and present-day ecotourists, of which I have been one several times, who visit the islands. Granted, significant differences exist, specifically regarding the collecting of animals and plants, but the multiple similarities deserve to be mentioned. The parallels provide a connection to history rarely experienced in our modern world of travel and tourism.

Nearly all of the places where the *Academy* anchored and where the men went ashore to explore and collect have been designated as visitor sites by the Galápagos National Park Service. The Park Service does excellent work preserving the integrity of the 97 percent of the archipelago that falls within its jurisdiction. Thus, visitors today can experience a Galápagos visit similar to the one the men on the *Academy* had, or even to Darwin's own experience in 1835. In many ways the islands have changed so little. The volcanic geography retains the same sense of place. A modern visitor stepping ashore, away from the few populated towns, is immersed in the sights, sounds, and smells that also greeted Charles Darwin, Rollo Beck, and all the others. Galápagos visitors can experience firsthand both the conservation ethos and the natural selection pathos.

Getting ashore has itself historical parallels with the landings of Darwin and the 1905–1906 expedition. Today, skilled boat drivers ferry visitors from their yacht or cruise ship to shore in a small "panga," the Spanish word used in the islands to describe anything from an inflatable Zodiac to a rigid-hulled fiberglass skiff or launch. When the Academy collectors went ashore that first day at Gardner Bay, and everywhere else they landed, they used an engineless panga that they rowed, albeit one made of wood (see figure 8.1). Slevin wrote, "While pulling in for the beach many seals followed the skiff, one large bull coming particularly close as if to make a lunge at the steering oar." Today, these dominant, mature, and sometimes aggressive "beach master" male sea lions control and patrol a section of beach

FIGURE 8.1 **Collecting Birds.** The field collectors often spent the day ashore collecting in their specialty and would rendezvous back at the shore to row back out to the schooner, where specimens were stuffed or preserved, before moving on to the next island. *Archives, California Academy of Sciences.*

with a harem of females. These alpha males frequently chase younger male interlopers for hundreds of yards, porpoising frantically off toward the horizon. Pursuer and pursued play out ancient evolutionary struggles for territorial domination and getting their genes into the next generation.

The Galápagos model of tourism reflects another larger-scale historical parallel dating to the 1970s and continuing to the present. In stark contrast is the "hotel model" that sprang up on all of the larger Hawaiian Islands and elsewhere. In the Galápagos, more than 95 percent of visitors follow directly in the footsteps of Darwin and Beck. More specifically, they follow in the wake of the *Beagle* and the *Academy* in that the Galápagos tourism model has substituted live-aboard boats for the omnipresent resort hotels of Hawaii.

Darwin had a rough time with seasickness aboard HMS *Beagle*. Today's visitors can battle that malady with medication, but the seagoing experience living aboard a vessel is historically parallel. Although Galápagos tourists fly to the islands in jet planes, they quickly step back in time once they set foot aboard their cruise ship, yacht, or sailboat. The deck might be made of fiberglass rather than wood, and the propulsion might be diesel rather than canvas, but the connection to maritime history is just as strong. Living aboard a vessel, whether it is the *Beagle*, the *Academy*, or any of the vessels plying Galápagos waters today, brings along the sights, smells, and sensations experienced by generations of seafarers. For the next week or two, Galápagos ecotourists, whether they know it or not, rarely go ashore at a place *not* visited by the crews of the *Academy* or the *Beagle*. Darwin's visit lasted only five weeks, compared with the "year and a day" the *Academy* and her crew spent in the islands. Consequently, the *Beagle* and Darwin did not visit as many places as the *Academy*'s crew. Regardless of which location you explore in Galápagos, history is all around you, both on land and in the water.

Natural selection reeks of death, especially when the losers of battles in the struggle for existence rot slowly in the equatorial sun. Stepping ashore during an El Niño year at Punta Suárez on Hood Island, I witnessed firsthand evidence of the ideas that make the archipelago so famous. In the Galápagos, even a revolting stench can provide scientific insight.

Galápagos has a dark side, because natural selection always involves death, and Charles Darwin knew it when he first described the process to the world in 1859 in his book *On the Origin of Species*. Darwin wrote that certain organisms "will have a better chance of surviving, and thus be *naturally selected*." We can still observe this process in the islands today.

The process of evolution contains a duality. The duality is a simple fork in the road and all organisms take one branch or the other. Going down one branch of the fork, some individual organisms survive by finding enough food, evading predators, attracting a mate, and reproducing, thus passing on their individual genetic material to at least one generation into the future. Others do not, and end up on the other branch of the fork. Some organisms live and some organisms die. Actually, all organisms eventually die, but what matters in evolution, by definition, is whether an individual organism passes along its genetic material. When thinking about evolution, we can look at individual organisms and ask how well *that* finch, or tortoise, or iguana is doing in the process of genetic survival, which is really genetic transmission to the next generation. Evolution can include a literal death and also a genetic death. When genes are passed down the generations, there is a kind of genetic immortality, until one day that, too, ends, when a species becomes extinct.

Evolution and natural selection also operate at two levels or scales: the individual scale of life and death and the scale of the genetic makeup of the entire population of a species. Natural selection itself operates on individuals, because individuals live or die, individuals reproduce or don't. So we can say that individual organisms are the unit of natural selection. One of the subtleties of evolution is that once an organism has reproduced, its genes have been passed along and its inevitable death does not cancel out those already transferred genes. The individual is dead, but its genes live on. Turning this around from the perspective of the Galápagos visitor, what are we to make of a dead organism we might see washed up on the beach or on a trail?

We can ask two questions. The first question is, has the dead individual organism already reproduced? If the answer is yes, then that organism's death is *not* an example of natural selection. Its genes were passed down. Its offspring are out there somewhere, carrying around the parental genes. Our dead organism is, at least while its offspring survive, in the realm of genetic immortality.

The second question is, was the dead animal somehow *different* from other members of the same species? In other words, did it die because it was different? Did it have faulty hearing? Substandard eyesight? Bad judgment? Or was the dead organism no different from any other member of the species and just suffered plain old bad luck? Was it bad genes or bad luck? Natural selection works only on bad genes, not on bad luck.

For the death of any individual we might see in the Galápagos to be an actual example of natural selection, we need to be able to answer no to the first question and yes to the second. No, the animal must not have reproduced, and, yes, the dead animal must somehow be different from other members of its species. Otherwise, the death might be tragic, but it is not an example of natural selection. These criteria for natural selection apply equally to organisms we see in the Galápagos and to road kill we might see while driving on roads back home. Death is not a synonym for natural selection.

Natural selection operates on individuals, but it is entire populations that evolve over time. Individual organisms do not themselves evolve. Individuals live or die according to events in their lifetime: finding food, avoiding predators, and surviving storms. Evolution and adaptation take place when a population changes genetically over time through numerous events of individual natural selection, sexual selection, random drift, mutation, or migration. A male blue-footed booby mates because it has successfully displayed to females feet that are "blue enough" or a specific sky-pointing behavioral repertoire expected by females. But that male bird we might stop to photograph at a visitor site does not itself evolve. Over time, the population could accumulate the genes for bluer and bluer feet, which likely accounts for their striking blue feet. The population could also accumulate the genes for more and more complex mating rituals expected by females and performed with precision by males. But individual males are stuck in their moment in time with set physical and behavioral equipment. Thus, natural selection operates at the level of individuals, and populations evolve genetically over time. Separating out natural selection from evolution is something visitors to the Galápagos can easily do by looking at organisms in the right way and asking the right questions. We can be citizen-scientists in the Darwinian tradition.

Death in the islands often occurs in plain sight, reminding Galápagos visitors that the islands are not Disneyland. The Galápagos archipelago might not be the "happiest place on Earth," but it is one of the most important and interesting places. Flies, maggots, blood, and the smell of death cannot be ignored. Nothing in the Galápagos is staged or cleaned up or sanitized for visitors who walk with their naturalist guide on the carefully prescribed sand and gravel trails set up by the National Park Service. Death is just a part of the landscape, at times an attraction as compelling as the living animals and included in the price of admission. Few visitors would actually pay to explicitly see or smell such a spectacle of death. It certainly is not what one signs up for in a Galápagos vacation. But if you can think of it as an unspoken event of insightful enlightenment, then the natural world of the Galápagos unfolding before us becomes less mysterious, less enigmatic, and more tangible.

Anyone interested in Darwin and his mechanism of evolution might like to see the process in person. And where better than the Galápagos? In the islands we walk in Darwin's footsteps, we see what Darwin saw, and we internalize natural selection by seeing, hearing, and smelling it. The unending and unrelenting danger of death in the mechanism of natural selection is as much on display as are the living endemic birds, iguanas, and sea lions. A shock to the senses can occur when one sees these very same charismatic animals dead along the visitor trails after having seen them alive elsewhere during the day. The realist understands that is why we come to the Galápagos. The combination of firsthand experience and equatorial geography amalgamates knowledge, like coalescing lava flows that intermingle to produce an island. Familiarity does not breed contempt; it breeds insight, cognition, and wisdom.

After the initial shock of seeing a dead iconic animal, never featured in tourism brochures and rarely featured in natural history documentaries, we should scientifically ask: Was this death due to bad genes or just bad luck? Was it tragic or simply tragically inevitable? Was it a *local* predator swimming, flying, or walking nearby that did the killing? Or was the agent of death a *regional* El Niño weather pattern causing shifts in food availability for marine organisms in the islands, or possibly even more widespread *global* climate change? Was that individual's death an example of *evolution* in action? In Galápagos and broadly in the eastern Pacific Ocean, 1905 and 1906 were El Niño years, so the men on the *Academy* expedition fully experienced the now-well-known regional weather phenomenon. Even the years that follow an El Niño cycle, called La Niña, can see a switch from increased death in the ocean to increased death on land. Both El Niño and La Niña are agents of selection, causing individuals to thrive or to die.

The scale of individual death, from local to global, should be considered any time we encounter a dead animal. Regardless of the cause or scale, and no matter how unpleasant, death is a necessary and fundamental part of Darwin's mechanism of natural selection. Avoiding the dirty details of death and natural selection in textbooks or natural history guidebooks removes the most important factor in the Darwinian equation. In any sanitized version, animals that die in this Darwinian process are euphemistically "selected against" or "selected out of the population." But in this case, "selected" means to die, not to live. Nature is doing the selecting in Darwin's natural selection, which stands in stark contrast to humans doing the selecting by means of artificial selection that leads to fancy dog breeds, faster race-horses, or improved vegetable crops. The ultimate cause of death could be local, regional, or global. In the Galápagos, these deaths just plain stink.

ENTERING THE LABORATORY OF EVOLUTION

The islands are rightly known as Darwin's living outdoor laboratory of evolution. What does this mean, and how can Darwin's evolutionary ideas be observed and understood? And do you need a white coat to enter this famous laboratory?

After all, back home we drive right past and avert our eyes when we see a deer lying on the side of the road or, worse, the tragedy of a domestic pet's death. But the Galápagos experience gets us out of our car and onto the trail, where we find ourselves face-to-face with horrific corpses, unable to avoid the smell of death as the tropical breeze swirls it around us. Hold your nose if you must; the smell is still there. The Galápagos experience is supposed to be about life, and the islands certainly possess life in abundance. And the experience is also about Darwin, and that's where things can get messy. So we find ourselves staring instead of averting our eyes, because it is OK to stare in the Galápagos, where animals are fearless and telephoto lenses are almost not needed. Confronting death up close, we might even start speculating about the cause of death. We can easily advance forensic theories like a CSI field technician, postulating starvation, or physical combat, or plain bad luck. And we're OK with the answer, because out here death is part of the landscape, like the lava rocks, beach sand, and the ocean. We embrace raw death and take its picture. And if we think long and hard, we come to understand that death makes this place what it is and always has been: the crucible of evolution and the cradle of new species. And those new species descended from their ancestors in the harsh reality of this volcanic archipelago. Today the islands are seen as the volcanic anvil on which new species are pounded out by the hammer of natural selection.

Immersed in Galápagos nature, often literally while snorkeling, we become Darwin, if Darwin could have snorkeled, which I'm sure he would have enjoyed. Our version of being

Darwin has us observing, speculating, and predicting. Our vision changes in the Galápagos, as it should when we visit.

Darwin was here on this spot on this island, he saw a particular organism, and now you're here seeing the same thing. This makes a Galápagos natural history trip superior to almost any other, and certainly better than a zoo or museum. Few Galápagos travelers leave unchanged.

So, yes, natural selection really stinks. Try standing right next to a three-hundred-pound adult sea lion that's been dead for about ten days and see if you disagree. To the extent that you agree, you form a bond with Darwin through time. The Galápagos presents us with sights, sounds, and smells that allow us to understand the mechanism of evolution in the very place where that mechanism, natural selection, got its start.

CHANGE OVER TIME—OR DEATH OVER TIME

While death is still an important factor for evolutionary biologists who extend Darwin's legacy, the timing of the death is also important. Death in the Galápagos can be easily observed by casual visitors and scientists alike—before, after, and while it occurs. The average tourist visit to Yosemite, or Yellowstone, or even Hawaii yields no chance to see even a fraction of the death, and thus the evolutionary machinations, that one sees, hears, and smells daily in the Galápagos.

Darwin envisioned nature as doing the crucial selecting and eliminated from the equation all traces of human breeders of plants or animals who desire certain profitable traits in, for example, thoroughbred racehorses, breeds of dogs, or the wool of sheep. Practically every domestic plant and animal we know of today has gone through numerous generations of artificial selection. And almost always a financial goal lurked behind those selective changes. Darwin called them domestic productions. The obvious differences between broccoli and brussels sprouts are real, and you might not like the way they taste, but they exemplify the success of selective breeding and artificial selection. A more uniform head of broccoli commands a higher price at the market, so the farmer selects for consistency.

Other examples of animals successfully modified by artificial selection are dachshunds, German shepherds, and toy poodles. Any sort of artificial selection to produce these domestic breeds, or Darwin's domestic productions, involves a series of value judgments by the breeder and a desired physical outcome over many generations, usually with an economic objective. From his mother's raising of pigeons, Darwin acquired a lifelong appreciation of how pigeon breeders could highly modify the basic bird in a few generations through selective breeding for competitive exhibitions.

Yet Darwin removed both human involvement and value judgments from consideration when he coined the term "natural selection." It is significant that, for Darwin, natural selection was unconsciously mediated by purely natural forces in a way that was directly analogous to artificial selection conducted consciously by humans. The selecting and the death must continue out there in nature for the Darwinian natural mechanism to work. Without these deaths, Darwin's celebrated process of evolutionary change would stop, like a ship suddenly running aground. For Darwin's gradual, stepwise, and sequential events of evolutionary change to work, some organisms live and others die, depending on their inherited genetic traits or sometimes simply on chance or luck. In the human world of plant and animal breeding, the organisms with desired traits are allowed to live and breed so that those traits— for example, fast running speed, ability to win races and prizes, or enhanced flavor of fruits or vegetables, or subjective aesthetic beauty in ornamental garden plants—are passed on to future generations.

To badly paraphrase a common aphorism, the wise organism chooses its parents well. Darwin's idea of natural selection requires the death of individuals that cannot surmount the numerous daily challenges presented by the natural environment, the crucial arbiter of life and death. Think of an English sheep dog separating individuals from the flock: except in a Darwinian process, the dog is blind and the sheep die if selected. The natural environment does not take sides during selection, does not judge good or bad, right or wrong. The natural world is neither merciful nor merciless. There is no emotion. But neither is there cruelty, apathy, or callousness. The natural environment simply separates fit from unfit, and the unfit are the ones that die. The environmental challenge that will bring death to the unfit if it is not met might be getting sufficient food, escaping predators, blending in well enough with the foliage, or withstanding a drought. The reason for the selecting does not matter, only the death matters. And death before reproduction means that no genes will be passed to a new generation.

The causes of death during natural selection can be as subtle as a finch not possessing a beak strong enough to crack the hardest seeds scattered in the dry soil beneath a sprawling plant (imagine trying to open Brazil nuts as a child), or as painfully slow and obvious as two beach master male sea lions pounding away for brutal hours at each other's blubbery necks in their game of thrones, or as violent as a bloody death over a harem of females waiting insouciantly nearby on the beach. Death in the Galápagos can be excruciatingly slow or surprisingly fast. It is slow, for example, for male sea lions whose long canine teeth do some real damage to a rival but are not quite long enough to quickly penetrate inches of blubber and deliver a definitive coup de grâce. Such short teeth and correspondingly blubbery necks necessitate a slug fest that usually results in a draw between two exhausted competitors, but sometimes it continues until one male is left bloodied, battered, and dead. Death can also come quickly, as when an agile frigatebird with a six-foot wingspan swoops and spins and maneuvers like its namesake sailing vessel to snatch up a chick of another species that might have a behavioral quirk that causes it, or allows it, to stray momentarily out of its nest and away from parental protection: bad move by the chick, good move by the frigatebird. But death is death, whether fast or slow, and selection is selection. Without death, selection cannot operate, and without selection, evolution is not Darwinian.

Death in a place like the Galápagos can also be the result of inherited bad genes—or, simply and ironically, the result of bad luck. Having bad genes usually implies a lineage of inferior fitness extending back several generations. Bad luck is just bad luck, like catching a red light when you're in a hurry. Bad genes are the raw material of selection in the production of new species by descent with modification; bad luck is simply a random, unavoidable tragedy. The most well-adapted fish still dies if the pond dries up.

SIBLICIDE: NATURE RED IN TOOTH AND "BEAK"

Death could also come from fighting a one-sided battle with your older, stronger sibling in the hot sun, right in the nest where you were born, where you might otherwise be safe and secure in the shadow created by your parents against the tropical sun. But in this case your parents watch and do nothing. This easily observed version of natural selection can be harsh and quickly interpreted as cruel, and it can certainly be personal on an eyeball-to-eyeball level. Like someone in a knife fight in a phone booth, the younger chick has nowhere to hide.

Darwinian struggle with the environment can entail any physical factor, such as food, climate, and space to live, but extends to other organisms in the environment as well. Other animals can mix things up violently in a variety of ways and get into serious conflict. When that conflict becomes deadly, when tragedy pulls the biological trigger, when survival depends on fitness, these are precisely the circumstances in which Darwin's mechanism is working the way he originally envisioned. Most Galápagos visitors would find such deadly scenarios abhorrent. A visit to the islands is more about biological conservation, preservation, and ecological restoration. But these harsh, natural, Darwinian scenarios play out daily in the islands during the breeding season of a particular bird, the Nazca booby. The islands' unique organisms that so profoundly move visitors are the result of these innumerable deadly conflicts, usually unseen, that extend back through all the past generations. The islands are a Darwinian showcase, and sometimes a Darwinian conflict can be difficult to stomach. But this family struggle ultimately makes Darwinian sense and ultimately vindicates Darwin, even if it seems contrary to expectation. This deadly conflict is called "obligate siblicide" in the Nazca booby in the Galápagos, formerly known as the masked booby. And a similar pattern is called "facultative siblicide" in the blue-footed booby.

Everyone has heard of homicide and suicide, and even fratricide, but siblicide? In the Galápagos, siblicide is not brotherly or sisterly love, but slow, remorseless, determined killing of a brother or sister under the hot equatorial sun. The Galápagos siblicide scenario starts innocently enough with mating parents and progresses over the next few weeks with nest building, then the laying of one egg, then another, each egg about twice the size of a chicken's egg. Visitors will see an obvious pattern of fishy-smelling white guano sprayed in a radiating star-shaped pattern in all directions for three feet around a Nazca booby nest, like a fecal supernova engulfing parent and offspring. The first egg laid hatches first, giving the chick a head start on strength and getting the drop on its sibling. That first wobbly chick with stubby white feathers starts eating regurgitated fish and squid, and steadily grows bigger and stronger and bolder by the day. Eventually an inclination—or, more likely, a genetically programed behavior extending back through many generations—will take hold in the older one's head.

By the time the second egg hatches, the battle is lost and the ending is a foregone conclusion. The first chick is already big enough and strong enough to forcefully shove the second chick right out of the guano-defined arena of life and into the unrelenting sun. Once there, the younger chick starves, dies, and dries up like an old leather glove.

But that's a brother or sister that has just been killed, and not some stranger or interloper. And the two parents, who equitably take turns incubating the eggs, fishing out at sea, then feeding any mouths that appear out of the eggs, take no action to help or hinder the second hatched chick as its older, stronger sibling unrelentingly shoves, nudges, and pokes it out of the nest. The pecking order has serious consequences in the Galápagos.

The outcome of the conflict is inevitable. If you are the younger sibling, you are going to die—it's just a matter of time. The parents remain as unperturbed by the imminent death of their smaller, younger chick as catatonic zombies. They go about their business as though they were indifferent agents of natural selection: neither merciful nor merciless. For the second hatched chick, life is short and life is brutal.

In these birds, natural selection favors the killing of one chick by another when parents produce more than one egg. But why two eggs in the first place? The second egg is an insurance policy in case the first egg fails to hatch or the first chick dies prematurely. If the parents were to raise both chicks, and thus expend a huge amount of energy to find food for both, then the parents would be more likely to die, basically from exhaustion. In the alternative scenario, the second chick hatches out successfully and is raised with no sibling rivalry or the dark, inevitable outcome.

The 1849 poem *In Memoriam A.H.H.* by the English poet Alfred, Lord Tennyson appeared some ten years before Charles Darwin published *On the Origin of Species.* One particular line in that long poem has been widely used to evoke the process of natural selection: "Nature, red in tooth and claw." As organisms interact, fight, and die, they often end up with bloody teeth and claws, like the Galápagos sea lion males that compete for exclusive mating rights with female harems. All that blood dripping down teeth and claws as animals viciously fight among themselves and compete in a dog eat dog world of struggle for dominance is pretty dark stuff, especially for anyone on a pleasant ecotourism vacation. But that's what Darwin had in mind when he envisioned natural selection. Animals struggle for anything that matters for their survival: mating privileges, food, and another day of life on a small patch of black lava on an island on the equator in the Pacific Ocean. If you don't have teeth, which modern birds do not, then your pointed beak is your weapon of choice. If you are a newly hatched and fluffy white Nazca booby, chances are your beak is going to get bloody, or else you will.

The field of sociobiology goes to great lengths to explain why you should protect your nearest relatives, especially siblings, who share as much as half of your genetic material. Those who are genetically close are much like you, or a mini-you. If you kill one-half of your own genetic material by killing your brother or sister, then isn't siblicide a bit like genetic suicide? Why does it make sense either genetically or evolutionarily to be a stone-cold killer in paradise, to be a murderer on display in Darwin's living outdoor laboratory of evolution?

The Nazca booby is involved in two of the strangest, but possibly most interesting behaviors in the Galápagos. One involves their drawing blood during siblicide; the other involves their blood being drawn by vampire finches. While these behaviors seem to fly in the face of Darwin, they actually vindicate Darwin.

How does siblicide make sense in the larger scheme of Darwinian evolution? Would it not make more sense for the parents to raise two chicks and double their reproductive success? Darwin described the struggle for existence as entailing "not only the life of the individual, but success in leaving progeny." Might the answer be counterintuitive? The answer, as it turns out, lies not in what happens right there and then in the nest with one chick living and one dying. The crucial effect lies instead in what happens to the parents the following year in the next breeding season. Common intuition might tell us that having more offspring is better, because you increase the amount of your genetic material in the world for future generations. And that is often true, but not with the Nazca booby. As it turns out, raising two chicks requires a tremendous amount of energy expenditure by both parents, with additional offshore fishing trips to supply both chicks with enough food. One or both parents usually die of exhaustion or starvation and don't perpetuate their genetic material to the same extent as the pairs whose younger chick dies.

So for Nazca boobies, it makes sense that natural selection has (1) shaped the biochemistry in their brains to allow the siblicide to run its "tragic" course; and (2) reinforced the advantage of the parental strategy of being a catatonic zombie.

An outcome of this family conflict is that each Nazca booby we see flying around, diving for fish, or nesting near a visitor trail is a murderer, a stone-cold killer that survived as a chick in an archipelago where Darwinian decisions are not always simple or pleasant. Death lurks in the background as a dominant force in the evolution of life. And murder lurks behind the eyes of every photogenic Nazca booby.

CAN I TAKE ONE HOME WITH ME?

Sea lions are a popular natural attraction in the Galápagos, both when lounging on land and when swimming gracefully at sea. They have twitchy noses and big, dark, limpid eyes that win over eager visitors, who can observe and photograph them from just a few feet away. And the pups take center stage in the impromptu show, either in the shallow tidal pools along the coast or on the sandy beaches where visitors often go ashore. With their beautiful furry skin and thick whiskers, they look like a sausage-shaped cross between a Labrador retriever and a house cat. But they, too, can present us with the face of death and tragedy.

Sea lion pups can sometimes be so emaciated that skin, rather than muscle or fat, drapes over their hipbones, ribs, and shoulder bones. We tend to be suckers for such a sad scenario. Often the pup's mother has not returned to feed it. The sea lion pup waiting desperately on the shore continues breathing slowly and intermittently, oblivious to other sea lions or the presence of human observers. The pup waits for nutrition it cannot acquire for itself—it cannot catch the fish or squid its mother hunts in deep water and will turn into nutritious and life-giving mammalian milk. At just a few pounds and perhaps just a few days old, before long the pup might not be breathing at all. When that happens, the concerned ecotourist walking by will see it on the rocks or on the trail, and smell it. The immediate thought might be what a shame that the tiny, helpless pup died. The next thought might be, is it possible to maneuver upwind to avoid the harsh stench of death and perhaps click a photo or two? Conservation and natural selection cannot always coexist.

All too commonly scattered along the white sand beaches and black volcanic shorelines are the stark white bones and folded brown furry skin of deceased sea lion pups. Perhaps their mothers never returned from an offshore fishing trip, or perhaps their mothers returned brimming with milk and optimism and could simply not locate their thirsty pups in time. Either way, the pups died. Sometimes on Galápagos shores, especially during El Niño years, visitors encounter larger sea lions washed up dead on the beach —ones that lived past the crucial stage of needing mother's milk but died for some other reason. The stark reality is that both types of carcasses stink, prompting observers to hasten down the trail or down the coast, preferably upwind. The Galápagos are not sanitized for visitors.

Darwin described how organisms change over time, how they descend from their ancestors with modification, becoming different from their ancestors over time. With enough time and enough change, a single type of organism in the islands could become two species—and two could become four species, and on and on, with one ancestral bird founding an entire lineage of birds. Before long, a place like the Galápagos Islands could be peppered with thirteen species of finches, or the Hawaiian Islands could be home to some fifty species of honeycreeper birds.

Darwin's mechanism of evolutionary change cannot sweep away its harsh results and replace them with family-friendly living museum displays, as in a zoo or public aquarium. In the Galápagos, the process of evolution looks horrendous and smells even worse.

Chances are the last thing anyone wants to think about while visiting Galápagos is the mathematics of probability. But math is the hottest topic in the ocean and on land, because evolution is all about probability. It is the probability of survival, the probability of mating, the probability of passing on genetic material from one generation to the next.

But evolution is not "random chance" like a winning lottery number or a lucky throw of the dice. The genetic variation on which natural selection works is randomly produced by mutations or recombination during cell division, but natural selection itself is *not* random. Natural selection is a process that weeds out unfavorable variations and greatly improves the likelihood of surviving, mating, and passing along genes. Walk the trails of the Galápagos and observe red throat pouches and blue feet and protean beaks, and listen to birds whistling, chirping, and squawking. All of these were shaped by natural selection as outlined by Darwin.

DARWIN'S LEGACY

Charles Darwin died in April 1882 before realizing his two greatest triumphs. First, the scientific community had not yet verified his mechanism of evolutionary change, called natural selection. The scientific experimental and theoretical verification of selection came in the 1930s from work on organisms such as fruit flies in glass tubes in laboratories in places like New York City, certainly far from the shores of the Galápagos. The eventual universal acceptance of natural selection vindicated Darwin. Scientists accomplished the vindication with many organisms in varied settings from laboratories in ivory tower institutions to muddy boots-on-the-ground fieldwork. An important part of that vindication was based on specimens collected in 1905–1906 and later studied, analyzed, and interpreted. Darwin's finches are the best example, but others include tortoises, land snails, plants, insects, lava lizards, mockingbirds, geckos, snakes, and iguanas. The menagerie was part and parcel of the vindication.

And second, the Galápagos Islands became more than just Darwin's namesake outdoor *laboratory* of evolution. Rather, the islands became a living outdoor *museum* of evolution, visited by more than two hundred thousand ecotourists each year. Designated the first UNESCO World Heritage Site in 1978, one of the largest marine reserves in the world, and a Biosphere Reserve in 1985, the islands warrant all the protection and scholarly study they can get.

NOTES

Page 81: *Many seals followed*: LSA, 35.
Page 82: *Five weeks*: Grant and Estes (2009, 7).

Page 83: *Better chance of surviving*: Darwin (1859, 3).

Page 83: *Evolution and natural selection*: Selection can operate at other scales as well—for example, sexual selection, group selection, genic selection, random drift, mutation, or migration. Natural selection is just one means through which evolution can occur, but it is the one that leads to adaptation.

Page 85: *1905 and 1906 El Niño years:* http://www.atmos.washington.edu/~mantua/ TABLES2.html, accessed July 18, 2015.

Page 88: *Obligate siblicide*: Anderson (1990a, b).

Page 89: *Raise both chicks*: Anderson (1995).

The voyage of the *Beagle* has been by far the most important
event in my life and has determined my whole career.

CHARLES DARWIN (1809–1882),
Autobiography

To young men contemplating a voyage I would say go.

JOSHUA SLOCUM (1844–1909),
Sailing Alone Around the World

9

Down to the Sea in Ships

SHIP'S LOG, 1905

DURING ITS SEVENTEEN-MONTH voyage out of San Francisco, the schooner *Academy*
crossed the equator only twice, despite spending a year and a day in an archipelago that strad-
dles the equator. The first crossing occurred when the schooner sailed south from Cocos
Island toward the Galápagos. The approach to the equator brought the vessel dangerously
close to the Ecuadorian mainland. The schooner was so close to shore that several conical
peaks of the Andes were visible. The *Academy* and her crew "crossed the line" of the equator
on September 18, 1905, with the South American continent to their left, just twenty-nine
miles off Ecuador's Cape San Lorenzo (figure 6.2). The wide expanse of the Pacific to their
right, off the starboard beam, held the crew's biological fortune.

Commenting on his own experience of crossing the line, Mark Twain made several humor-
ous observations in his 1897 book *Following the Equator*. "In the distance," Twain wryly
noted, "it looked like a blue ribbon stretched across the ocean. Several passengers kodak'd
it." Twain also related an anecdote about the "slow" progress of the sailing ship he was on: "A
sailor explained to a young girl that the ship's speed is poor because we are climbing up the
bulge toward the center of the globe; but that when we should once get over, at the equator,
and start down-hill, we should fly." As if that were true, after crossing the line, the *Academy*
built up momentum and soon flew toward Darwin's Islands.

The *Academy's* actual crossing set in motion the traditional ceremony whereby old salts
initiate first-time crossers. Beck noted the boisterous shenanigans the crew engaged in, stating in his field notes, "Neptune came aboard but was overpowered & ray necks kept from
getting shaved. I got a cut on edge of eye & others minor bruises but all right." This rowdy
ritual of "crossing the line" on the *Academy* took place only eight years after Mark Twain
described a similar crossing. Either disappointed or relieved that his own ship did not engage
in the solemn ritual, Twain wrote, "We had no fool ceremonies, no fantastics, no horse play.
All that sort of thing has gone out." Playing the role of historian, Twain observed, "In old
times, a sailor dressed as Neptune used to come in over the bows . . . and lather up and shave

everybody who was crossing the equator for the first time, and then cleanse these unfortu-
nates by swinging them from the yard-arm and ducking them three times in the sea." Gifford
wrote in his personal diary, "Beck, Nelson, White, and Slevin [the group of experienced
'Shellbacks' on the *Academy*] try to shave remainder of crew but failed." It sounds as though
the "Pollywogs" offered stiff resistance.

Despite the apparent cruelty with which some of the experienced Shellbacks might
attempt to treat the uninitiated Pollywogs, Twain thought the ceremony took on a level of
humor at sea that might seem out of place on dry land. After having passed the initiation, the
new Shellback is welcomed to the order by his Royal Highness, Neptunus Rex, and allowed
to pass to the other side of the equator.

The Academy men went ashore for the first time on September 25, 1905, seventy years almost
to the day after Darwin's first experience. During that first full day ashore on Hood Island at
Gardner Bay, they discovered an array of the "indigenous productions" of Darwin's "mystery
of mysteries"—the origin of species. Once in the islands, the men were free of seasickness and
liberated from their restricted diet of canned salmon, hardtack, and coffee. More important,
they had clear instructions from Director Loomis back in San Francisco about the business
they needed to attend to.

All hands turned out for breakfast at 6:00 a.m., and a landing party of Beck, Ochsner,
Gifford, and White, the cook, went ashore. Later, in the afternoon, Beck, Slevin, Gifford,
King, Stewart, and Hunter rowed to the south end of the island to visit the albatross colony. The botanist, Stewart, commented years later, "When one lands for the first time on
almost any of the islands, one is immediately struck with the great abundance of lichens," and
"Lichens often lend a striking appearance to the vegetation." Any visitor to the islands will
appreciate Stewart's further observation about lichens: "They often cover the lava-boulders
on the lower parts of the islands to such an extent as to give them, when seen from a short
distance, the appearance of being covered with paint."

The men on the *Academy* happened to arrive in Galápagos in September 1905, eight months
after Manuel J. Cobos and the territorial head of Galápagos, Leonardo Reyna, were killed by
angry and justifiably disgruntled "workers" on January 15, 1905. The much-despised Cobos died
from two gunshot wounds, as well as an unplanned swan dive out a second-floor window, after
being beaten with rifle butts and kicked by his assailants. His death appears to parallel the way
he treated some of his beleaguered workers, who functioned in many ways like slaves. Señor
Reyna, on the other hand, was killed after his police escort fled out the back door and he was
shot once in the neck and stabbed in the stomach, all the while pleading in vain for his life.

With murder still fresh in the air, the plantation manager took precautions with the crew
of the *Academy*. Slevin reported that he pointedly "asked us to be very careful to see that
no firearms got into the possession of the natives." The gringos were welcome to collect the
products of evolution, but not to foment revolution.

In addition to the sea lions sent rushing from the beach "clumsily into the water" with
a slap or two when woken by the men, they saw their first marine iguanas on the basalt
shoreline. The men marveled at them just as Darwin had in his time. The iguanas were scattered lazily about on the warm rocks and "brilliantly marked with large blotches of red and
green" in the characteristic coloration of the Hood Island variety. Enticing the iguanas into
the water was difficult, despite their unique seaweed diet. The crew attributed the iguanas'

reluctance to a fear of sharks lurking near shore. Standing on the beach looking seaward, the men saw several large sharks and manta rays at the edge of the breaker line. Later that day, after retrieving the shore party in the *Academy*'s rowboat Nelson, the mate, hooked three sharks from the deck of the schooner.

The extreme fearlessness of Galápagos birds impressed Darwin and the *Academy* crew alike and continues to draw birdwatchers to the islands today. "Tameness," as the Academy collectors called it, is an elusive quality in otherwise wild animals, difficult to grasp, with a puzzling origin. The expedition's entomologist, Williams, knocked down a heron with a stone, and Hunter watched the endemic waved albatrosses as the birds ignored him. Upon seeing his first albatross at sea in the southern ocean, the ornithologist Robert Cushman Murphy (1887–1973) wrote home to his wife, "I now belong to a higher cult of mortals, for I have seen the albatross!" Hunter recorded his first meeting with the mythical albatross in his field notes with less exultation. The Galápagos waved albatrosses on Hood Island exhibited unusual behavior, which Hunter documented in his field notes. "When disturbed," Hunter observed, "they would go through rather slight fencing movements with each other," emitting "a loud hoarse squawk." Having arrived at the albatross colony at the end of the breeding season when adult birds were common and "rotten eggs are equally so," Beck and Hunter wandered about the open area where pairs of birds sat quietly on nests (figure 9.1). "When mad," Hunter continued, "the feathers over the eye are slightly raised giving considerable expression to the face." As fascinating and exotic as the birds were for any of the wide-eyed Californians, the objective was to acquire museum specimens. Moreover, a large quantity was needed. To this end, Hunter reported, "about 25 old birds were taken."

FIGURE 9.1 **Bird Eggs.** Rollo Beck (right, with shotgun) and Joseph S. Hunter collecting birds and bird eggs. *Archives, California Academy of Sciences.*

The Galápagos albatross colony on Hood Island was in pristine condition in 1905 and has remained largely so in the hundred years since this expedition, although with fluctuating numbers of birds in recent years, possibly due to by-catch in long-line fishing.

The Academy collectors observed these active and inquisitive birds on many occasions, but never as intimately as on Hood Island. Galápagos mockingbirds provided Darwin with arguably his best example of species differentiation on separate islands. The Hood mocking-birds were particularly aggressive and predatory toward each other, while being trusting and inquisitive about humans, as the men soon learned. While on Hood, Slevin noted that they were "particularly trustful and hopped about the ground within a few inches of one's hand while we were engaged in turning over pieces of lava in search of insects." Their tameness was further demonstrated at lunch when "they would even alight on one's knee and pick up the crumbs dropped." Hunter noted, "Mockers will eat almost anything in the shape of meat. One carried away a lizard that had been killed that was nearly as large at itself." The essayist Barry Lopez had a similar experience nearly one hundred years later when he wrote, "Mockingbirds snatch at your hair and worry your shoelaces—you are to them but some odd amalgamation of nesting materials."

Darwin was an experienced collector and knew how to use his shotgun to get bird specimens. He was impressed by the tameness of Galápagos birds and commented at length about this unusual phenomenon. "I will conclude my description of the natural history of these islands," Darwin wrote ending the Galápagos section of *The Voyage of the Beagle*, "by giving an account of the extreme tameness of the birds." Darwin found that the land birds "often approached sufficiently near to be killed with a switch, and sometimes, as I myself tried, with a cap or hat." For Darwin, a bird in the hat was worth two in the bush, and he further noted, "A gun is here almost superfluous; for with the muzzle I pushed a hawk off the branch of a tree." However, Darwin believed the birds' increasing shyness over time was due to excessive killing by bucca-neers and whalers, because "sailors, wandering through the woods in search of tortoises, always take cruel delight in knocking down the little birds." With the ongoing and increased "persecu-tion" of these birds, Darwin was impressed that they hadn't become more fearful.

The finches could "in many cases be caught with a butterfly net," while ground doves could be "knocked over with a stick." At the top of Hood Island, Beck "killed a hawk with a stick" and found centipedes in its stomach. Elsewhere in the Pacific, a similar phenomenon was noted in 1903 and 1911 on Laysan Island in the northwest Hawaiian Islands by the zoologist W. A. Bryan, who found that "there is not a species on Laysan that can not be caught in the hands with a little care and patience." On Laysan, the tame birds impressed Bryan all the more because hundreds of thousands of birds had been slaughtered there by poachers. Even in the face of extermination, the Laysan birds remained predictably tame.

The *Academy*'s herpetologist, Slevin, extended his observations of tame species to include the reptiles. "What is said in regard to the tameness of the birds can also be said of the liz-ards." Slevin found that the little lava lizards in the genus *Tropidurus* could be "killed with a switch and, like the birds, would come and pick up crumbs that dropped to the ground under one's feet." Hunter observed that after he severed the tail of a lava lizard with his pock-etknife, the lizard promptly spun around and ate the wriggling remnant, "no doubt think-ing it was some insect." In addition to all the other specimens the men brought back to the schooner, a couple of sacks of live marine iguanas were placed in a cage prepared for them.

Here they scrambled around for a few days, but they traveled back to San Francisco in alcohol. Catching the elusive marine iguanas was tricky, until you got the hang of it. Slevin found them to "cover the ground fairly fast when chased rapidly," and his strategy was to "approach them slowly, and run when nearly on top of them." He would have gladly collected additional marine iguanas, except that he had "no barrels handy and no place to put them." As a compromise, he vowed to "lay in a few skulls as soon as possible."

All this tameness allowed for productive collecting, and "everybody enjoyed their first day ashore and came back laden with specimens." That first day's bounty kept the men busy on the schooner well into the evening, skinning birds, preserving lizards, and pressing plants. Loomis would have been proud of his industrious young men.

SEPTEMBER 26 TO OCTOBER 2, 1905

For the next seven days, the schooner remained anchored in the protection of Gardner Bay while the men worked ashore on their specialties or stayed on board to finish up the latest round of specimens. The men took occasional side trips to nearby Gardner Island, just north of Gardner Bay. A pattern of daily activities developed, with the members of the expedition assuming their assigned roles, focusing on their particular group of organisms, and finding activities that suited them, both on land and on the schooner. At times, everyone pitched in and collected whatever crucial specimens were readily at hand. Stewart remained an exception. He rarely engaged in collecting anything but plants, somewhat to the chagrin of his more flexible shipmates. Beck and Gifford not only were active collectors of birds and reptiles, but also specialized in photography.

Slevin added to the collection of small reptiles that John Van Denburgh, the physician-curator who had stayed behind in San Francisco, subsequently described. Slevin's specimens of lizards, snakes, and geckos found by "splitting open the trunks of dead trees" formed the basis of three papers published by Van Denburgh between 1912 and 1913. By the second day on Hood Island, Slevin had collected seventy-five lava lizards using a stick, "which was the weapon for collecting." Undoubtedly due to Darwin's influence, he also started noticing inter-island differences in the lizards. The females on little Gardner Island were smaller than those on larger Hood Island, and "the coloring under the throat a little darker color, almost chocolate." Evolution was all around. All the men had to do was look.

Centipedes and land snails gathered by White and Ochsner became part of the invertebrate bounty. The vegetation of Gardner Island was less dense than that on Hood, but the fauna was about the same, minus the goats. At the summit of Hood, Rollo Beck reported finding a few old tortoise bones, but no live animals. Yet the fast-disappearing tortoises were one of the main reasons for this expedition, and consequently everyone was on the lookout.

Nelson continued the repairs and maintenance on the schooner by tarring the rigging. At night, Parker kept anchor watch to make sure they did not drag their anchor and drift perilously close to shore. Later in the expedition, staying off the rocks would not be so easy.

The first mosquitoes of the expedition, but certainly not the last, made their appearance that first week. The temperate weather continued, with fine, clear evenings. The weather was perfect for collecting during the day and working on specimens under the yellow glow of a kerosene lamp in the evening.

Foraging for food on land was a daily event and a necessity if the men wanted diversity in their diet. Goats remained the most common source of meat, and pigs were added if they could be found. Writing about the 1905–1906 expedition eighty years later in a biographical sketch of entomologist Francis Xavier Williams, Howard Evans wrote, "They often 'lived off the land,' shooting and eating the feral goats and pigs as well as devouring native fish, doves, flamingos, ducks, and tortoise livers." While they were on Hood Island, a particular gastronomic specialty appeared on the menu for the first time. The men went ashore and killed numerous Galápagos doves with a firm swat from a stick. Thus chosen by means of "natural selection," these specimens were brought back in a burlap bag and turned into dove pie by White, the cook. In fact, the men killed so many doves in the islands during the year that Hunter mused in his bird notes that it was a wonder they didn't make them extinct. The occasion of a warm "dove pie supper" appeared to have been a rare and welcome treat for the men. Although far from home, they still needed their comfort food!

Without refrigeration on the *Academy*, the men had to hunt almost every day. As soon as they landed on Hood Island, Ochsner set off in search of goats, and he quickly returned with "two kids and two big bucks." Galápagos goat meat was "a most acceptable addition to the bill of fare," because no fresh meat had been served on the schooner since she had left Cocos Island three weeks before. "We now have goat meat every night for supper," Slevin reveled, "and cold goat for lunch."

Interactions with the wildlife continued on Hood Island whenever the men went ashore. Williams and Ochsner encountered a large bull sea lion reminiscent of the ones that had formed a similar welcoming party on their first day at Gardner Bay. While sitting in the stern of the skiff, which they had run up on the beach, the dominant "beach master" rushed to within a foot or two of the stern in an obvious display of macho territoriality. Williams hit him over the head with an oar, causing the startled sea lion to dash for the water. Even mother sea lions valiantly defended their pups by snarling and making short dashes at the intruders. Tameness went only so far, even in the Galápagos.

During their yearlong visit in Galápagos, all eleven men on the *Academy* worked extremely hard, at a level that would exhaust ordinary men. Through all of their hardship and strenuous collecting trips ashore, the men complained remarkably little in their letters home to friends and family, field notes, or personal diaries. They knew they had a job to do for Loomis and the museum, and they did just that.

Their weekly reprieve from collecting specimens came on Sundays, a day reserved during the expedition for catching up on all things unattended to during the week, including sleep. Whenever the *Academy* was anchored on a Sunday, and not actively under sail, the crew took it easy. This weekly reprieve probably had little to do with the religious significance of Sunday, even though the expedition leader, Rollo Beck, was a devout Methodist, and more to do with the tradition of resting on Sunday. During the "work week," their physical exertion in the equatorial sun sometimes brought them to the brink of exhaustion. At times they were probably so tired that they forgot their names. If a Sunday reprieve was missed because they were under way (and obviously completely occupied with sailing the schooner), they did *not* automatically take a day of rest at the next anchorage. Instead, they kept up their rigorous pace of collecting ashore, sometimes also skipping the following Sunday's rest if they were heading toward their next destination. So they rested only on Sundays, when it was convenient, rather than arbitrarily during the week. Yet Sundays were not sacred, either.

As with other aspects of the expedition, the men quickly developed a pattern during their well-deserved Sundays. In his published log, Slevin frequently noted that the men were "variously employed" on Sundays, usually aboard the schooner, but sometimes also ashore, although specimen collecting was kept to a minimum. Almost without exception, they had a late, leisurely breakfast on Sunday. The remainder of the day they variously spent overhauling their collecting gear, repairing shoes that had been torn up on the sharp lava rock, and "putting up" the remaining specimens from the past week.

Beyond the general routine, certain members of the expedition fell into personal routines each Sunday. Very commonly, Rollo Beck and Edward Gifford went ashore and took photographs. Williams fished, usually successfully, from the decks of the schooner. Hunter became the crew's shoe repair specialist. Parker, Nelson, and Ochsner frequently hunted ashore for goats or pigs, and they were as successful as Williams at fishing. Sometimes Nelson, White, or Parker killed a sea lion or two and rendered their oil.

Sunday meals were often more elaborate than those on the other days, especially if fresh meat or fish was available. Following their late breakfast, they often had "Sunday dinner at 3:00 p.m. and supper at 8:30" in the evening.

Without an auxiliary engine, the schooner was a slave to the wind. Transiting from one island to the next had to be coordinated with favorable breezes. Numerous poorly known anchorages in the islands could potentially turn disastrous if wind and swell conspired against them. The schooner had twice previously suffered the indignity of being battered by the wind. When still named the *Earnest*, she sank in shallow water off Isle au Haut, Maine, in 1876, and was tossed unceremoniously aground on Goat Island in San Francisco Bay in 1904 (figure 9.2). In the Galápagos her crew needed to be hypervigilant. The lack of wind often proved a greater problem than excess wind. Poorly charted shallow "reefs" had to be deftly negotiated. When sailing at night, the men quickly learned that an entire island could loom up dead ahead in the darkness, unheralded except for the absence of the stars on the horizon that were blocked out.

Depending on the wind, which governed so much of their activity in the islands and the distance to be traveled between anchorages, the collectors' departure from one island might be preceded by fieldwork ashore for a few hours if time, tide, and wind allowed. Under favorable conditions, they hoisted anchor and set sail at or just before dawn, or even in the early evening if they planned to sail at night.

For example, their departure from Hood Island on October 2, 1905, was preceded by a full day ashore collecting. Williams got his "usual catch of insects," along with a snake and some geckos for the herpetologists. Others rowed a skiff to the albatross colony near Punta Cevallos. Beck provided fresh meat by shooting two more goats.

Nelson and Parker took down the canvas awning they had constructed for shade and generally busied themselves as they "made the vessel ready for sea." Leaving Gardner Bay, the crew raised the anchor at 8:30 p.m. and the schooner "shaped course" for a second Gardner Island off the southeast coast of Charles (Floreana) Island. During the night they safely made the passage across the channel between Hood Island and Charles Island. But their island goal was sighted at 2:30 a.m. Not wanting to get too close in the dark, they had to "tack the ship and hove to" a safe distance offshore until daybreak. From 7:00 a.m. to 9:30 a.m. they made their way up to Gardner, then anchored and landed the shore party for fieldwork. Nelson manned one of the schooner's skiffs, or "pangas" as they are called in the

FIGURE 9.2 **Grounded Like a Beached Whale.** The schooner *Earnest* went hard aground on Goat Island, now Yerba Buena Island, in San Francisco Bay in April 1904. Damage from this grounding caused the US Navy to put the vessel up for sale, and she was purchased by the California Academy of Sciences. *US Navy Historical Center, Washington, DC.*

islands today, and got Beck, Williams, Hunter, Ochsner, King, and Slevin ashore, but not without incident.

What occurred next, while not resulting in serious injury, was one of the worst "wet landings" the men made in the islands. They found that a "rather high surf [was] running" that day as they prepared to reconnoiter the new island. Carefully approaching the wave-swept shore, all the men scrambled out of the skiff onto a narrow ledge, and Nelson held the small boat just offshore to await their return. The narrow ledge was at the base of a sheer cliff, which they followed a short distance to a steep gully that appeared to be the route up to the top of the island. Joe Hunter watched as King and Williams took a higher path, then followed the others to the gully. While waiting for them to scramble up the gully ahead of him, Hunter was grabbed by the ocean. "An extra heavy swell came in reaching to my knees," Hunter painfully recalled. "It was followed by another still higher, the first one reached above my knees and the second reached to my arm pits."

Unable to hold on as the forceful water receded, Hunter was at the mercy of the sea. "It carried me off my feet and I started to slide on the slippery rocks back into the ocean. I tried to get a hold with my fingers but the water kept dragging me back. Just as I yelled to the other fellows that I was gone, my toes caught in a crack and I stopped sliding." More embarrassed than hurt, Hunter took only a few seconds to crawl up the slippery rocks to a safe place. But the damage was done, and when the water receded, Hunter realized he had lost both his prized gun and a smaller gun called an "aux." Ochsner nearly experienced the same fate, but he was a little farther along the cliff and managed to grab a projecting rock that kept him from being carried into the water by the surging swells after he was swept off his feet. The others got thoroughly soaked, but stayed on their feet.

A quart low on pride, and without his gun and aux, Hunter managed to get to the top of the island with King after scaling the cliffs rather than taking the chance of getting back in the skiff. The others still needed to land at a safe place on the island, but first they had to get off the island and back into the skiff to make their way around. This retreating maneuver also nearly ended in disaster. Getting into the skiff, Beck almost caused it to capsize when the keel caught on a rock. Luckily the boat righted itself as the water rose with the next swell.

Rowed by Nelson in the skiff, Beck, Williams, Ochsner, and Slevin landed safely on the northeast side and made it to the island's summit without incident. As they climbed upward, they found the north side covered with cacti and brush, along with abundant frigatebirds, doves, mockingbirds, and ground finches. Hunter's knowledge of the mockingbirds of this tiny island revealed much about the changes taking place in the islands, even then. After just two weeks in the islands, Hunter was starting to make some inter-island comparisons. "The mocker here," Hunter noted, "is not as inquisitive as the species on Hood." On a theme that would be echoed by conservationists a century later, Hunter presciently noted, "It is the species that formerly was common on Charles Island but has been practically exterminated on that island." These were insightful comments from someone who had not yet even visited Charles Island. The knowledge possessed by these intrepid collectors regarding rare and dwindling species shows a depth of understanding not usually attributed to the members of this expedition. Reading the published literature and talking with Beck about his three previous Galápagos trips constituted the men's preparation as they moved from island to island. Also on Gardner Island, Slevin managed to collect a snake and "the usual number of lizards" before all hands returned to the schooner at 1:30 p.m.

The next destination was the even smaller Champion Island off the northeast coast of Charles. Without a suitable anchorage, the schooner "lay to" a short distance from the island while a landing party went ashore. While Williams stayed on board, Slevin collected some beetles, spiders, and centipedes for him, in addition to his own lizards. Beck and Gifford brought back a few young blue-footed boobies. After an hour ashore, the landing party rowed back to the schooner, and the crew "shipped course" at 5:00 p.m. for Post Office Bay on Charles Island, arriving there an hour later.

From their anchorage in four and a half fathoms of water, the schooner was cradled by the gently curving coastline of Post Office Bay. From the deck of the schooner, the men looked northward and saw several of the larger islands in the archipelago. They would reach all those islands in due time, but for now they had letters to mail.

OCTOBER 4–6, 1905

After an early breakfast, Beck and Nelson rowed off to visit a saline lagoon on the north-east end of the bay, at Punta Cormorant. The rest of the party landed directly opposite the anchorage and walked straight inland across the beach for an encounter with history. Although their mission was scientific, here they intercepted the long and complex human history of the Galápagos. Human footprints on this remote beach presented a clear message: wind and tide had not yet erased the footprints of recent visitors. Just behind the narrow beach by some thirty meters, but back far enough not to be immediately visible, was the old barrel post office for which the bay was named.

Ranking among the most unusual "post offices" in the world, this sunbaked letter repository consisted of nothing more than "a barrel erected on top of a post and painted red." Upon seeing this unique service some 14 minutes of latitude south of the equator, Slevin began what would become a lifelong fascination with the history of the "Barrel Post Office of the Galápagos." This stalwart repository was not diminished by rain or wind or dark of night, and most distinctively it did not require numismatic attachments of any sort. By the time Slevin's *Log of the Schooner Academy* appeared in 1931, he had thoroughly investigated the barrel. During their visit in 1905, the men of the schooner *Academy* not only were surrounded by maritime history, but were making their own mark on it. An inscription on the upright post holding the barrel read: "Erected by H.M.S. *Leander*." Slevin wrote in 1931, "Crew of various vessels calling at this anchorage had painted or carved the names of their vessels on the post or barrel." The men found the names of His Majesty's ships *Amphion* and *Virago*, the French cruiser *Protet*, and the USS *Oregon*.

Among the scientific predecessors of the schooner *Academy* that had left their mark at Post Office Bay was the US Fish Commission steamer *Albatross*, which had "joined the hunt for Galápagos specimens" with official visits in 1888 and 1891. Alexander Agassiz, the son of the anti-evolutionist Louis Agassiz, and successor to his father at Harvard's Museum of Comparative Zoology, led the 1891 expedition.

The idea of the post office barrel went back to whaling days when ships visiting these waters made voyages of a year or more. Any homeward-bound ship picked up the letters left in the barrel by other vessels and marked them as "ships mail" for delivery in the United States without postage stamps being attached. Edward Gifford put this tradition to the test. He composed a letter to his father in Alameda, California (just across the bay from San Francisco), and placed it confidently in the barrel. Only after the return of the *Academy* to San Francisco did Gifford and the other men learn of the circuitous route the letter had taken. Picked up by the British yacht *Deerhound*, the letter was forwarded to Washington, DC. But rust from the iron barrel hoops had obliterated part of the address so that only "Gifford, Alameda" was visible. The conscientious workers of the US Post Office traced the addressee and delivered the letter, but also wanted to know the letter's origin. They had to wait until the schooner returned to learn all the details, which Gifford duly provided.

The highly modified ecosystem of Charles Island in 1905 became evident as the men climbed a six-hundred-foot hill and saw cat tracks and wild cow pies and heard the harsh braying of a donkey. The negative effect of introducing animals to the island was clearly visible. "The cats have about exterminated the *Tropiduri* [lava lizards]," lamented Slevin, "and the wild dogs the sea iguanas on Charles Island." A flock of skittish wild goats that Slevin thought had been "driven away from the fertile part of the island to this dry and barren portion" constituted further evidence of the changes brought to Charles by exotic species. Humans brought their animals, and these animals changed the islands. The forces of deadly competition in Darwin's "struggle for existence" had been partially transferred away from the natural inhabitants. Moreover, these recent domestic arrivals now played out the evolutionary drama of "the survival of the fittest." Despite the natural or human-induced rarity of certain native species in the islands, they remained fair game for the collectors. Cats had apparently decimated the lava lizard population on Charles, but Slevin nevertheless collected as many specimens as possible. "I also got five lizards," he recalled after a beautiful calm morning of collecting at the saline lagoon at Post Office Bay, "good luck as they are very rare on Charles Island." For Slevin, this chance encounter with a rare species meant an opportunity to enhance the museum collection back in San Francisco. Even though they were rare, his job was to collect them anyway.

Back at the beach, all hands enjoyed a swim after "a hard day bucking through the brush." At least two of the men could not swim, however. It was very likely that Rollo Beck stayed in the shallows, along with anyone else uncomfortable in deeper water.

The intrinsic beauty of an organism did not keep it from appearing on the dinner table. During their collecting trip to the small saline lagoon at Punta Cormorant, Beck and Nelson shot seven flamingoes. Hunter commented on this exotic gastronomic treat by stating, "Had fried flamingo for supper. It has a flavor much like wild duck but stronger." While flying over the schooner, the flamingoes let out a honk that "sounds much like a Canada goose." Slevin could not resist bagging a bird to add to the collection. He found that "the ducks are very tame and I killed one myself with a 22 rifle and dust shot cartridge."

Hunter continued having bad luck with guns. "I did not shoot any small birds on account of losing my aux." Beck became exasperated by Hunter's clumsiness and let off steam in his field notes, writing, "Hunter lost another aux today, making three he's lost this week!"

The inherent beauty and extreme tameness of the Galápagos fauna did not deter the men from their single-minded task of collecting. They dissociated intrinsic beauty from scientific necessity. They easily spent a few minutes carefully observing a living animal and its complex behavior, then dispassionately shooting it, adding it to their collecting bag, and moving on down the trail. Joe Hunter's comments about ducks and flamingoes at Cormorant Bay on Charles Island were typical: "There were a nice bunch of ducks around. I killed eight of them, two got away." For Hunter, "nice" had greater numerical significance than an aesthetic one. Likewise, regarding the long-legged pink flamingoes still easily seen today in the lagoon, Hunter wrote, "A flock of the bright plumaged birds feeding in the lagoon is a very pretty sight." Hunter got within seventy-five yards of the flamingoes before he "killed two of them with the first barrel and the third with the second." Modern readers, sensitized to the plight of species everywhere, might cringe when reading Hunter's comment: "The flamingo is a slow and graceful bird under ordinary conditions but when forced to run its legs get tangled up in the most amusing manner."

Having finished their collecting in the vicinity of Post Office Bay and Cormorant Bay by late morning on October 6, 1905, all the men returned aboard; they weighed anchor at 11:30 and set sail for the short passage to Black Beach Anchorage. The men sailed the schooner "by the wind" on the west side of Charles Island. Once at Black Beach, they intersected the tragic trajectories of many facets of the human history of the islands.

After a three-hour sail they arrived at the Black Beach Anchorage or Black Beach Roads, as it is sometimes called. The men anchored the schooner in eleven fathoms along with a small vessel described as a sloop from Chatham that had brought over a hunting party to kill wild cattle and "jerk the beef" for the settlement. The temporary hunter's camp, with its ragtag group of men drying meat in the sun and hoping for whiskey, grew into the small village of Puerto Velasco Ibarra, as it is known today. The Black Beach Anchorage was adequate in calm conditions, but as it was "merely an open roadstead," heavy winds caused problems with anchor dragging.

At the north end of the volcanic black sand beach, a wooden cross was firmly planted in the sand. The cross bore no visible inscription, and the men wondered about its significance. Was it a grave marker, or did it have some other meaning? Near the cross began a trail leading to the interior of the island. Once ashore, Beck, Ochsner, and Nelson did not let their lack of knowledge of Spanish get in the way of interacting with the locals in the hunting camp. Discerning that the word "whiskey" was the extent of the hunters' English vocabulary, Ochsner produced some for the thirsty men from the schooner's "medicine chest," which the hunters "seemed to relish beyond everything." In exchange, the hunters provided the *Academy* with juicy oranges, directions to the freshwater springs located farther inland, and a penguin stuffed with sand. Having mastered the art of making friends and influencing people, the scientific collectors planned to stay on the island for several days. Further diplomatic relations occurred when the hunting party's "major domo" came out to the schooner with two sailors bearing fresh beef.

OCTOBER 7, 1905

Despite the highly altered ecosystem of Charles Island, which had been inhabited since the early 1830s, Ochsner collected numerous land snails that covered the trunks of small trees. All the men enjoyed "a feast of oranges, which were sweet and juicy." While eating lunch in a grove of lemon trees, Slevin delighted in "seeing the many little birds hopping around in the nearby trees, without the slightest fear, and eyeing us with the greatest curiosity." The birds' fearlessness made them easy candidates for inclusion in the growing museum collection, as Hunter noted: "The two little flycatchers are common. They are so tame that they can be killed with short sticks." Even without his gun and aux, Hunter was able to add to the collection.

In addition to the island's fauna, Hunter observed the pattern of altitudinal vegetation changes: brown and dry near the coast, wetter and greener in the highlands. The expedition's botanist, Alban Stewart, was the first to publish on this connection between rainfall and island vegetation. Hunter wrote of these early observations: "There is a belt of vegetation from the shoreline to an elevation of about 600 feet of cactus and thorn legume trees. Above this to 1200 feet is a belt of different vegetation and still higher to 1500 to 1600 still another and then from about that elevation to the top of the island there is a grassy belt. The top is 1750 feet."

Good collecting during the day sometimes led to good eating at supper, and such was the case on board the schooner at Black Beach. With all hands aboard at 6:00 p.m. for a big Saturday evening meal, the cook served up "beef soup, teal duck, flamingo, corn-starch pudding and ripe oranges." This spread was "quite a variety for a desert island," and Slevin further commented, "Flamingo tastes somewhat like duck but is a little drier and more oily."

The Galápagos might have been a collector's paradise, but they were also witness to a complex human drama. Over Sunday dinner, the Academy men learned from the chief of the local hunting party just how violent these islands were. The hunters informed them that Manuel J. Cobos (whom Rollo Beck had first met on his 1897–1898 trip to the islands), cruel owner of the veritable slave colony called "Hacienda El Progreso" on Chatham, had been brutally killed in a revolt on January 15, 1904. As it turned out, extending the squalid equatorial human drama of survival of the fittest, one of the men responsible for killing Cobos was on the beach just then with the hunting party. The crew of the *Academy* concluded from the looks of the hunters that they were "a bunch of desperados." Upon hearing that one of the hunters had been killed in a squabble a month ago, the collectors felt justified in their conclusion. The only woman among the hunters was the camp cook, who looked "like she could cut anybody's throat with pleasure."

Having no news yet from San Francisco, a few of the men on the *Academy* wrote letters, which were delivered by Nelson to the "Chatham sloop," to be taken to Guayaquil when the trading schooner left for the mainland. This set in motion a series of events and a change in plans for the *Academy*. The hunters returned to Chatham in their sailboat, and two days later they returned to the Black Beach Anchorage with the secretary of the governor on Chatham and a small escort of soldiers. The secretary brought the men of the *Academy* letters from home and newspapers from San Francisco. It was the first news from home they had received in 107 days.

With a Jamaican serving as translator, the secretary and the captain of the guard were entertained aboard the schooner at dinner on October 12, 1905, and "one would think, judging from their appetites, that they had never anything to eat in their lives before." Beck received a letter written in Spanish stating that it was customary for vessels visiting the islands to call at Chatham Island first. After trying and failing to devise a means to circumvent this bureaucratic request, Beck, a reluctant diplomat, decided an appearance with the governor at Chatham was in order. Early in the expedition, the schooner had visited the coastal town of Ensenada in Mexico to go through the proper channels and obtain permission to collect on the offshore Mexican islands, but apparently the men felt no compunction to do likewise in the Galápagos. They either did not understand or did not respect Ecuadorian sovereignty over the islands. Perhaps Beck's three previous collecting trips in the islands led him to believe that seeking permission was not necessary. Or on his previous trips there was no authority figure, such as the governor, present in the islands to demand that the collectors seek permission before starting their work.

OCTOBER 13, 1905

Beck decided after going ashore and talking again with the secretary that it would be prudent to go directly to Chatham Island as requested by the governor rather than to continue on to Albemarle (Isabela) Island as planned. Committed to this detour, the men made ready for sea. They took aboard the entire group from Chatham, including the secretary, captain, interpreter, and five soldiers. Slevin facetiously noted that the soldiers "made a fine looking army. One had an old French army hat about the vintage of 1880, some had no hats nor sandals, but were the proud possessors of undershirts and trousers, which seemed to be the uniform of the day. Their firearms were about the same period as the uniform hat and were old bolt-action rifles. One had a double-barrel shotgun, the firing pins of which were rusted fast. No ammunition at all was in evidence. No doubt it is too dangerous to let the soldiers get hold of any cartridges." The Galápagos were so far off the beaten path that the Ecuadorian government did not invest in the military force stationed there. The islands were, for organisms as well as soldiers, "cursed by God and Nature." Slevin had more respect, perhaps facetiously, for their commanding officer. "What the army lacked," Slevin continued, "the commanding officer supplied. He was dressed in a light blue uniform of French design, with red epaulets, red stripe down his trousers, a huge red pompon on his hat, and a cavalry saber that, for size and polish, was the last word." Yet in Beck's view, this military puffery was merely an impediment to his single-minded mission of collecting specimens.

After breakfast with the secretary and army captain on board the schooner, the crew brought the remainder of the "army" aboard, weighed anchor, and set sail for their begrudged diplomatic rendezvous on Chatham. Nelson and King turned over their "stateroom" to the secretary and army captain, who were sarcastically wished a good night, "trusting the bedbugs would not drive them out before daybreak." Fickle winds prevented a direct northeast passage from Charles Island to Chatham Island, forcing those on board to spend most of that night and the next day drifting just south of their destination. They made no headway against the steady current and decided to tack the ship away from land to await the wind. At sunrise, light airs and calms continued to retard the schooner's headway, and Beck put out the skiff and shot about thirty shearwaters. Calm weather was the optimal condition for collecting more of the offshore tubinares for Loomis to study back in San Francisco. Eventually, a breeze sprang up, and the schooner made her way up to the anchorage at the ominously named Wreck Bay and a settlement called Puerto Chico, which is known today as Puerto Baquerizo Moreno.

After the schooner was anchored and he rowed ashore, Beck jumped through the necessary hoops to secure permission by paying an official call on the governor. Beck chafed at the idea of kowtowing to the major domo of the islands and planned to make this visit as perfunctory as possible. Yet ironically, both Beck and the collectors befriended the governor and others on this island, or vice versa, and pleasant return visits to Chatham took place several times over the coming year. The men found the coastal settlement at Wreck Bay to be a humble collection of buildings consisting of nothing more than "a warehouse, the plantation manager's house, several native huts and the light keeper's house." The men on the schooner *Academy* were mariners dependent on any available aids to navigation. For them, the Wreck Bay light was a disappointment. As important as the light was for safe navigation, and given the name of the bay it marked, they found little more than "a lantern placed on the top of a pole." It was a simple structure in an archipelago where everything was simple. How this "beacon" helped guide a vessel safely into the bay remained a mystery. From their anchorage in the bay, the crew saw the sugar plantation and upper settlement called El Progreso, located about five miles inland.

OCTOBER 15, 1905

All hands were up late for Sunday breakfast. Beck and Williams went ashore to speak with the Italian plantation manager, who spoke in French with Williams. After lunch with the manager and some of the engineers from the sugar mill, the whole group migrated back to the schooner for a visit. One of the engineers, a German named Sing from Guayaquil, befriended the collectors. The manager was very cordial, giving permission to the *Academy* crewmembers to hunt on the island and inviting them to visit the upper settlement and sugar mill at El Progreso.

Beck, Williams, and the manager rode up to the hacienda on horseback to visit the island's governor, while Slevin, Gifford, and White slogged up through the mud. En route to El Progreso, they noted green and luxuriant vegetation. The warm ocean and tropical atmosphere had conspired recently to bring plenty of rain. Although not yet part of the common scientific lexicon, an El Niño event of moderate intensity had influenced the islands in 1904–1905, and the effect on the vegetation was still evident. The luxuriant growth of Galápagos wild cotton, which was in full bloom with "large yellow flowers not unlike those of the Mariposa Lily," was very noticeable. The group also got their first exposure to the distinctive *garúa* season that normally dominates Galápagos weather from July to December. "A misty rain began falling at about 500 feet," Hunter wrote in his field notes after his excursion to El Progreso, "and continued until we reached the settlement."

The walking group reached the late Manuel Cobos's erstwhile slave colony of El Progreso after an hour and a half. The settlement consisted of several corrugated iron buildings along with the sugar mill as its prominent centerpiece. The evidence of "progress" at El Progreso was in short supply, given the simple grass huts of the plantation laborers and the abundant stories fresh in the minds of its inhabitants of the years of tyranny perpetuated by Cobos. The men learned that while the sugar mill had a monthly capacity of fifteen hundred bags weighing a hundred pounds each, a total of thirty-six hundred bags had been shipped to Guayaquil the month before.

Although Manuel Cobos was dead, an escort took the men of the *Academy* to the office of another, more important figure in Ecuadorian history. The governor of the archipelago who had summoned them was none other than General Leónidas Plaza Gutiérrez, who had been president of Ecuador from 1901 to 1905. Although relegated to political exile in the islands in late 1905, General Plaza would regain the presidency in 1912, only to be replaced in 1916 by the namesake of the town (formerly Puerto Chico) located at Wreck Bay, Alfredo Baquerizo Moreno. The name "Plaza" is remembered today in the islands mainly for two small islands off the west coast of Indefatigable (Santa Cruz) Island. General Plaza named Isla Plaza Norte and Isla Plaza Sur in honor of his two daughters.

Although General Plaza was accustomed to the splendor of the presidential palace in Quito, his new office was "in an old shack of a building constructed of corrugated iron" and his desk nothing more than an old kitchen table with the drawer missing. In tropical fashion and in a style befitting a former president, he wore a white duck suit and a navy watch cap, likely donated by a sailor off a visiting warship. General Plaza treated the collectors with great respect. After conversing with them through the Jamaican interpreter, he readily granted them the permission they sought to continue their work in the islands. The twenty or so soldiers at El Progreso, no two wearing the same uniform or cap, were turned out barefoot to drill for the benefit of the Academy group. After Beck took two photos of the military parade in the drizzly rain, he was impatient to get back to collecting. Beck kowtowed only when necessary; in this case it meant free access to the biological bounty of the islands.

General Plaza left a strong and lasting impression on the young collectors by giving them a generous sample of his very strong home brew, or *aguardiente* as it is known in Spanish. As Plaza observed the tears streaming down their cheeks, he remarked that the brew might be a "little strong" for inexperienced drinkers. "Not only was it a little too strong," the twenty-four-year old Slevin painfully recalled, "but much too strong for any human being." Concerned for the longevity of his tongue, esophagus, and stomach, Slevin took particular pains not to be included in any more official visits with General Plaza and celebratory toasts with his firewater. "Even now, 25 years afterwards," Slevin wrote in his 1931 *Log of the Schooner "Academy,"* "whenever I think of General Plaza I can almost feel my insides on fire."

While not impressed by the drinking ability of the men, General Plaza was very supportive of the expedition's scientific goals. The Jamaican interpreter informed Beck and the men that "the General says he can tell from your looks that you are gentlemen." This remark flattered the group of young Californians who had been without a shave for 110 days and without a bath since their stop at Cocos Island more than a month before.

OCTOBER 16, 1905

During the entire voyage, everyone on board the *Academy* came to appreciate mate Frederick Nelson's ingenuity and maritime skills. Nelson devised a way to make more substantial side trips from the schooner by constructing a "spritsail" rig and mast for the ship's boat, a longer and more substantial auxiliary vessel than the other two rowing skiffs. With sail power added to their repertoire of skiffs that otherwise had to be rowed, overnight trips to more distant locations became possible. The schooner could be anchored in one location while the ship's boat sailed elsewhere on a tangential collecting trip. Giving his nautical handiwork a trial run, Nelson tried out the new sail and found it satisfactory. He picked up the shore party from Chatham at the end of the day and sailed everyone back to the schooner.

Injuries to the collectors were either rare or rarely reported. For example, Beck recorded in his field notes on October 17, 1905, "Stewart not collecting as sore toe prevails." Stewart's own field notes from the expedition are apparently lost, and one cannot verify his version of the injury.

While walking along the dirt road from Puerto Chico to the hacienda at El Progreso and in the wooded areas near the two settlements, Hunter noted an avian disease that perhaps accounted for the anomalous scarcity of birds. He noted that "many, in fact nearly all, have the peculiar disease of the feet that I have observed in the States. Mockers are very badly affected."

Hunter had mixed success interacting with the residents of the island. He struck out when he "tried to buy some things at the store . . . they do not feel like doing business." At least he was welcome at the hacienda, where he had dinner with the engineer, Mr. Sing. Beck also left several photographic plates to be developed at the hacienda.

Human impacts on native species of Chatham took several forms, in addition to the diseased bird feet observed by Hunter. After sailing along the coast for some distance looking for marine iguanas, Slevin "was informed by a negro working on the plantation that the natives used to eat them, and even the eggs, and that he never saw any on the island."

OCTOBER 18, 1905

Strong winds kicked up in the afternoon and caused the schooner to drag her anchor in Wreck Bay, something that is never a positive development for a vessel dependent on the wind and waves. Not wanting to become another victim of this ill-named bay, the men became acutely aware that they had to replace the anchor and ten fathoms of chain lost at Socorro Island off Mexico in late July 1905. Their spare anchor was much too heavy to handle, and their kedge anchor (despite some chain wrapped around the shank to give it weight) was much too light to hold the schooner, even under ordinary conditions. So the men entered the world of Galápagos commerce. Beck arranged with General Plaza to buy an old anchor for $22, but they had to get the schooner up to the wharf to take it aboard. The schooner's mainsail and jib propelled them from the anchorage up to the wooden wharf, but while the vessel's headway was interrupted by the trusty kedge anchor, one of the flukes broke off. (Although damaged, the kedge anchor, a vital piece of equipment, continued to see heavy use throughout the voyage, especially anytime the *Academy* went aground.) The immediate danger was trying to keep the schooner off the beach. To prevent the *Academy* from being washed ashore by the heavy groundswell that was running that day, the crewmembers quickly ran one line directly to the wharf and another over to the beach to hold the schooner off. They accomplished their mission when they took the "new" anchor on board and shackled it to the starboard chain. Then they let go all lines and sailed safely out of Wreck Bay.

OCTOBER 19, 1905

The schooner's next destination, and the first island the collectors visited with the official sanction of General Plaza, was the nearest island due east of Wreck Bay, Barrington (Santa Fe) Island. Despite its proximity, the journey took thirty-six hours. Although they came tantalizingly close to Barrington after only four hours, the wind dropped and the ocean current swept them steadily away.

The trip to Barrington highlighted how frustrating life could be during the age of sail, and without the benefit of an auxiliary engine. Each fresh breeze brought the sailors and the schooner to life. Then, both the men and the vessel would settle back down as the wind did the same. Herman Melville wrote in his 1856 book *The Encantadas*, "Nowhere is the wind so light, baffling, and every way unreliable, and so given to perplexing calms, as the Encantadas."

Multiple tacking maneuvers zigzagged the *Academy* from port to starboard and back again, all through the night. The schooner repeatedly "beat up towards the island," only to drift away each time the wind fell. By daylight on the nineteenth they had drifted ten miles north of Barrington and spent the remainder of the day trying to get the schooner opposite the anchorage. Steady winds at night were not of much use, except for getting within sight of the island, because the crew constantly needed to keep a safe distance offshore until sunrise, when the tricky Barrington Anchorage could be approached. Melville captured the sense of frustration and futility of counting on the wind and current when he wrote about the islands, "Nigh a month has been spent by a ship going from one isle to another, though by ninety miles between; for owing to the force of the current, the boats employed to tow barely suffice to keep the craft from sweeping upon the cliffs, but do nothing towards accelerating her voyage."

The Academy men were abundantly rewarded when they finally anchored the schooner in the sheltered and extremely picturesque bay on the northeast corner of Barrington Island. Despite its small size, however, this island had a disproportionately large human impact. When the men landed on the fine white sand beach, they were met by numerous sea lions. Yet they also found wooden pegs pounded into the ground that had been used recently by hunters for stretching goat hides. Feral goats were abundant on Barrington. Walking about the island, the men saw several large herds. Beck shot two goats in short order, and the meat was "always a welcome addition to our larder." Not only did the native plants and animals exhibit differences from island to island, so did the introduced species. "The Barrington goats are not nearly so fat and healthy looking as those on Hood," Slevin wrote.

Hiking up to the plateau near the top of Barrington, Beck, Slevin, Williams, and King found a large and thriving colony of land iguanas. The abundance of the iguanas discovered there by Slevin recalled Darwin's observation seventy years earlier about a similar breeding colony on James (Santiago) Island. "I cannot give a more forcible proof of their numbers," Darwin wrote in *Voyage of the Beagle*, "than by stating that when we were left at James Island, we could not for some time find a spot free from their burrows on which to pitch our single tent." For the men on the *Academy*, this was *not* to be a night of the iguanas as it had been for Darwin and his assistant, Sims Covington, on their shore leave from the *Beagle*. But the museum in San Francisco needed specimens of this distinctive Galápagos reptile that had evolved in such splendid isolation from its nearest relatives.

Generally, the men found land iguanas sitting at the mouth of their burrows, retreating inward when approached. Darwin had watched patiently while a land iguana carefully dug a shallow-angle burrow until half its body was underground. "I then walked up and pulled it by the tail," Darwin confessed. "At this it was greatly astonished, and soon shuffled up to see what was the matter; and then stared at me in the face, as much as to say, 'What made you pull my tail?'"

Collecting on Barrington was relatively easy, if you had a couple of assistants, because although the iguanas were abundant, they were also quick and elusive (figure 9.3). Slevin worked in concert with King, Beck, and Williams to outsmart the speedy cold bloods. "It takes about two or three people to catch them," Slevin wrote. "They seem to lose their heads when chased by several persons and don't make for their holes, but run around and get caught in the brush where they are easily captured by the tails." The men darted around like young boys playing tag in a schoolyard, stirring up the loose volcanic soil, eventually getting

FIGURE 9.3 **Collecting Marine Iguanas.** Rollo Beck added several marine iguanas to the collection. *Archives, California Academy of Sciences.*

the iguanas into burlap bags. Bringing them back to the schooner was hard work, though. Slevin commented, "About six is all a person cares to carry." Under Slevin's direction, during three days the men collected forty individuals for skins and skulls or to be preserved whole in alcohol or formalin. These forty iguanas made up a small fraction of the individuals in the entire colony, and their removal likely had little effect on the overall breeding success of the Barrington population.

On Barrington, the iguanas' claws dug easily into the ground, and eggs were clearly being deposited in the burrows that resembled those of a ground squirrel. The Barrington Island species of land iguana has the two-part Latin name *Conolophus pallidus* and is unique, or endemic, to this postage stamp–sized piece of equatorial real estate. This single-island version of the Galápagos land iguana is generally larger and has skin and scales with a more pale yellow coloration than its widespread sister species, *Conolophus subcristatus*, which occurs on several other islands in the archipelago.

In the fall of 1905, at about three hundred feet above sea level, the Barrington land iguanas were busily preoccupied with burrowing and nesting in the ancient red volcanic soil. Darwin found the breeding ground on James Island to be so riddled by the "lizard-warrens" that the ground was constantly caving in, "much to the annoyance of the tired walker." Darwin managed to avoid twisting his ankles too badly and went on to comment on the iguana's behavior. "They are not at all timorous . . . raising themselves on their front legs, nod their heads vertically, with a quick movement, and try to look very fierce." Slevin discovered that the iguanas did more than just look fierce. Their extremely aggressive behavior on Barrington brought home Tennyson's succinct Darwinian maxim "Nature, red in tooth and claw." Slevin witnessed the iguanas being "very vicious to one another, grabbing each other by the jaws and drawing the blood." Their aggressiveness became extreme when one captured iguana "tore the whole lower jaw off another." Darwin's paradise could be hell if you trespassed on your neighbor's territory. But an even darker shadow soon passed over this breeding colony.

When the schooner returned to Barrington eight months later, on July 9, 1906, the men found that the iguana colony had been "nearly exterminated by some natives from Chatham" who had come over on a fishing sloop. Although curious from a modern conservation perspective, this fact did not prevent the Academy men from collecting yet more specimens, despite the devastation of this previously thriving breeding population by the locals. Slevin wrote that the locals had "cleaned out the entire iguana colony." But the small number of iguanas remaining meant only that Slevin and the others collected fewer. "We saw a few however scattered around," Slevin observed. "We brought three back with us." Rather than leave those three individuals on the island to rebuild the population, they opted to kill them. Fortunately, the land iguanas of Barrington continue to exist today, to form the basis of numerous comparative studies with the other, more common species of land iguana. Further searching by Slevin and the others in July 1906 revealed "numbers of iguanas scattered all over but none in colonies." Fortunately, Slevin wrote, they were "still plentiful on the island despite the visits of the natives." The land iguanas had dodged the bullet of extermination by the local residents from Chatham and held onto the island with their sharp claws. The resilience and tenacity of some species in the face of obliteration is refreshing.

Tiny Barrington Island serves as a showcase, or microcosm, of the larger problems facing Galápagos organisms in 1905 when the schooner *Academy* and her crew visited. It also exemplifies the serious conservation problems that continue today. The tortoises of Barrington tell a parallel story to that of the devastated land iguana colony. During the expedition, Slevin and the other men on the *Academy* met a fellow named Captain Thomas Levick who ran a schooner between the islands. Levick frankly informed the collectors that "30 years ago tortoises were found scattered all over Barrington and that he had taken them off of there." Rarely does one meet the actual person responsible for the extinction of a species, but the *Academy* crew seems to have crossed paths with the decimator of the Barrington giant tortoise. Levick informed Slevin that he "doubted very much whether we would find any as he said they were all killed off long ago."

Even this pessimistic news did not deter the eager collectors as they "landed after dinner [= lunch] and proceeded on a tortoise hunt" on their second Barrington visit in early July 1906. After finding some initial signs of tortoises, just some old bones and broken eggs, Slevin was convinced they would have "good luck" in finding live tortoises on Barrington and proving Levick wrong. But further searching by Beck the next day revealed only "more old bones" and "a very old piece of dung." Slevin and field assistant King carefully searched the valleys on the north coast but found "no signs of tortoise whatever." The men discussed the situation and concluded that the tantalizing egg fragments from the day before were nothing more than very old eggs that had been unearthed by land iguanas digging nests. Levick was sadly correct about the devastation he had brought to Barrington, and Barrington tortoises are now permanently extinct there. Rothschild's sad prophecy of a "damnable shame" had supporting evidence.

Captain Levick may well have been following the lead, or the orders, of the much-reviled Manuel Cobos of Chatham Island. The Ecuadorian historian Octavio Latorre attributed the extermination of tortoises on Barrington and Charles Islands and the population on southern Chatham Island to Señor Cobos and the men working for him. "Another item of intense exploitation was turtle oil. Cobos' teams of 'oilers' quickly exterminated the species of tortoises native to southern Chatham. They then turned to the tortoise population on Charles, which met the same fate. Finally, they went on to the tortoises of Albemarle." Levick appeared to have been the agent of extinction of the Barrington tortoise population, mirroring Cobos on the other islands. "No one will ever know," Latorre lamented, "how many tortoises were slaughtered but it is sure to have been several thousand." Latorre insightfully noted that the "accomplishments" of Cobos during his years in the Galápagos "have been ephemeral and only ruins remain." Moreover, "the extinction of the tortoises on Chatham Island and Charles Island [and] the destruction and reversals of the ecology on his island have been permanent."

Darwin himself participated, in a meager way, in the tortoise demise of James [Santiago] Island. Eating and enjoying tortoises was an obvious parallel between the crew of the *Academy* and that of the *Beagle*. While staying in the upper region of James Island for a few days with his assistant, Sims Covington, Darwin wrote, "We lived entirely upon tortoise-meat." No complaints were lodged about the food, and Darwin fell back on his recent South American experience when he wrote, "The breast-plate [was] roasted (as the Gauchos do *carne con cuero*), with the flesh on it [and it] is very good." Darwin had a further impact on the tortoise population when he noted that "the young tortoises make excellent soup; but otherwise the meat to my taste is indifferent."

OCTOBER 24, 1905

The men wrapped up their first collecting visit to little Barrington on Tuesday, October 24, 1905, without actually finding a tortoise. In fact, they had yet to find their first tortoise. On the Sunday before their departure from the island, they had relaxed on board the schooner, "keeping the Sabbath" in the island's scenic anchorage. Joe Hunter "stayed on board, loaded shells and spent the remainder of the day reading." He would be ready to collect when Monday rolled around. Nelson went out fishing in one of the smaller skiffs and caught four large fish known locally in the islands as *bacalau*, or trout grouper. Beck also stayed aboard in order to "take things easy."

The next destination was the uninhabited Indefatigable (Santa Cruz) Island. On this island the expedition members made their mark on geographic history and did significant collecting as well. After a full morning of collecting ashore on Barrington, they weighed anchor on October 24, 1905, and shaped course for the archipelago's central island, plainly visible from Barrington. They made good time crossing the twelve-mile channel and reached the southeast coast of Indefatigable by 3:00 p.m. Their first anchorage was a mile offshore from a bay not often visited today by ecotourism boats. They dropped anchor in six fathoms at Puerto de la Aguada, a Galápagos geographic name that has fallen between the cracks of history. Rollo Beck visited this same bay in October 1897 during his first trip to the islands on the Webster–Harris expedition on the schooner *Lila and Mattie*. Today this small, scenic bay is a visitor site in Galápagos National Park. It remains remote and unchanged. The bay can be reached on land from Puerto Ayora, most of the way by vehicle and then on a short footpath. Its present name, El Garrapatero, was coined in reference to the incorrect assumption of a tick-eating habit of the introduced ani bird.

An adventurous and potentially dangerous trip ashore was in store for Beck, Nelson, Ochsner, and Williams. They pulled on the oars of the ship's larger of two rowboats as they headed toward the sandy beach at Puerto de la Aguada to look for the supposedly plentiful water that the bay's name suggested. Heavy swells outside the bay made the schooner roll considerably, and near the beach, submerged reefs caused sizable breakers to go rolling in. The party in the ship's skiff had a rough trip trying to avoid the largest of the breaking waves. Parker and Nelson needed to be vigilant at night to keep the schooner safe at this exposed anchorage. Undoubtedly, a stomach or two became queasy.

Once ashore the collectors found a nice white sand beach protected by outside reefs. About a hundred yards back from the beach was a half-dried-up lagoon that provided enough water to fill half a small cask, but it was brackish and undrinkable and suitable only for washing. With the sun riding low on the horizon, the shore party returned to the schooner, but "had a hard pull back owing to the heavy swells, which sometimes broke close on to the skiff." They were out in the surf, which for them was not a good place to be. Staying both dry and in the skiff was a high priority for Beck, who could not swim, and they managed both. Staying alive was also a priority, but not something to dwell on.

The next morning started with near tragedy, greater in magnitude than just the risk of capsizing a skiff. Their exposed anchorage fairly close to shore on the weather side of the island would have made any mariner nervous. This was not the place to anchor a vessel without an auxiliary engine. Long, heavy swells came up from the south and were so large that, in the words of Hunter, "it seems as if they must go over the ship." At 5:30 a.m., Parker rushed below from his anchor watch and urgently called to Nelson to let go the heavier port anchor because they had started to drag the newly purchased and lighter starboard anchor. With the crisis averted, the landing party went ashore with both skiffs at 6:30 a.m.

They pulled the skiffs safely above the high-tide line, and the men fanned out in the direction of a small hill about two miles distant, in what appeared to be "the best looking tortoise country." Beck used his knowledge of this corner of the island from his previous collecting trips for Lord Rothschild and was able to guide the men up into the territory where tortoises lived. Cactus and thorn bushes made the going rough and slow in the morning sun. But nothing could impede the men in their search for tortoises. Numerous tame mockingbirds and doves accompanied them in the thick brush, their beauty perhaps offsetting the collectors' newly scraped skin and torn clothing. While looking for live tortoises that day in 1905, Beck found the skeleton of another individual that he had killed the last time he was at this spot, in 1902. Although the shell and bones were bleached white by the sun, the skull was still in good condition, so he added it to the collection. Reaching the base of the hill behind the beach, Beck found a small live tortoise, the historic first specimen of some 266 tortoises eventually brought back by the expedition.

Finding and returning to San Francisco with a diverse collection of giant tortoises was the raison d'être of this expedition. It was such an important goal that the entire crew

FIGURE 9.4 **Collecting Giant Tortoises.** The extreme difficulty of collecting giant Galápagos tortoises cannot be overestimated. A live tortoise, before being skinned, could weigh as much as a full refrigerator. The field collectors used poles cut from small trees, with the ends wrapped with blankets to provide padding for their shoulders. The rough volcanic terrain, completely devoid of the groomed tourist trails we associate with visitor sites today, and nearly impenetrable vegetation made transporting each tortoise down to the coast a remarkable accomplishment. *Archives, California Academy of Sciences.*

would periodically set aside all other interests or duties to lend a hand, and a shoulder or back, to retrieve these tremendously heavy animals (figure 9.4). The average tortoise weighed one hundred to two hundred pounds. Shortly Beck, Williams, and Slevin found two more tortoises after they "fell upon a fresh track." They followed the track for about ten minutes and soon came upon a big male tortoise walking slowly through the brush. While taking a look at their prize, they heard a noise in the brush to one side, and turning around they saw a large female heading in the direction they had just come. The male weighed five hundred pounds, as much as a fully loaded refrigerator, and the female weighed two hundred. They killed the female on the spot and cut out her liver for dinner, but as it was getting too late to do anything with the male, they left him turned over on his back with his massive feet tied securely with lashings to a tree "so he would not travel inland and make us pack him further on the following day." But the wily male would not give up the ghost so easily.

The next day, now that they knew they were in tortoise country, the men "scattered out and followed different trails." The game was on. These trails were from three to five feet wide and crisscrossed the area, often degenerating into water-filled wallows. This was ideal tortoise country. A tortoise could be anywhere out there, and the men looked carefully for clues to help them zero in on their quarry. Two of the best signs were fresh wallows and fresh dung, indicating a tortoise had recently passed through the area.

After young Ernest King and the experienced Joseph Slevin skinned out Beck's first small tortoise in the highlands above Puerto de la Aguada, King and Hunter packed it out to the coast and the waiting schooner. Despite its "small" size, this was no easy task. Hunter wrote that he and King "had a rough time" because they "missed the trail and had to force our way through the thick brush." Getting just this one small tortoise back to the beach took the two young men three hot, sweaty hours. After another hour, along with the rest of the shore party, they were back on the schooner. Slevin acknowledged that this first day of tortoise hunting had been "a hard day's work, but nothing to what the morrow had in store for us." Bigger tortoises posed bigger problems for the men to solve. Luckily, they got some practice with a small one.

Back on the schooner that night, the men ate the liver of the two-hundred-pound female tortoise, but without Chianti or fava beans. Ironically, during the night while the men slept on the schooner anchored offshore, the five-hundred-pound male tortoise outsmarted the collectors. Early the next morning on October 26, 1905, when they went ashore to continue the work of the day before, Slevin and the others discovered that "in his struggles for freedom, he had dug a miniature crater in the ground, broken all of the lashings, the ends of which were still fast to the trees, and had traveled a considerable distance up the hill, where we found him after following his trail through the tangled undergrowth." This leviathan had worked through the night to break free of his lashings (four fishing lines) and made a break for it, moving at the deliberate, slow-footed pace these giant reptiles are known for. In his bid for freedom, his leaden pace got him only fifty yards from the spot where he had been restrained. His hard-won freedom was short-lived. "All hands started skinning operations at once." It took nearly six hours to finish both the male and a female they also collected.

Collecting tortoises in the field was not for the faint-hearted or for those with a weak back. Beck had done it many times and knew that it entailed entrails. The other men were novices, but they soon gained experience to rival Beck's. These reptiles were obviously so heavy that the larger ones simply could not be transported the several miles from where the men found them, all the way back to the beach, and then rowed in a skiff through the pounding surf to be lifted aboard the schooner. For the larger tortoises, and sometimes for the smaller ones, the only practical way to proceed was to kill them on the spot and partially skin out the shell, quickly removing the heaviest of the internal organs and some of the muscles. Frequently, the men left the leathery neck and legs attached, curling them out of the way and back into the concave shell pockets from which they had protruded in life.

Collecting and then preparing a tortoise specimen by mummifying the remains required several days of work. Back at the schooner, after additional meat was sliced off with sharp knives, the entire tortoise shell and remaining carcass were submerged for several hours or days in large wooden "pickle tubs" constructed on deck for this purpose. The tubs were filled with a "bath" of salt, alum, and arsenic that quickly and effectively desiccated any remaining tissue to permanently stop bacterial decay and thus embalm the tortoise's legacy forever. After the pickling fluid rendered the tissue biologically inert, the tortoise was removed from the wooden vat and allowed to drip-dry. Before drying was complete and while the flesh was still pliable, the neck and appendages were neatly tucked back into the shell to make a streamlined package to be stored below decks. When fully dried out and desiccated, the remaining skin, tissue, and muscle had the consistency of rather toxic beef jerky. Smaller tortoises sometimes were brought aboard alive, then skinned and pickled as time allowed, often while the schooner was transiting under sail to the next island.

In the coming months, the men would repeat this process many times on the other islands in the archipelago. The task of skinning was necessary and part of their field work. Collecting tortoises was, after all, the main reason for this expedition. As unpleasant as the task of skinning a tortoise may seem to modern readers, let there be no illusion that it was a simple one. These efforts were only a prelude to the more strenuous task of getting the silenced behemoths back to the beach.

Each tortoise was far too heavy for any one man to carry alone, so the men rigged up a carrying device to get the job done. They lashed each tortoise with manila rope to a sturdy branch about ten feet long and five inches in diameter. Wool blankets covered the ends of the poles and served as shoulder pads (figure 9.5). Without the blankets, the pain of the poles digging into the men's shoulders was unbearable. With two men to a tortoise, they walked back to the coast over the rough lava and through the heavy, thorny underbrush. They looked like big-game hunters in Africa. When they came across impenetrable underbrush, the men used machetes to hack out a trail. They advanced slowly and with great effort, taking three hours to reach the beach in the hottest part of the day. The Sabbath was six days away. None of them knew if they would get a day of rest. If walking over the rough volcanic landscape and dodging thorny plants left and right was difficult on the way in to look for tortoises, it was a nightmare on the way out. On days like this one they certainly earned the $1.50 a day Loomis was paying them. These were the dog days that added new meaning to the phrase so

FIGURE 9.5 **A Live Tortoise.** Most often the tortoises being collected were brought down to the coast alive, to be skinned and preserved later on board the schooner. Here, from one of Beck's earlier Galápagos expeditions, two men carry a medium-sized tortoise on a pole with blankets for shoulder padding. *Archives, California Academy of Sciences.*

commonly written in the Academy collectors' field notes, "All hands observing the Sabbath." Despite saying nothing about how tired the men were after a full day or a full week of tortoise collecting, those five words summarized their state of physical exhaustion.

Carrying the tortoises was so difficult and exhausting, despite "all the heavy meat" having been cut out, that the men did it for no more than twenty minutes at a time over the rough lava and for thirty minutes over the smoother parts. Two pairs of men assigned themselves to each pole. Those not actively carrying the tortoise took a short respite or hacked a trail ahead, while the others trudged on with their reptilian burden. In one of the many understatements of the expedition, Slevin wrote that the trip down to the beach with the two large tortoises was a "long and hard journey." As a herpetologist, Slevin was also concerned with collecting other reptiles besides the tortoises. Once, while shouldering the dreaded padded pole, he saw a snake he wanted to collect, but had to keep moving.

Back on the schooner, the collectors had tortoise liver for dinner again. Yet this time they found it to be "a little oily and of a peculiar taste, but not at all bad when one becomes accustomed to it." Slevin described the significance of this new addition to the menu, stating, "During future times on the expedition we always lived on tortoise liver when obtainable, with the exception of that of the Duncan Island tortoises, which we found to be dark and tasteless. The liver of the Indefatigable tortoise is rich yellow, as is that of the tortoises of all the other islands except Duncan." Anyone who is not a fan of eating liver (and I'm certainly not) can be glad this internal organ now usually stays where it belongs.

In addition to the tortoises themselves, the men collected tortoise eggs from nests and sometimes from gravid, or pregnant, females. The two-hundred-pound female whose liver was eaten that first night after tortoise hunting had begun "had eggs in her ovaries at different stages of development," according to Slevin, who dissected her reproductive system in the field. Some of the eggs contained embryos with hard shells, while others were still in the yolk stage. Eggs that had already been laid left vacant spaces farther down on the uterus. This female tortoise (and many others like her taken during the expedition) had been actively reproducing. Thus, removing her from the population had a negative impact on the species as a whole. The gene pool can shrink only so much before it dries up completely.

The men also found two tortoise nests on one of the trails and opened them up. Inside each nest Beck found about a dozen eggs placed in the shallow round depression. Hardened soil covered a hole eight inches in diameter. The hardened soil gave way to soft soil a foot down, where the eggs had been laid. Beck speculated that the female urinated on the nest to moisten the soil, then circled around to thoroughly pack the upper six inches of soil. The tortoise eggs were white and spherical, measured about the size of a billiard ball, and had a hard, thick shell. The men brought both sets of eggs down to the schooner, where they planned to "blow them" to remove the contents (much as one does with chicken or other bird eggs with two small holes on either end). Slevin managed to get "the embryos out of four eggs and preserved them in alcohol." The miniature fetal tortoises would join their gigantic counterparts in the permanent research collections of the California Academy of Sciences.

Although the men skinned the large tortoises for several hours in the field to make them lighter for the long walk back to the beach, one man still needed several hours' work to get a single tortoise ready for the pickle vats.

OCTOBER 27–30, 1905

On October 27, Slevin "stayed aboard the schooner all day skinning tortoises." By the end of a full day's work, he had "two females in pickle and the big male nearly ready." Slevin hoped that he would "have the male in pickle tomorrow and try and blow some eggs." Collecting giant Galápagos tortoises was neither easy nor quick.

Before leaving Puerto de la Aguada, Beck went ashore and returned to the tortoise grounds on October 27, 1905. Beck's remarkable toughness as a veteran field collector clearly distanced him from his younger and less experienced compatriots. He sneered at anyone who could not keep up with him. In his notes Beck wrote that "others worked beach till noon being too tired after yesterday's work to do much!" (figure 9.6). He had plenty of energy left and while in hot pursuit of a "little female" tortoise, he entered a low "tunnel" in the vegetation, just the height made by a tortoise bulldozing through the brush. A rain shower made the ground wet, and Beck noticed fresh tortoise footprints in the soil. He promptly "got on hands and knees and followed [the trail] for 20 yards and saw chiquita headed up hill." He tied her to a tree as a temporary measure while he crawled farther down the tortoise tunnel, but without any luck. The absence of trails leading into the area where *chiquita* was found puzzled Beck, who wrote, "Can't determine yet how these tortoises come into this place."

The last tortoise from this area was a "fair sized" female. Beck wrote that she was drinking at a small water hole "when I found her [and] tied her up." The next day, October 28, 1905, Beck, Williams, and Slevin went inland and skinned and packed her out. Slevin described

FIGURE 9.6 **Loading a Tortoise into the Skiff.** The difficulty of transporting tortoises, even relatively small ones, is apparent here: two men on an earlier Rollo Beck expedition are transferring a tortoise from the shore to a skiff, hopefully avoiding being capsized. *Archives, California Academy of Sciences.*

what all the field collectors, except Beck, were experiencing when he wrote, "Our shoulders are getting sore now after the past few days and packing is becoming somewhat painful." Despite their protestations, Beck whipped the novices into shape before long. After collecting these initial 5 tortoises, they added another 261 to their collection before heading home. The men had a long way to go and their tortoise training was just beginning. Soon, they would be experts.

The next day, October 29, 1905, was a Sunday. Despite having worked extremely hard for the past few days, and with some of the men suffering from tortoise-induced painful shoulders, they got up at 5:00 a.m. and had the schooner under way at 6:00 a.m., shaping course for their first visit to the largest island in the chain, Albemarle. Resting that Sunday would have been a tactical mistake, and they wisely set sail on a fine day for the next island. After a slow start under foggy skies and with only a mild southwest breeze, at 9:00 a.m. the wind "hauled around" to the southeast, filling the *Academy*'s sails, and the weather was fine for the rest of the day. For the next ten and a half hours the schooner was as alive as it had ever been. She was kept steadily on a port tack throughout the day in a scene that would have made William E. Woodall, the schooner's builder, proud. Wind can be visualized as an arrow, and therefore is always described by the direction it comes *from* rather than the direction it moves *toward*. With the wind coming up from the southeast and passing over the port, or left, side of the schooner, the men had an entire Sunday at sea under perfect conditions.

It was too late in the day when they reached the coast of Albemarle, at 7:30 p.m., to make a safe anchorage. So they kept "beating back and forth" throughout the night under light winds and clear weather. At 8:00 a.m. the next day, October 30, they "hove to off the west side of Brattle and landed a shore party" on the small crescent-shaped islet. Hunter complained in his field notes that they spent excessive time "rowing about a mile and a half on account of Parker's caution." Complaints about Parker, such as this one, show up periodically in the field notes and diaries of the eight young men. The complaints foreshadow a deepening dissatisfaction with Parker as navigator and shipmate.

For its diminutive size, Brattle (Tortuga) Island proved to be quite interesting both geologically and biologically. The island is a broken-down volcanic tuff cone, eroded away completely on the south side, leaving just the northern crescent of a formerly circular crater. The formation of Brattle was probably identical to the volcanic formation of similar islands, such as tiny Molokini off the southwest coast of Maui or the crescent-shaped Hanauma Bay on Oahu. The open bay of Brattle Island faces due south, about a mile from point to point, with two small rocks protruding nearby. The crater's rim has steep sides, no sand beaches, and poor landing places. Getting close to the north side in the skiff, the men were able to find a protruding rock they could jump ashore on. If executed well, it would be a dramatic dry landing—if botched, an embarrassing wet one. Their landing was dry, and once ashore, the men scrambled up the steep slope covered with loose, sliding rocks to the top.

On Brattle, they collected marine iguanas, lava lizards, geckos, snakes, beetles, scorpions, and a big black and red grasshopper: not a bad haul for a tiny island with precipitous cliffs plunging straight down to the water's edge. They found very few plants, and the island was "almost destitute of vegetation." The poor soil cover supported only about twenty species of plants that were scattered about quite widely, not providing much cover for birds. Aside from a wide assortment of seabirds (several of which were breeding or had just finished),

Hunter noted just three species of land birds: the large-billed flycatcher, the yellow warbler, and one of Darwin's finches, the small ground finch.

After two hours of collecting, the collectors' one and only visit to Brattle was completed. The men signaled the schooner, which ran down before the wind and picked up the shore party. Another two hours of sailing brought them to the protected anchorage of Turtle Cove, near the small town of Vilamil. Outlying reefs protected the anchorage, which was suitable only for small vessels. They dropped anchor in just two fathoms of water. Out of diplomatic respect, they dipped their colors to the Ecuadorian flag on shore and cleared up the decks, making the ship ready for port, after which three island men and another Jamaican interpreter boarded the schooner. Here on Albemarle the men intersected the complicated human history of the Galápagos, just as they had done on their earlier visit to Chatham.

They met Don Antonio Gil, the most well known person on the island. He had established the colony he called Vilamil in 1893, and his descendants still live in the town today. The *Academy* crew got along famously with Señor Gil, and they returned several times during the expedition to enjoy his hospitality. The relationship got off to a good start when Don Antonio gave them a quarter side of beef and two watermelons. He knew how to make friends, and food was certainly the ticket with the hungry men of the *Academy*. The men also wanted to know if they could find a source of fresh water, and Don Antonio told them that "there is plenty, and good drinking water." When Beck, Nelson, Ochsner, and White returned to the schooner that afternoon, they reported that "the water is hardly fit to drink but good enough to wash clothes in." Perhaps Señor Gil had different standards for drinking water, and perhaps he had other reasons for wanting the men to like him and his island.

The island was alive with sexual energy. Hunter noted, "Warblers are chasing one another as if they are mating," which they probably were. After shooting birds, he dissected them in the process of preserving them as museum specimens and observed their reproductive anatomy, finding that the mockingbirds had "testes and ovaries well developed." Male birds of several species sang loudly to attract mates, and the males of Darwin's finches had "large testes and good plumage." Another fifty years had to pass before DNA would be discovered. Yet DNA was being exchanged in abundance in the bushes of Albemarle.

Vilamil was a veritable Shangri-La for the men of the schooner *Academy*. Unlike their experience at Puerto Chico on Chatham, despite a warm and encouraging welcome from General Plaza, they felt at home in Vilamil. They took on water, bartered for fruits and vegetables, and collected specimens. Later in the trip they sang and danced and drank wine late into the night with their Vilamil compatriots.

They began their day by taking on fresh water and attending to the mundane but necessary task of scraping barnacles from the schooner's bottom. Watering the ship was a slow process, because water had to be carried in small quantities by mule in wooden casks down to the ship. At the rate of two small casks of water per mule, the men brought on board 150 gallons during the day, which they had earlier bartered for 100 pounds of flour. In a separate deal, a box of hardtack and 50 pounds of Mr. Spreckles's Hawaiian sugar netted them several hundred of the local sweet potatoes they called *atoyas* (*otoys*), "a sort of coarse potato with rather a sweet taste," and other sweet potatoes called *camotes*. With fresh water more readily available, "all hands took advantage of the opportunity to wash clothes this morning."

In the afternoon, the men cleaned the schooner's bottom for the first time since leaving San Francisco. They swung the main boom out to port in order to list the vessel to one side for scraping and painting. Parker and Hunter applied the elbow grease, with Hunter's sarcastic statement, "Helped Dodo scrape side of the ship," revealing his continued and growing dissatisfaction with the expedition's navigator. The next day, they scraped and painted the other side of the ship. Parker and Nelson did the work while the men collected specimens. The schooner resembled a beehive with all hands actively working, not to mention swatting the numerous mosquitoes buzzing about.

NOVEMBER 1, 1905

The collecting began in earnest up and down the coast and on the mountain above the town of Vilamil. The small settlement of Santo Tomás on the slopes of Sierra Negra volcano was the destination of Williams, Ochsner, and Slevin. They walked up the road to reach the second village established by Don Antonio for mining sulfur from the steaming volcanic fumaroles on the inner slopes of the volcano. Don Antonio considered the town of Vilamil to be the island's "industrial center," where uplifted coral formations were mined and burned to produce lime. The economy was diversified by fishing and cattle ranching on the moist upland slopes of Sierra Negra, which makes up the bulk of southern Albemarle Island.

Williams, Ochsner, and Slevin had a pleasant experience walking up the road to Santo Tomás. "A native who had a small garden which we passed presented us with a watermelon and a papaya, both of which tasted very good and were quite a luxury to us, as we had not tasted fruit for some little time." Back at the beach they enjoyed a nice visit with Don Antonio, and "the natives were quite interested in looking over our catch and gave Ochsner's pistol a minute examination."

All of the men were actively engaged in fieldwork and they "found collecting in our various lines quite productive." Hunter went with Gifford along the coast to the southwest, where they found a large mangrove-lined lagoon whose "bottom [was] soft mud and quicksand." Avoiding the hazards, they noted the abundance of the small ground finch, *Geospiza fulginosa*, as well as numerous other birds. Despite the human presence in the area, the ecosystem seemed very healthy.

The men also found tortoises on the southwest side of the island, and adventures were to be had! A mini-expedition of Beck, King, Williams, and Slevin set out in the smaller, fourteen-foot skiff, and they pulled steadily down the coast about seven miles to the southwest from Vilamil to the vicinity of a place they called "Iguana Cove" on Cerro Azul. They found a fine stretch of sand on which to land and beach the skiff, "back of which was a lagoon surrounded by a dense growth of trees." Beck knew this spot from a previous collecting trip, again proving the sagacity of Loomis in appointing him head of the expedition. Beck's experience was invaluable (figure 9.7).

A few wild cattle were scattered around the lagoon, probably strays from Don Antonio's herd. They showed no fear or interest in the men. Walking just a half mile inland, the party quickly found five tortoises on the cattle trails and in shallow burrows in the bushes. One of them was promptly skinned because it was the farthest away and had to be made lighter for the walk back. The other four tortoises got a ride back to the schooner alive. Using the same technique as before, the men securely lashed a tortoise to a pole and two men carried it to the beach. With five tortoises and four men, they had to make two complete round trips. It was "a good, hard job," in Slevin's understated words.

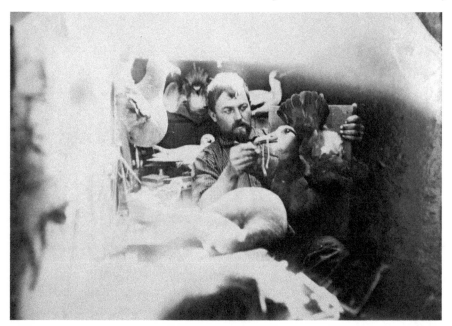

FIGURE 9.7 **Rollo Beck, Master Taxidermist.** Rollo Beck on an earlier collecting expedition below deck stuffing an albatross. Beck was renowned in the ornithological community for stuffing birds so exactly and repetitively that when laid out in a museum drawer the bird skins resembled cigarettes lined up. *Archives, California Academy of Sciences.*

The next challenge facing the four men was to get all five tortoises and themselves into the small skiff and safely back to the schooner. First, they launched the skiff into shallow water and carried the tortoises down the beach and into waist-deep water. Next, they hoisted the tortoises up over the gunwales and let them tumble heavily into the skiff. Last, all four men hopped aboard. With such a heavy load, the skiff had precious little freeboard (the distance between the water and the gunwales) and even less margin of error for capsizing. They worked as a team, though, and made it happen without incident. Williams and Slevin provided propulsion, taking measured strokes with an oar in each hand. Their backs were turned to the bow and their feet firmly planted among the tortoises. Beck sat in the stern, as befitting the chief of the expedition, handling the steering oar and watching for breakers. Before long, they crossed through the breaker line and maintained a delicate balance with each passing wave. At one point they "shipped a sea which very nearly swamped us." King got busy right away bailing out the skiff, and with a few more strokes on the oars they were safely outside the breakers and rowing steadily back to Vilamil.

The men spent the next three hours pulling back up the coast to the anchored schooner. They settled into a steady rhythm of rowing, bailing, and steering. As soon as they got back to the schooner and went aboard near sundown, all four men "immediately shifted into dry clothing, as we got a good soaking getting through the surf." Dog tired and chilled to the bone, they were proud of their success. During the trip down the coast and back, their balance and strength greatly exceeded their freeboard, and their stamina got them and the five tortoises safely back to the schooner (figure 9.8).

FIGURE 9.8 **Live Tortoises on the Beach.** After the tremendous expenditure of energy to get large specimens of giant tortoises down the shore from the highlands, each tortoise had to be transported by rowing offshore to the waiting schooner, which had to anchor out beyond the near-shore surf. *Archives, California Academy of Sciences.*

Meanwhile, the other men had spent the day less dramatically but no less productively. Walking inland, Hunter saw and collected more birds. He made two interesting observations: a specimen of small ground finch, *Geospiza fulginosa*, was different from the others taken (it had light feathers and a dark beak) and also fed among the bushes, "more like a certhidae than a geospiza." The other observation Hunter made concerned the vegetarian finch, *Playspiza crassirostris*, "feeding on a small wild tomato." The vegetarian finch brought a berry from the inside of the tomato bush and crushed it against a twig before swallowing the squashed berry.

The collectors spent the last day of this first visit to Vilamil ashore collecting near the coast and along the muddy road to Santo Tomás. They received watermelons and a chicken from people they met during the day, increasing the sense of hospitality they felt toward the place. Water barrels filled one of the skiffs to the gunwales, and it was towed behind the other skiff by Nelson's reliable pulling at the oars, all the way out to the waiting schooner. They also took on more *atoyas*, which the men substituted for "the potatoes long since used up." When the schooner departed from Vilamil the next morning, the men did not miss the troublesome mosquitoes that had tormented them during the long night.

NOVEMBER 4–6, 1905

They could not wait to get an early start as they hauled the skiffs on board, weighed anchor, and set sail for a return visit to Indefatigable Island. But weak winds kept them from even getting past Brattle Island until the afternoon. Nelson continued his maintenance work on the equipment, this time repairing and painting the ship's boats, their lifelines to their island collecting. During the day King and Slevin constructed a sturdy tortoise pen on deck between the main hatch and the foremast to contain the live tortoises until they could be skinned at "odd times." This reptilian purgatory was filled and emptied several times during the voyage. The other men worked with the numerous specimens just secured on Albemarle, stuffing, skinning, and preserving what they had worked so diligently to collect.

At 8:00 p.m. the schooner was "hauled on the wind for the night" only about five miles off the west coast of Indefatigable. This maneuver reduced their forward progress. They had crossed the channel in good time from west to east, improving on their time in the other direction. But at 4:00 the next morning, they discovered they had drifted to within two miles of Barrington Island, so they used the light southwest wind to sail on a port tack back toward their objective.

After a fine, clear night of sailing across the channel between Albemarle and Indefatigable Islands, the crew of the schooner sought anchorage on the south shore of Indefatigable Island in an unnamed, uninhabited bay and promptly ran aground. They struck the nemesis of mariners and boaters worldwide: underwater rocks hidden from view. It was Sunday morning, November 5, 1905, six weeks after the schooner and crew had arrived for a year's worth of field work in the Galápagos Archipelago.

The men knew this was an "unnamed bay," but it would not remain that way for long. Their first experience with the bay could easily have been their last. Slevin wrote, "On nearing the spot where we were going to let go the anchor, we struck on a submerged reef just about the middle of the bay." The schooner went aground under misty and drizzly skies, but that did not dampen their enthusiasm for this "good bay." When a sailing vessel hits a rocky bottom, the unnatural contact between wood and rock can be heard as well as felt through the soles of the feet. The shock of grinding sounds of a boat keel on a submerged reef brings a sinking feeling to the stomach of any sailor or mariner. The first job when a vessel goes aground is to reverse direction as quickly as possible. But without an engine to slam into reverse, the men on the *Academy* had to rely on pure muscle power.

The stress level, tension, and confusion among the men would have increased immediately. How could this have happened? Who was to blame? What should they do next? Was the entire expedition in jeopardy?

Without knowing how serious the grounding was, they sprang into action. The fate of the entire expedition depended on it. Two men scrambled to row out a short distance in one of the skiffs to drop the kedge anchor with a heavy chain and a thick line back to the schooner. They used the anchor as a stable point to pull against with a hand winch in order to work the schooner free of the rocky grip where its keel rested on the bottom. As soon as the kedge anchor was on the bottom and its tines had dug in, those still on the schooner used muscle power to winch the anchor line to break the schooner free of the rocky grip. This is called "kedging off" and is a common technique for freeing a grounded vessel even today.

None of them knew what had previously happened to this schooner on October 15, 1876, when she sank in shallow, frigid water off Isle au Haut in Maine. But even had they known, they still would have worked frantically to prevent a similar disaster from happening. The men learned an important lesson about this Galápagos bay: low tide exposed the top of the reef they had just struck. The reef definitely had to be avoided on future visits. Moreover, its location had to be noted carefully on the navigation chart.

Fortunately, no physical damage was done to the hull, and the crew got the schooner free. Safely off the rocks, they quickly dropped a heavier anchor in fifteen feet of water "with fine, sandy bottom." The damage done in grounding the schooner, if any, was another intangible black mark on the record of the navigator, Parker. These rocks were something to avoid on future visits during the expedition and to note carefully on the navigation chart. From the deck of the schooner, Hunter saw two pelican nests in the distance and noted that they were anchored "about 200 yards from land." Rolling with the swells with enough water under their keel, the men were embraced by the curving arc of a bay they had just named Academy Bay and propelled into history.

The men stayed on Indefatigable Island for the next eleven days, making several overnight excursions inland to capture the island's bounty and to explore its impenetrable vegetation. Beck's Sunday reconnaissance revealed a source of borderline fresh water just behind a small sandy beach. "It is possible to drink this water without getting sick," Slevin wrote after braving a swallow or two of the brackish water; however, "it is not to be recommended except as a last resort."

The men of the schooner *Academy* intersected the human history of the island once more when they found the remains of a grass hut near the water hole. Not knowing anything about the hut at the time, Slevin reported in hindsight that it "was used by a negro who had been marooned on the island by Manuel Cobos, the owner of the plantation on Chatham." The Academy men actually met this marooned man, rescued only when a passing vessel brought him to Vilamil. He told them "of being on the island for a whole year, living on raw tortoise meat and what shellfish he could gather." This unwilling Man Friday in a cruelly vindictive Cobos version of *Robinson Crusoe* was one of the few to survive such punishment by Chatham's late tyrant. Rollo Beck met Cobos back in October 1897 when he sailed among the islands on the schooner *Lila and Mattie* for Lord Walter Rothschild on the Webster–Harris Galápagos Expedition.

On November 6, 1905, all the collectors went ashore at Academy Bay after breakfast, except Slevin and King, who stayed aboard to work on the Albemarle tortoises. Not a soul was in sight, but there was evidence that residents from other islands had been getting salt out of several lagoons near the anchorage.

COLLECTING CONTINUES

Williams reported finding an old trail heading inland from the beach up through the cactus and thorn bushes. But he lost the trail after a mile or so where the green zone began. Any trail that existed in the past was soon overgrown by thick vegetation. Despite this hindrance, Beck went up the trail and found his first tortoise from this side of the island. He killed a large female tortoise and left her on the trail that they had cut into the interior of the island. As elevation and moisture increased, the plant life became more luxuriant, with larger trees, creeper vines, ferns, moss-covered rocks, and the first orchids, "one of which has a pretty pink flower, put in an appearance."

The next day, November 7, 1905, Beck and Slevin's overnight inland excursion began. They took provisions, blankets, and water for a few days of camping. King and Williams hiked along with them up the trail for the day and planned to carry out yesterday's tortoise. After lunch, King and Williams spent the next five hours walking back to the coast with the heavy tortoise strung under a stout pole. Beck and Slevin set up a makeshift camp, then pushed farther up the green mountain. After hunting all day, they came across three more tortoises (a large male and two smaller females), which they tied to trees before returning to camp for the evening.

The night was cool enough to use the wool blankets they brought along, a pleasant change from the clammy nights below deck on the rocking schooner. Up high on the mountain as they tried to settle in for the night, Beck and Slevin were visited by aliens, not the outer-space variety, but the mundane rat, ant, and mosquito variety. "There were so many rats running about," Slevin complained, "that we had to tie our provisions up in the trees with rope yarns." In the morning they found that the rats had feasted on the handles of their skinning knives, "which were well soaked with tortoise fat." But the rats were unable to gnaw away at the men's enthusiasm for collecting. Despite the additional, insidious torment of crawling ants and flying mosquitoes that "made sleeping rather impossible," Beck and Slevin carried forth, with or without sleep, in pursuit of their lumbering, antediluvian reptilian prey.

The large male tortoise they had tied up the day before was the picture of island tranquility when they found it "grazing along in the grass similar to a cow or horse." Upon approach, they thought the animal was "perfectly deaf, as we went right up and yelled." The only reaction they got was when the tortoise actually saw the men and quickly drew his head in with a loud hiss.

King hiked back up to the tortoise camp the next morning, November 8, 1905, to help Beck and Slevin skin and pack out the three tortoises, including the big "deaf" male that had fruitlessly sought protection in its scientifically valuable shell. Beck hunted about the area for more tortoises, hacking a trail up the mountain through thick vegetation. Using a barometer, the men determined that the productive "tortoise belt" on Indefatigable Island was about two hundred feet above sea level and filled with cactus and brush that made finding tortoises, well, thorny.

The men on the schooner organized a second mini-expedition and set out that same morning. The group consisted of Williams, Ochsner, Stewart, Gifford, and Nelson. They started "up country," hoping to make the summit of the island. Each man was loaded down with food, blankets, a canteen, and a gallon molasses tin full of water—enough provisions for three days. They followed the regular tortoise-packing trail up as high as possible until the morning glory vines impeded their progress and "it became necessary to cut a trail every inch of the way." Up there in the moist zone, the men learned how thick the vegetation could grow. They faced an impenetrable green wall, every bit the dense, tropical rainforest or the "tangled bank" that Darwin referred to in the last paragraph of *On the Origin of Species*, his exquisitely envisioned metaphor for the complexity of nature. Darwin wrote, "It is interesting to contemplate a tangled bank, clothed with many plants of many kinds, with birds singing on the bushes, with various insects flitting about, and with worms crawling through the damp earth, and to reflect that these elaborately constructed forms, so different from each other, and dependent on each other in so complex a manner, have all been produced by laws acting around us."

High on the mountain in their Darwinian struggle, the men made camp in a small clear-ing and learned how difficult it was to start a fire in such moist country. After a lunch of "beans, coffee and canned fruit," they collected various specimens before bedding down for the night, only to be tormented by ants and mosquitoes. "To make things more interesting," Slevin recalled after having exited to the coast with a tortoise for the comparative luxury of the schooner, "it began raining at dark and a tarpaulin was rigged to keep bedding as dry as possible and to catch rain water for the canteens." After such a miserable night, it is not surprising that "all hands arose early," either to continue slowing hacking their way up the mountain or to turn back to the coast. For Ochsner and Stewart, there was nothing quite like sleep deprivation to change their minds about hacking through impenetrable vegetation. The geologist and the botanist packed up their things and headed downhill.

Meanwhile, on Beck and Slevin's overnight inland excursion, only Beck stood his ground against rain, rats, insects, and the impenetrable tangled bank, the green wall. He was now camped alone in the tortoise country, not letting the pests or the weather dampen his inde-fatigable collecting spirit. Any hour of the day or night could find Beck collecting. Of the three barn owls he shot on November 9, 1905, even he was impressed by "one killed at 3:30 a.m.!" The intrepid group of "mountain climbers," as they called themselves, was in level country that had exceedingly thick vegetation. They reported back to Slevin when they regrouped that the green wall was "even worse than that of Cocos Island, if such could be possible." Trudging up the wet slopes of Indefatigable, they ran into patches of nettles, which became "very thick and troublesome," and the undergrowth reached seven or eight feet in height. Adding to their misery, several heavy rain showers passed over, soaking all the men to the bone. They made upward progress on the mountain at a rate of less than half a mile per hour.

Despite their strong desire to reach the top of the island, discretion proved the better part of valor, and they decided to call it quits just before noon. With provisions running low and progress so painfully slow, they prudently decided to retrace their steps back down to the campsite of the night before. Rain pounded them again that night, compounding their dis-comfort. Williams cleverly used a bottle of preserving alcohol to keep the fire going to cook dinner. They could not get down the mountain soon enough.

Back at Academy Bay, Parker had to deal with the schooner's touching bottom again at low tide when she swung on her anchor chain. Hitting the rocks this time was more severe than when they first came in to anchor, but again, fortunately, no damage was done. Several of the men secured a deeper and safer anchorage by raising the main anchor and then haul-ing the schooner a little farther out with the kedge as they had done before. Their first days in Academy Bay made the men acutely aware of the dangers that surrounded them, both on land and in the sea.

The defeated and retreating mountain climbers headed back to the coast on November 10, 1905, collecting specimens along the way. More showers drenched them thoroughly and made traveling and collecting "rather uncomfortable." To make things worse, Williams and Nelson developed sore feet. Blisters and wet socks often go hand in hand, and the men were all "pretty well tired out" by the time they reached the coast just after noon. For these men, the understated phrases "rather uncomfortable" and "pretty well tired out" would translate today into much more wretched descriptions of misery and physical exhaustion.

The rest of the ship's company was ashore packing out tortoises using the technique of two men carrying a long pole. It was a long, incredibly arduous job. The tortoises lived far inland and the country was very rough (see figure 9.4). Once the men had transported the tortoises down to the coast and ferried them out to the schooner, Slevin was usually in charge of getting them fully skinned out and into the pickle tanks. Whenever they were in good tortoise country, Slevin had to work long and hard to finish up the backlog. On November 11, 1905, the shore party brought back "one of the largest tortoises taken so far." Even the navigator, Parker, went inland and helped carry the tortoise rather than stay with the schooner as he usually did.

All hands took a well-deserved rest on Sunday, November 12, 1905. The extreme physical exertion required for tortoise packing outdistanced the effort expended in any other aspect of collecting by the men in their respective specialties. On this day, though, the exhausted men relaxed as they fell into their typical Sunday routines. Some of them indulged in fishing over the side and repairing badly worn shoes. Others did light collecting, as did Williams and King, who ventured into one of the shallow lagoons at the head of the bay for seashells. Gifford continued his photographic documentation of the expedition and the islands themselves. Occasionally Beck also took photographs, as he did on November 11, 1905, of the group carrying a tortoise down a cliff that they had earlier climbed to get into the interior of the island.

When he was not helping carry the tortoises, Hunter searched tirelessly for birds. His astute observations, carefully recorded in his field notes, are among the most detailed from the expedition. Fortunately, Hunter included more than just geographic locations and numbers of species seen or collected. For example, when he witnessed a group of six mockingbirds fighting, all he could see was "a ball of wings, tails and feet." This territorial fighting in Galápagos mockingbirds has become known today as a "flick fight." To Hunter, who stood and watched the birds fighting for several minutes, this aggression appeared to have no visible cause other than "the scrappy disposition of the species."

Ashore in the shallow lagoons around Academy Bay, Hunter observed and collected more birds. He saw a flock of five ducks and commented, "They were very unsophisticated swimming toward me to learn what new kind of beast I was. [But] once shot at they soon learn to keep away." Back on the schooner, Hunter's dissection of the birds revealed that they were "near the breeding time."

The vegetarian finch was one of many tame birds of the archipelago Hunter observed while collecting. Hunter found a small flock of this species of Darwin's finches scattered around a deep water hole, feeding both on the ground and "on the flowers of the wild passion vine that grows commonly here." In the trees along the lagoon, Hunter found pelicans, plumbeous herons, and great blue herons, all of which "uttered hoarse squawks and croaks as we came past and interrupted their dreams."

Hunter shot a cactus finch, with the scientific name *Geospiza scandens*, while it was "eating the unripe bud of an opuntia" cactus. A hawk he shot near the beach contained in its stomach "the remains of a centipede and the feet of a duck. All the remainder of the duck was digested. The feet were swallowed whole and were not injured to any extent." Hunter captured another hawk alive when he found it near the carcass of a big tortoise, eating the meat the men had carved off. They kept the hawk for some time, and "it quickly became accustomed to being handled and did not offer to bite or claw."

Even the schooner's mate, Frederick Nelson, frequently helped the collectors ashore when his shipboard duties permitted. On the last day at Academy Bay, Nelson went out in the ship's boat (which he had rigged with a mast and sail) to visit "a big rock off the main island." Although unnamed back in 1905, and for many years after, this small island is easily visible from Academy Bay and is today called Isla Caamaño. When he reached the tiny island, Nelson found sea lions and a healthy population of marine iguanas. Although these common species did not warrant a collecting visit by the other men, Nelson contributed, in a small way, to the overall picture of animal distribution in the archipelago. Even though Nelson did not keep field notes, each time any of the men on the *Academy* made an observation, either simple or profound, their written record provides us with an invaluable snapshot of Galápagos life more than a hundred years ago.

After eleven days of productive collecting at Academy Bay and with fifteen more tortoises (and numerous other specimens) added to their growing reptilian collection, the men weighed anchor on November 17, 1905, and set sail at 6:30 a.m. to head north for a trip along the eastern shore of Indefatigable Island. Right away they were plagued by wind trouble. They raised sails and tried to beat out of the anchorage, but the wind was too light and headway was impossible. After two hours of patiently waiting for the wind, they were forced to drop anchor to keep the schooner from drifting onto the rocks. The wind freshened a little at 10:00 a.m., and it took them until 11:30 a.m. just to get clear of the mouth of Academy Bay. Their immediate goal was the vicinity of Gordon Rocks, just north of the two small, then-unnamed Plaza Islands near the red-colored volcanic cinder cone today called Cerro Colorado. Fortunately, the wind picked up considerably in the afternoon and there was "a fine sailing breeze" the remainder of the day.

The location of the next anchorage was never precisely described, but can be closely approximated from descriptions in several volumes of the expedition's field notes. Beck clearly stated that the schooner was headed for Gordon Rocks and "anchored [a] few miles north of Gordon Rocks." Slevin described the location in somewhat clearer terms when he wrote, "We dropped anchor in 15 fathoms of water in a little cove on the northeast side of Indefatigable." Although there were two potential anchorages in the protected water between the two Plaza Islands and in the small bay near the prominent volcanic tuff cone of Cerro Colorado, it appears that the crew stopped farther north. Their anchorage was most likely about three to four miles north of Gordon Rocks.

They stayed at this location for two and a half days despite the fact that they "found a very strong current running and schooner riding to a taut chain." Strong southerly winds over several days caused the schooner to drag its anchor, "making our navigator [Parker] quite ill at ease."

Hunter called their new anchorage "Fossil Cove" because Ochsner, the geologist and paleontologist, found "two beds of fossils both near shore about 10 feet above sea level." Slevin also reported that "Ochsner found some good fossil collecting in the cliffs close to the anchorage." Beck indicated in his field notes that Ochsner found about twenty-five species of shells in the fossil ledge. Hunter also stated that he "went to the fossil outcrop and from there to nearly opposite Gordon Rocks," and his excursion was almost certainly to the south of the anchorage. Layers of fossiliferous limestone also occur at Cerro Colorado, and the "fossil ledge" discovered by Ochsner extends for several miles along the northeast coast of Indefatigable.

While Parker fretted, and rightly so, about the safety of the schooner at this unprotected anchorage, the men were preoccupied with collecting and exploring. To those standing on deck with turtles and sharks swimming around the schooner, Indefatigable appeared "very barren and the shore line is marked by several steep cliffs extending to the water's edge." One of the first orders of business was more shoe repairing, and Beck reported that "Hunter fixed my shoes," as Hunter also did for some of the others. Beck was remarkably expeditious at stuffing birds, and before going ashore that day he not only "put up a box of birds," but also fixed a place in the hold for the tortoises, making room for the fifteen new specimens from Indefatigable, including four "big ones."

The next day, November 18, 1905, Beck went ashore and walked inland "to examine the country" and to look into the possibility of hiking to the top of the island from the east side after the failure of the mountain climbers to reach the summit from Academy Bay. But Beck wrote in his characteristically truncated style, "country not good for going to top," and no attempt was made. Just to reach the wooded portion of the island, where the impenetrable green wall of vegetation might thwart their upward progress, would have required traversing the very rough and long, gradual slope of the volcano. The lower elevations were "extremely desolate and covered with cacti and dried grass" and riddled with abandoned land iguana burrows. Beck knew that Darwin had described in *The Voyage of the Beagle* the difficulty "walking over these lizard-warrens, [where] the soil is constantly giving way, much to the annoyance of the tired walker."

Slevin kept up his shipboard tortoise skinning and pickling work as a counterbalance to Beck's work on shore finding tortoises. On November 18, 1905, Slevin wrote, "Stayed on board all day and fixed tortoises. I am unable to wire the plastrons as they take up too much room in the hold."

The next day was another typical Sunday, with "all hands taking a day of rest." Nelson and Ochsner, who remained close friends for many years after the expedition, went fishing in the sailboat and caught eleven of the good eating groupers today called *bacalau* in Spanish, the largest measuring two feet long and weighing twelve pounds. Gifford went ashore to take some photographs of the fossil clams and snails embedded in the cliffs. Williams tried his luck at shark fishing with a large baited hook over the side and caught a seven-footer.

The men spent a half day collecting on November 20, 1905, which yielded a variety of specimens before they left Indefatigable. Slevin got twenty-nine lava lizards but found them to be scarcer and more skittish than elsewhere because "the country is open and the hawks probably chase them." Nelson and King gathered some driftwood from the shore and brought it on board the schooner for the galley stove.

By 11:30 a.m. all hands were back on board, the sails were hoisted, and the crew began to weigh anchor. Normally the process of lifting the anchor went smoothly, but this time the men ran into trouble. They spent nearly an hour of hard work to get the anchor and thirty fathoms of chain off the bottom. They set a course to navigate around the north end of North Seymour Island and made good time of seven knots by sailing "wing and wing" with a following wind. By early afternoon they rounded the tip of North Seymour with a good breeze, but then the wind died down and they were almost becalmed.

At 4:30 p.m., they let go the anchor in "a large shallow bay on the southwest side of South Seymour [= Baltra] in six fathoms of water." This bay was perhaps what is today called Caleta Aeolian. Slevin, Beck, and Ochsner went ashore and Beck shot two goats for fresh meat, while Slevin and Ochsner got a large sea turtle on the very wide sand beach. Two men grabbed a second turtle by the hind flippers, but it was too close to the water and "gave us a fine bath and some good, hard slaps with its flippers." Turtles were common here in the shallow water, and Slevin saw at least a dozen.

The next day, November 21, 1905, all hands went ashore on South Seymour, except Beck, who stayed on board to skin birds. Slevin quickly sized up the geology and noted that "the island, as far as I can see, consists of three elevated plateaus and two wide valleys, all covered with boulders of lava of a dark brick red." They found fossil marine mollusks "scattered all over this part of the island," and Ochsner brought back a large collection. Slevin found that lava lizards were common, as were land iguanas, but the iguanas were scattered all over rather than in colonies as he had seen earlier on Barrington. The land iguanas, Slevin noted, were also less active and fewer lived in burrows. Here they seemed to prefer living among the rocks. Perhaps influenced by Darwin, Slevin also noted size and coloration differences with the Barrington land iguanas, and Nelson reported that "one came and drank the blood of a goat he [Nelson] shot." Although "vampire finches" are known from two islands in the archipelago (usually drinking blood from an open wound on a booby), this is the first, and only, reference to vampire iguanas. Slevin collected several land iguanas from the island, "putting them in sacks and packing them down to the landing place," where Nelson had beached one of the skiffs for repairs and new bottom paint.

Slevin noted that northwest of their anchorage there was "a fine lagoon, separated from the ocean by a long sand beach on which the surf broke furiously." When Slevin returned to South Seymour twenty-two years later, in December 1927, aboard Captain Allan Hancock's research vessel *Oaxaca*, this fine lagoon had completely disappeared. Not only were living species dynamic in the islands, but so was the landscape.

On their last day at South Seymour, November 22, 1905, Gifford and Hunter went to a lagoon in the first bay south of where the schooner was anchored. Gifford patiently took photographs of the dozen ducks swimming in the lagoon, and Hunter noted, as only a true museum collector could, "They were as usual very tame, swimming up to look at the camera. I shot 10." Back on the schooner and upon inspection of the ducks' well-developed ovaries and testes, Hunter concluded, "Nesting time must be rather near." But the ducks learned quickly and reverted to being skittish, staying "out of range after a few shots."

The tortoise pickling tubs on deck saw heavy use, and Slevin and King stayed aboard to scrape and paint the tubs. Meanwhile, Williams and Beck hiked to the south end of South Seymour Island, where they stood on the sea cliffs for a panoramic view of the narrow and shallow strait separating South Seymour from Indefatigable, which is today called the Itabaca Channel. The specimens they gathered while "general collecting" were also brought aboard, along with five goats shot for food. The men stayed on the island until 10:00 that night, and at the end of a long day they brought back the big skiff, which had sat on the beach all day, bottom up for the paint to dry.

Early the next morning, November 23, 1905, the men weighed anchor and sailed for nearby Daphne Major Island. Slevin described this tiny speck in the archipelago as "an almost perfect crater some 250 feet in height." Like researchers visiting Daphne today, Nelson, the mate, was able to land the men in a skiff at only one spot on the island's south coast because "overhanging ledges seemed to extend around most of the lee side of the island." Going ashore at the only accessible place led them inevitably upward toward the crater's rim, then down to explore the interior of the crater, which "was found to have a fine floor of white sand about on the same level as the sea."

Hunter wandered around the island, making observations that would ring true years later when the finches of Daphne came under the evolutionary microscope of Peter and Rosemary Grant of Princeton University. Hunter moved freely among the ancestors of the Grants' study subjects, stating that "in places Geospizas were common." These finches included the species Hunter accurately identified as "*scandens, fulginosa,* and *fortis.*" Unbeknownst to Hunter, the descendants of these very same birds would reveal much about the process of Darwinian natural selection. Their nondescript black beaks would be precisely measured between the jaws of vernier calipers, the birds held gently in one hand by the Princeton biologists, while the calipers documented a measured evolutionary response to changing food supplies following a particularly intense El Niño phenomenon.

Hunter also noted firsthand the other half of the Daphne evolution story the Grants would study, namely, increased finch reproduction in response to changing food availability following El Niño. Hunter recorded that "Geospiza nests were numerous in cactus trees." Hunter was standing in the middle of an ongoing evolutionary experiment. In an entire archipelago of islands known today as Darwin's living outdoor laboratory of evolution, Daphne Major was arguably the best island for studying the effects of environmental change on bird beaks. But the Academy collectors were focused on the *products* of evolution rather than in the *process* of evolution. Hunter and his colleagues were *collecting* the products of evolutionary change, not documenting it.

Among many other birds seen and collected on Daphne, Hunter collected tropic birds from their nesting holes in rocks. He accurately observed, "When pulled from the hole their shrill cry is almost earsplitting." His honesty in depicting the actual events of field collecting, as unpleasant as they might seem to us today, distinguishes his field notes from those of all the other collectors. Hunter's attitude and honesty about pulling struggling, squawking birds from their nests affords us insight into the daily activities of this yearlong expedition.

In addition to providing valuable and important specimens, Daphne Island afforded the men another maritime misadventure. At one point while Hunter was alone at the rim of the island where he had a panoramic view, he noticed the skiff, empty, drifting out to sea. He quickly raised the alarm and all hands raced to the landing place. Slevin and Hunter hailed Nelson, who answered back that the skiff had accidentally drifted away while he was collecting marine shells at the tide line. Besides being responsible for the drifting skiff, Nelson was now stranded above high tide, but with several sharks circling in the water below him. Nelson was facing his own Charybdis. The overhanging ledge that encircles the island had him pinned down, with a sheer cliff above and shark-infested waters below. Nelson calmly chose neither to scramble up the cliff above nor to swim in the dangerous waters below. Wisely, he decided to stay put; the other men would pick him up later.

Stranded on the tiny island, all hands gathered at the rocky landing place and fired their guns into the air, frantically waving their hats to get Parker's attention as the schooner sailed by. With the schooner hove to, and after a half-hour wait for the men ashore, Ochsner and White rowed the small skiff from the schooner to shore to pick up Beck, Williams, and Slevin. Then Gifford, Hunter, and Slevin rowed back for the stranded Nelson, who faced a serious challenge. The skiff could not approach close enough to the rocks for Nelson to hop aboard, because the periodic wave surge could have damaged the dinghy, or worse. Had the dinghy gotten close enough for Nelson to hop in, it would have been a classic "dry land-ing"—the phrase used by today's ecotourism operators to describe running a skiff or panga up close enough to the rocks for passengers to nimbly jump ashore, or back aboard, without getting their feet wet. In the situation Nelson faced, he ended up getting more than just his feet wet. The men manning the dinghy's oars had to stay a couple of boat lengths away from the rocky shore, and even that was a challenge as they rowed hard to maintain a safe distance. With no other option available to Nelson, he made a fully clothed dive into the shark-infested water and the skiff quickly backed in to pick him up.

Returning quickly to the schooner, the men boarded and ran down before the wind to chase after the wayward skiff, which had now drifted far to leeward. Their misadventures were not yet over. "As the schooner approached the skiff," Slevin wrote, "a swell suddenly caught it just right, turning it turtle, the oars falling out and floating away." To rescue the capsized skiff, three men got into the sailboat, which was being towed astern of the schooner. Once they reached the capsized skiff, Nelson, Ochsner, and Williams managed to gather up the floating oars, right the skiff, bail it out, and head back to the schooner. But they were not home free yet. In the process of approaching for the pickup of the two small boats, Parker, the navigator, tried some fancy seamanship that caused the schooner's boom tackle to get fouled up in the mast of the small sailboat, swamping it and unceremoniously "throwing the occupants into the water." Ochsner and Williams treaded water, the unfortunate victims of Parker's less than adequate seamanship. Hunter expressed his continuing disdain for Parker by describing this ill-fated maneuver as "several tacks and a display of seamanship by Dodo." Some of the other men shared Hunter's sentiment, and the negative opinion of Parker con-tinued to build.

Nelson, still wet to the bone from his ocean plunge, rowed up from behind in the skiff and rescued Ochsner and Williams. No permanent damage was done to men or equipment, but in the excitement, the cook, Jim White, "burned up the potatoes." In addition, the sacks of sand used for ballast in the small sailboat went overboard when it capsized. With the sailboat bailed out and the situation under control, resentment toward Parker continued to fester. The soaked men headed below deck for dry clothes, while the schooner hove to the wind with both small boats safely taken in tow.

Not willing to let a little maritime accident derail their collecting, the men sailed the schooner back up to Daphne Island and a party landed at 2:30 p.m. for another two hours of collecting. Later that day, the schooner headed back across the channel to Indefatigable Island to anchor in Murex Cove on the north coast near the large, mangrove-lined lagoon now called Black Turtle Cove.

JUMPING THE SHARK

The following morning, November 24, 1905, some of the men went ashore on Indefatigable Island while Hunter stayed on board all day to skin birds. Reaching the shallow lagoon that extends more than a half mile inland, the men quickly saw, as ecotourists do today, "many turtles, mantas and sharks." In his field notes Rollo Beck called the rays, which had white spots on their backs, "very pretty," a rare opinion expressed by the otherwise laconic chief of the expedition. When a shark approached Nelson while he was in the water gathering shells, the eighteen-year-old field assistant, King, "struck the shark with an oar and sent it heading full speed in the opposite direction." Beck had his own shark encounter in the shallow lagoon. While he waded around its edge looking for turtles, the sharks swam right up to him and he poked them with a stick to keep them away. He continued to wade around the lagoon, with sharks slicing in on graceful trajectories. Upon returning to the schooner, Beck proudly noted his passage through the gauntlet of teeth and jaws, where he had "managed to keep from getting bitten."

The men saw an abundance of sea turtles feeding on the young shoots of the mangroves that lined the lagoon. When sliced open by the men, the turtles' stomach contents confirmed their dietary preferences. Out of three turtles taken the first day, the men got two at once by grabbing a copulating pair. In general, the turtles were "very wild," except when otherwise distracted by copulation, in which case the men could approach them easily. In a mating pair, though, the female seemed more likely to notice danger and swim away. To avoid losing the pair of specimens, the men used a gaff to hook the female first and then proceeded to haul the oblivious male off her back, as an easy bonus in their biological "twofer."

After a longer visit to the lagoon the following day, November 25, 1905, the men proclaimed it to be "a wonderful place for turtles, shells, fish, sea urchins, oysters and various forms of marine life." They brought back three more marine turtles, but the additional weight made their skiff ride low in the water. Additionally, a choppy sea and a stiff headwind made it "a long, hard pull back to the schooner." Beck returned from a separate seven-mile excursion inland with thirty-five doves in hand, for a dove pie dinner. He had struck down all the doves with either a stick or with rocks, because no one wanted buckshot in his dove pie. Also, the birds were so tame that guns were superfluous. The proximity of the lagoon, with its mating turtles and circling sharks, to the anchored schooner had one drawback, however: mosquitoes. They arrived in force that night.

If they didn't sleep well because of their insect visitors, the men rested the next day, Sunday, November 26, 1905, by doing odd jobs about the ship. Before breakfast they even indulged in a bit of hair cutting and beard trimming. A group of three men planned to trek far inland the next morning, and they spent that Sunday getting ready for an early start. The evening meal, which capped their day of relaxation, consisted of fresh, roasted doves, "a welcome addition to the bill of fare."

For the next three days, from November 27 to 29, the men fanned out in small groups in several directions, using the schooner as a base of operations. Slevin and King stayed on board to skin the growing collection of sea turtles. Gifford and Nelson traveled the farthest. First they went back to Daphne in the sailboat, or gig, as they called it, where they collected a dozen tropicbirds. On another day, they started out for Daphne, but the wind slackened and they made for South Seymour instead, where they shot three goats for fresh meat. With no giant land tortoises in the area, their cousins, the giant sea turtles, became the main objective of the collectors.

Three separate groups of men rowed back to the protected lagoon, mainly to catch more turtles. They had already secured six turtles from the lagoon, when Beck and Parker caught another four, followed by Stewart and Nelson's catch of two more. Finally Beck, Ochsner, and Williams added another eight, raising the total to twenty. Slevin was busy from sun up to sun down skinning and pickling the turtles. He wrote, "We have quite a collection of them on deck now." The excess turtle fat thrown overboard presented a new collecting opportunity, allowing Beck to shoot eight Grace's petrels that had come in to feed.

The group that headed inland to explore and collect consisted of Hunter, Ochsner, and Williams. They were loaded down with provisions for two or three days' camping in the hills. Their goal was to reach the green zone to prospect for tortoises, but this proved difficult. After they had slowly traveled through the hot lowland country for seven miles, the same distance traveled by Beck two days earlier along the same route, the barometer indicated an elevation of 1,160 feet. Along the way, the men found the vegetation monotonous and the birds scarce, except for doves and hawks. In the evening after they had lit the campfire, a group of short-eared owls joined them. All four were shot. Their dinner consisted of roasted doves they had collected. But when the sun set, they "slept rather cold" and "spent the greater part of the night keeping up the fire." Their misery continued the following day.

During the morning, Hunter, Ochsner, and Williams found it difficult to push through the dense brush and over the rough lava flows, making their forward progress "anything but fast." Beyond their campsite, the terrain continued upward and unchanging. Except for thicker brush, it was "all very dry, rocky and rough." The heat caused them to drink water faster than expected, and their three-day supply of two gallons each ran low after only a day and a half. By midday on November 28, 1905, with their water supply too low for comfort, they abandoned their quest and concluded that "apparently it is not possible to reach the green zone from this side." Defeated, the three men headed back to the coast.

Parched and tired, Hunter, Ochsner, and Williams got to the coast at 5:00 p.m. and lit a signal fire on the beach to herald their early return. Their water supply had barely lasted them back to the beach, which put them in danger of serious dehydration. Slevin rowed out from the schooner to take them back on board.

The next day, Thursday, November 30, 1905, was Thanksgiving Day. It was the only Thanksgiving the men celebrated in the islands. They maintained continuity with their lives back in California by preparing a suitable feast, tailored to the fauna of the islands. While roasted turkeys were eaten all across America on this most American of holidays, the men of the *Academy* set their sights on a smaller, more exotic bird. All hands turned out at sunrise for a 6:30 a.m. breakfast, then went ashore to hunt for their Thanksgiving dinner: Galápagos doves. Their cooperative effort was quite productive, although a bit hard on the local dove population. Their hunt yielded one hundred of the tasty morsels in three hours. Beck seemed to have been particularly motivated and efficient at dove swatting. He remarked that he "killed 41 from 7:45 [a.m.] to 9:10 [a.m.]," which is a dove every two minutes, on average. The men mustered together at the landing place on the beach to pluck and clean the doves before turning them over to the cook. This was a special occasion, regardless of the hemisphere they had traveled to, so White gave the big dishpan an extra cleaning before preparing their Thanksgiving dove pie. At 2:00 p.m. all the men sat down to enjoy a meal consisting of the dove pie main dish and a coconut pie for desert. They celebrated much as they would have with their families back in California.

After sharing the communal meal aboard the schooner and commenting on the unique use of the big dishpan, they gave thanks for the continued success of the collecting expedition. The tranquility of the day shifted dramatically when Williams put the shark hook overboard. He hooked a nine-foot tiger shark that put up quite a fight before being pulled to the surface and "dispatched with a bullet." The folks back home didn't have nearly this much excitement on Thanksgiving. Once on deck, the shark's stomach contents revealed a goat hide and several balloon fish. The waters of Galápagos really *were* shark infested.

Thanksgiving Day was finished off several hours later with a second feast of turtle egg cake, cold goat meat, and local fruit, a uniquely Galápagean meal. With all the men in a satisfied and contented mood, Ochsner gave a relaxing flute concert before all the men turned in for the night.

December 1, 1905, was a full sailing day, although progress was slow at times. After an early breakfast, the crew hoisted the sails and shortened the anchor chain in anticipation of a strengthening sailing breeze. Settling for a light breeze that got them under way at 8:00 a.m., they shaped course for nearby Duncan (Pinzón) Island. They spent the early part of the day under light winds before a good sailing breeze sprang up in the afternoon, filling the sails and raising the men's spirits. The wind carried them steadily south to Duncan. When they reached the anchorage on the northeast side, they used the skiff to set out an anchoring buoy on the sandy bottom near shore. They dropped their main anchor just before sundown.

Rollo Beck had visited Duncan Island three times during his earlier collecting trips. He knew the island and its resident tortoise population fairly well, or at least he thought he did. He wrote about it in a paper he published in the 1903 annual report of the New York Zoological Society. Beck and others who had collected there thought the Duncan Island tortoise race to be nearly extinct. In a letter home to his brother in 1897, the leader of the 1897–1898 Rothschild Galápagos expedition wrote, "Twenty-nine of our tortoises were taken from Duncan Island, where it was supposed they were extinct. They were at the top of the island, and in the bottom of an immense crater."

The first tortoise in 1905 from Duncan was a small one. Hunter had found it on a reconnaissance trip when he climbed to the top of the island's caldera, then headed down "inside the old crater." Hunter's experience of the island and its vegetation was typical of what the other men experienced. Although the island was rocky and the brush not very thick, the dense growth of thorn bushes made traveling difficult. Near the top of the steep-sided caldera the brush became thicker, but not green, as it was on several other islands. Two species of delicate orchids and an abundance of ferns, lichens, and moss hanging from the bushes revealed the moistness of the air.

Upon reaching the rim of the crater, Hunter descended into the extinct volcanic caldera, where he found his tortoise quarry. He immediately proceeded to skin out the very fat female, which had several eggs still in the yolk stage. Two Galápagos hawks flew in and fed on the scraps of tortoise. Focusing on retracing his steps back to the shore with the still-heavy tortoise, Hunter did not pay his usual attention to birds, although a hawk nearly landed on his hat. One interacts with Galápagos wildlife just as one interacts with the water in a swimming pool: it surrounds you and you become part of it while you are there. A skinned tortoise on his back and a hawk swooping down to his hat while he pushed through heavy thorn bushes was complete immersion for Hunter.

Beck spent the day on the southwest side of Duncan Island, where he found twelve tortoises, which he tied up securely. Along the way, Beck had an interesting tortoise encounter in which "one very old flat backed one chased me for 10 feet before I stopped him." Beck was accustomed from his previous visits to finding tortoises on Duncan on the west and south sides. Thus, he was surprised to find them so far from their usual habitat. A heavy growth of lichen, which served as the tortoise's food, covered the bushes in this area. Beck noted that the tortoises here had very little cactus to eat, and the moisture they acquired came from eating the "weeds." A plan was now set for the men to return to skin out the tortoises and carry them back to the ship. But the tortoises had to endure their lashings for at least twenty-four more hours, because the following day was a Sunday and all the men stayed aboard the schooner to rest with "all aboard taking things easy."

On Monday, December 4, 1905, the men looked forward to more serious collecting on Duncan, where the dozen tortoises and more were waiting. This small island, so heavily collected in past years, yielded a disproportionately large number of the total land tortoises collected on this expedition. Fully one-third of the 256 tortoises collected during the expedition came from Duncan, an island thought to have virtually no tortoises remaining.

The tortoise collecting on Duncan became a multi-day relay, shifting men and effort from the anchored schooner to the upland base camp and back again. Camping inland for up to six days at a time, the men increased their overall collecting efficiency. Other men hiked in daily to retrieve the skinned, or sometimes still-living, animals from a common rendezvous point. Everyone worked hard, and the tortoise collection grew by leaps and bounds.

Duncan Island proved to be, in several ways, the ideal place for the men to strike it rich in the tortoise department. The small island offered only so many places for tortoises to hide, although they had eluded ardent collectors in previous years. Camping on the slope of the Duncan caldera was pleasant for the men, because there were no mosquitoes or rats to prevent them from getting a good night's sleep. The distance the men had to travel from shore to camp and back again with tortoises was manageable. Of greatest biological interest, the tortoises of Duncan were noticeably small, which allowed the men to easily carry them down to the schooner. Because of their small size, many of the Duncan tortoises were still alive when they were carried aboard the ship. In fact, they remained alive on deck until it was their turn to be skinned out and pickled.

The men spent a total of sixteen days on Duncan, focusing largely on collecting tortoises. Occasionally they gathered species in their own specialties, but tortoises took precedence on most days. One of their first tasks, though, consisted of establishing a comfortable camp in a small valley near the top of the island. Beck and Nelson selected this strategically located spot for their camp, and they were the first to use it. Several others joined Beck and Nelson during the following weeks.

The men built their comfortable field beds out of tall grass growing abundantly near the camp and slept well most nights. The shore party coming in each day brought additional food and water, and left with tortoises strapped to their backs. Slevin spent several days at a time at the inland camp. His pleasure was transparent when he wrote, "A plentiful supply of doves was always at hand and we all agreed that we had a first class camp." Moreover, Duncan had "nice and cool" weather. All these conditions contributed to the type of collecting they had come to the Galápagos for. Naturally, they made the best of a good opportunity.

Frederick Nelson came into his own during these weeks on Duncan Island. He reveled in his new role as camp cook. During this extended stay away from the schooner, Slevin reported that the job Nelson did "was a first class one," and a week later he gushed, "Nelson is a fine camp cook and we enjoy our dove suppers every night." Good food, plenty of water, a good night's sleep, and plenty of tortoises: what more could a group of young collectors want in the Galápagos?

Another efficient plan was to designate a common "meeting place" at a point where the main trail descended down the side of the island. The men brought recently skinned tortoises to this spot for the daily shore party going back to the schooner to pick them up "and carry down the day's kill." This procedure saved Beck, Williams, Slevin, and Nelson time. Instead of going back and forth from the camp to the schooner, they stayed in camp to continue hunting and skinning tortoises. By specializing for a few days, they maximized their collecting efficiency.

The problem on Duncan Island was not finding tortoises, but getting the large number back to the schooner for processing. "They are quite common," Slevin observed after a couple of days in camp, "and we have no difficulty in running across them just as fast as we can skin them." Contrary to previously published reports, they found the Duncan tortoises more abundant on the southern slopes of the island, where brush covered by thick moss provided ready food and cover. At any given time, Beck and the others staked out several tortoises around the island's valley and adjacent hillsides. Each day the collectors sent down four or five tortoises. The deck of the schooner was soon littered with the reptiles.

Even while ending the lives of so many tortoises, the men jotted down in their field notes occasional observations of living animals. The stomach contents of the Duncan tortoises revealed that they preferred to eat "cactus, grass, and moss." Anyone who has watched a Galápagos tortoise in a zoo, for example, or at the rock wall enclosures at the Charles Darwin Research Station, or on a Galápagos nature documentary can appreciate how the men of the *Academy* encountered these intimidating-looking giants on the slopes of Duncan. "Some of the old males taken would stretch out their necks," Slevin observed. Despite a fierce expression, the tortoises did not attempt to bite the collectors. The oldest tortoises had a thick layer of moss and lichen growing on their backs, which afforded them perfect camouflage, "and at a short distance, [they] looked exactly like the lava blocks which were covered with the same thing."

The men added a camp pet to the homeliness of their upland collecting outpost. On December 7, 1905, the group caught a hawk and "made it fast with a rope-yarn to a small dead tree alongside our camp fire" (figure 9.9). It stayed there for the duration of their camp and "probably never lived so high before in its life" on the abundant tortoise meat they fed it nightly. After nine days the hawk was "cut adrift" and its benevolent treatment had ended, leaving the hawk in a position where it would "have to work for a living." This benevolent attitude apparently did not extend to Hunter, who spotted the released hawk the next day. "While skinning [tortoises] a hawk came and lit a few feet away," Hunter recounted. "A rope yarn on the left foot showed that the fellows in camp had caught him." This was just another bird to Hunter. "I hit him with a stone and then caught him," he explained. "He was uninjured and allowed me to handle him without making any resistance." But even Hunter would not stoop so low as to add this loyal pet bird to the museum's growing collection. "I offered him several pieces of tortoise meat which he readily accepted." When the kindhearted Hunter finally let him go, "he was so full that he could hardly fly."

When rain fell during their time on Duncan, the men found it a mixed blessing. When they awoke one night to find a steady rain falling, they quickly spread the canvas tarpaulin over themselves on the ground in order to "sleep as best we could." As miserable as this unwanted rain left them feeling at dawn, it had the unforeseen benefit of bringing out the tortoises in even greater numbers. Slevin did not seem to mind at all that he had to "take a

FIGURE 9.9 **Upland Collecting Camp.** An efficient way to collect giant tortoises was to set off to the highlands of an island for several days, accumulating tortoises, then transporting them back to the shore and then to the schooner. The men had to deal with ticks, noisy rats, and rain, among other discomforts. In this photo, a Galápagos hawk, which kept the men company for several days, is perched on the right. *Archives, California Academy of Sciences.*

brisk walk down the valley to try and warm up." The weather eventually cleared up, and they took off their wet clothes and hung them up to dry.

Previously invisible tortoises came streaming out of the thorny brush to feed on the thick greenery and to drink water from puddles in the lava. They were soon added to the bounty on the schooner. After drinking from the water holes, the tortoises seemed to be filled to capacity with water. "Those that we found after this," Slevin observed, "were mostly filled with water which seemed to be all through the body and would come out as soon as the plastron was cut into." Thus, in addition to spilling their blood, the tortoises spilled their water.

The tortoise collecting on Duncan Island proved productive and prolific. Nevertheless, after five or six days of camping inland, the men chose to return to the schooner to observe their day of rest. Slevin described their retreat from the highlands to the schooner on December 9, 1905: "In order to take a day of rest and spend Sunday on the schooner," Slevin recounted, "we all left camp at four o'clock." On the way down to the coast, Beck had three small skinned tortoises in a sack, Nelson carried a medium-sized tortoise alive, Williams a small live one, and Slevin a large skinned one. After an hour and a half they reached the coast, where they all "took a wash for the first time in five days."

Despite his heavy load of tortoises, Beck managed to collect numerous doves on the way down for Sunday dinner. With twenty-nine tortoises packed out during the week and twelve more staked out in various places on Duncan Island, they retired to the schooner with "all hands aboard ship keeping the Sabbath" before returning on Monday for another six days of camping inland. After a week ashore they were rewarded with "doves and cocoanut pie for Sunday dinner." Williams added to the larder by catching twenty fish of various kinds.

Slevin was concerned about the growing backlog of tortoises that had accumulated on the schooner over the past two weeks. He knew he would have to spend a few days getting "things in the tortoise line fixed up." The tortoise catch on Duncan was unexpectedly good and somewhat overwhelming. Slevin noted his concern when he wrote, "with such a grand rush they could not very well be in the best condition." Although Beck spearheaded the tortoise collecting, Slevin took on the grizzlier shipboard task of getting them skinned out and pickled and the more meticulous task of getting them cataloged. Data collection on the tortoises was an important part of the fieldwork; without solid documentation, scientific studies of the museum specimens would be rendered nearly worthless.

Yet sometimes data collection under such circumstances proved haphazard. The meticulous Slevin lamented the sloppiness of some of the other men. They sometimes skipped entering the measurements Slevin normally took on each specimen, but only when Slevin happened to be "at work elsewhere" and unable to supervise data collection. But even Slevin was not perfect. Under the rigorous field conditions, some of the data "were lost from the field notebook on account of rain, and some from wear." Ironically, while a population of tortoises was being rubbed out, their measurement data were likewise reduced to a smudge in a field notebook.

Ashore on Duncan, the men continued to enjoy dove suppers every night, making a significant impact on the dove population. The noticeable change merited an insightful comment from Hunter: "The flock of doves in the crater shows the effect of our visit in its greatly reduced numbers." Beck was less pessimistic about their impact on the dove population, noting after two weeks that "doves [were] still plentiful." He "shot 10 for breakfast" and collected "40 for Sunday dinner tomorrow in ½ hour." Their daily consumption of doves was steady, with "12 to 16 every day as well as 120 for [the] ship."

The tortoise population suffered its own parallel diminution. The men carried eighty Duncan tortoises downhill to the schooner between December 2 and 17, 1905. Hunter reflected in his field notes that these tortoises were not even the last of the Duncan Island race. Although neither extermination nor extinction of species was a goal of the expedition, the men collected without regard for either of these possibilities. They simply collected because collecting was good. Of the eighty Duncan tortoises, Hunter speculated, "there are to my belief as many more on the island and likely more yet." Apparently the Duncan Island tortoise had a grip on the muddy slopes of that tiny island that would not be easily dislodged. Beck was less optimistic than Hunter about the remaining tortoise population and was surprised by how many tortoises they actually found. Beck believed that perhaps only forty to fifty remained when the expedition left Duncan, and he originally thought they would get no more than twenty specimens from the island. Yet Beck was incorrect on both points, which is a tribute to the health of the Duncan tortoise population.

Physical injuries to the men in the field were rare (or perhaps only rarely reported). Beck's notation on December 16, 1905, proved an exception. Toward the end of their stay on Duncan, he wrote, "Stewart in today as his twisted ankle is better." When and how Stewart, the botanist, injured his ankle remains unclear.

With the eighty newly collected tortoises from Duncan Island in various stages of skinning, pickling, and preservation, the expedition weighed anchor and set sail on

December 18, 1905, for Jervis Island (Rábida). This even smaller island, due north of Duncan, yielded at least one big payoff. Sailing north and east, the men passed close by a group of three nameless islands that were actually part of the side of an old volcanic crater. Yet they did not stop because the tiny islands were barren and did not appear very promising from the deck of the schooner. They also sailed past a small island to the east of Jervis, but likewise gave it a pass, as nothing of interest caught their attention. After a few hours of sailing they reached Jervis, dropped anchor in seven fathoms, and sent a collecting party ashore on the scenic "red lava sand" beach, still a popular ecotourism visitor site today.

Slevin, Nelson, and Williams, who broke out some provisions for storing in the ship's lazaret, attended to domestic duties on the schooner. They also stored about 45 of the newly collected marine turtles and land tortoises below deck in the hold, raising the trip total to 101 tortoises. After working all day on the tortoise backlog, Slevin managed to get "the mess straightened out as well as possible."

From their convenient anchorage at Jervis, the men directed their efforts over the next several days in two directions. While most of them went straight ashore on Jervis to explore and collect specimens, some of the men made two separate side trips in the sailboat to the nearby coast of James Island. Spreading out allowed them to effectively cover more territory without having to move the schooner and give up their safe anchorage.

Gifford's early reports indicated that Jervis Island looked promising. He reported, "Lots of old tortoise trails and old droppings." Just how promising the men would not know until they explored the island further. Once again the expedition intersected the human and scientific history of the archipelago. Slevin confirmed the tortoise potential on the island when he later went ashore on Jervis and hiked to the top of the island with Williams, determining with their barometer that the summit was 1,050 feet above sea level. "I saw old tortoise trails in the ashes," wrote Slevin, "and one trail ran right up a valley clear to the top." So far they had not actually seen any living tortoises on Jervis, but Slevin "saw lots of old dung that was apparently rained on as it was bleached out rather white." They had to look farther on the island to find the maker of the bleached dung.

Beck, Ochsner, Gifford, and Nelson made up the first group to cross the five-mile channel between Jervis and James in the small sailboat. While on James Island on December 19, 1905, Beck and the others hiked three miles over very rough lava of several kinds and textures to reach a small crater. No particularly noteworthy specimens rewarded the men's difficult trek across the lava. Moreover, their return back to the coast, and then all the way back across the channel to the schooner at Jervis, took much longer than anticipated.

The sailboat sustained slight damage getting out of the lagoon when its keel struck a submerged reef. Meanwhile back on the schooner, the other men became anxious when the small sailboat did not appear by dark, and dinner was delayed while they waited for Beck, Ochsner, Gifford, and Nelson's return. Progress in the sailboat remained agonizingly slow. Loaded down with four men, two sea turtles, and a marine iguana, the boat also had to buck a headwind all the way back across the channel. With the anchor light on the schooner thoughtfully raised high to serve as a homing beacon, the men in the small sailboat successfully navigated back to the anchorage by 9:00 p.m.

The following day, Nelson and Slevin beached the sailboat opposite the schooner's anchorage and repaired the damaged keel. Finishing the repairs by 11:00 a.m., they set off on a second trip to the same lagoon where Beck and the others had landed the day before. Many wild donkeys, both on the higher slopes and at the landing place, inhabited the coast of James Island where the men landed. That day, Slevin and Nelson's primary objective was the plentiful green sea turtles in the lagoon. They collected four turtles, one of which was the largest yet taken, from under the edges of the mangroves lining the lagoon. Slevin noted that they had not yet seen any hawksbill or loggerhead turtles in the islands. After a short trip ashore for lava lizards, Slevin and Nelson sailed back to Jervis without incident and in time for dinner.

While Slevin and Nelson hunted turtles over on James Island, the other men went ashore on Jervis and stumbled across the most historically interesting specimen of the trip so far. Back in 1891, while collecting tortoises and other specimens on twelve islands in support of his geological subsidence theory for the origin of the Galápagos organisms, the German-born and -educated Georg Baur had intentionally transplanted a tortoise from Albemarle to Jervis. Now fourteen years later, Rollo Beck found the 350-pound female tortoise "at the head of a slide of cinders about 500 feet up the hillside." What Georg Baur had left behind, Rollo Beck carried home. Georg Baur's experiment turned into Rollo Beck's trophy.

Beck decided on December 20, 1905, to bring the behemoth back to the schooner alive, which required several men to return ashore in the afternoon with lashings to pull it down the inclined slope. Once at the beach with the tortoise, they used their technique of wading out with the floating tortoise into waist-deep water before hoisting Baur's relict into a skiff for the short row back to the mother ship. The tortoise lived another fourteen days on the deck of the schooner, before it was killed on January 3, 1906, and assigned the anonymous museum number of 8134.

Hunter continued his active pursuit of birds and other animals for the two days they spent on Jervis Island. "In a cove on the East side of the landing place were a couple of seals," Hunter wrote in his field notes. "I thought they were fur seals and shot one with the rifle. It turned out to be a female hair seal." While the others were focused on their specialties or taking the sailboat over to James Island, Hunter walked the entire island from top to bottom in search of birds. Working around the shallow lagoon near the anchorage, Hunter "shot a pair of yellow crowned night herons in a tree near the lagoon." He remarked about the ease of their capture, stating, "They were very tame, not minding the shooting around them."

But not all the birds in the lagoon proved to be so tame or so easy to collect. Hunter had to be quick and resourceful with one particular duck. "It was," he observed, "for a Galapagos duck quite cautious, not coming within range of the gun. I surprised it in one corner of the lagoon and killed it." Mockingbirds were abundant on Jervis, and Hunter shot eleven, all of which were actively breeding, as revealed by the condition of their ovaries and testes. Few entries in his journal are more poignant, however, than the short line about Jervis mockingbirds that reads, "One pair were seen *in copulo* and were shot."

Before leaving Jervis after another half day of collecting on December 21, 1905, the young field assistant, Ernest King, went ashore and caught a few marine iguanas. Unlike the iguanas on some of the other islands, the Jervis iguanas were all small and rather timid, "so some [had] to be shot."

JAMES BAY: A HOTBED OF HISTORY

The next stop for the schooner *Academy* was the history-laden James Bay on the northwest coast of James Island (Isla Santiago), which they reached easily on December 21, 1905. This bay, with its coffee-colored beach, was a very well known "port of call." Charles Darwin disembarked there from the *Beagle* for an extended shore leave in October 1835. Darwin was equally interested in getting away for a few days from the source of his perennial seasickness as he was in exploring and collecting on the island. The Academy collectors had the luxury of spending two weeks at James Bay on this first visit, twice the time Darwin had spent ashore.

Another interesting episode in Galápagos history took place at James Bay in 1813, during the War of 1812. Buried in a makeshift grave on the beach at James Bay was Lieutenant John S. Cowan, who had been killed in a pistol duel with Lieutenant John M. Gamble on August 10, 1813. Both men served under Captain David Porter, who brought the thirty-two-gun US frigate *Essex* to James Bay during its voyage of destruction of the British whaling fleet off Galápagos waters. Cowan's final resting place went unnoticed by the Academy collectors. The exact location of Cowan's gravesite is currently unknown.

James Bay was the final anchorage of the schooner *Academy* in 1905. They spent the next two weeks anchored there and fanned out to make multiple trips ashore. The men celebrated both Christmas and the New Year in this history-laden bay, in a suitable, if low-key, Galápagos style. Over the next two weeks, the men variously collected onshore during the day, rowed offshore in a skiff for seabirds, or made extended camping trips inland to reach the more inaccessible parts of the island. Just to the south of their anchorage, they saw a large recent lava flow. They also noticed that the island was heavily wooded with "a fine green zone at the higher elevations." There was every indication that James Bay would be a productive locality worthy of an extended stay.

Their anchorage at James Bay appeared ideal at first, but in the coming days it revealed its subtle treachery. The men found out that a strong undertow existed and that heavy surf made it difficult to get on and off the beach in their skiffs. Danger also circled them below. Whenever they threw turtle or tortoise meat overboard while preparing specimens for the pickle tubs (figure 9.10), large sharks circled the schooner grabbing up the scraps. Scraps of turtle meat and fat too small to interest sharks drew in the seabirds. Beck and King rowed out a short distance from the schooner in a skiff to shoot petrels and capitalized on this bounty of floating flesh. Beck wrote that he "got 20 Graces petrels astern after fat of turtles." But he was disappointed that "no Aestrelata [had been] seen since leaving South Indefatigable."

Over the next few days, from their arrival on December 21 until Christmas Day, the men engaged in general collecting or staying aboard the schooner to finish off the backlog of specimens. For Nelson and Ochsner, one of the first orders of business meant going ashore and heading inland to look for fresh meat. On their first attempt, they saw one pig but returned empty handed. The following day they saw several and shot two. Williams and Gifford also went well up into the green zone and reported back that although the island was heavily

FIGURE 9.10 **Pickle Tubs.** After the giant tortoises were skinned, some tissue still adhered to the inside of the shell; often the legs and neck were not skinned and were left attached to the shell. With the neck and legs turned into the shell, the entire shell was placed for a couple of days into a pickle tub that contained a solution to eliminate the remaining moisture to prevent rotting. *Archives, California Academy of Sciences.*

wooded, the undergrowth was not as dense and impenetrable as it had been on Indefatigable Island. This was promising tortoise territory.

Hunter stayed busy in his enthusiastic pursuit of birds. He was one of the few men on the *Academy* whose field notes more precisely revealed the area of James Bay. He commented that "there is a fringe of green trees along the beach; behind this there is a 3 acre lagoon up to 2 feet deep." This was prime waterbird habitat, and Hunter wasted no time. He described the next sequence of events as he readied himself: "There were a number of ducks on the water. They were very inquisitive, all swimming toward me to see what I wanted." Hunter wasted no time demonstrating his intentions. He also wrote, "There was one flamingo [in the lagoon] which I secured after a little trouble. It was rather wild and would not allow close approach without trickery." Inquisitive or not, these birds were all fair game for Hunter.

Having demonstrated his resourcefulness with the timid flamingo, Hunter "then went after the ducks [and] secured a dozen, many with the aux." Whether the birds were timid or tame, Hunter was a persistent and thorough collector. But even he failed to get close enough to a diverse group of other shorebirds around the lagoon, all of which he wrote off simply as "wild" when they did not tolerate his approach. He grew frustrated in his attempts to shoot any of the black-necked stilts, turnstones, semipalmated plovers, or curlews. They quickly scattered whenever he approached. As much as Hunter preferred the "tame" birds and became frustrated by the "wild" ones, he also noticed that his actions influenced bird behavior. "Small birds were rather common early in the morning," Hunter observed, "but when I began to shoot they had thinned out so that there were not many."

He also noticed close to two dozen flamingo nests on the east side of the small lagoon. The nests were constructed as "circular platforms of mud about 8 inches high and having a slight cavity on the top with no lining." Although he saw no fresh eggs in the nests, several rotten eggs were lying around.

Hunter was determined to get a pig for Christmas dinner, despite the fact that Ochsner and Nelson had already shot two. He chose an area "behind the needle rock between the two lagoons" where he was sure there was "quite a stamping ground for pigs and burros." Following fresh trails and going past a "place where a burro had laid down a few hours before," he was in hot pursuit of dinner. "I jumped one pig near the needle rock, but it was too wild to get a shot at." But before long he "jumped another in a small cave" that turned out to be a 175-pound boar with white hair and black spots. Pork was now plentiful on the schooner, with at least three pigs having been shot over two days. All of the pig hunters proved themselves to be able providers, and the entire crew celebrated Christmas dinner, their 181st day out of San Francisco, with a "dinner of beans and wild boar" and the cook treated the men to a plumduff for dessert.

Because Christmas fell on a Monday in 1905, the men had the rare treat or perhaps burden of two days of rest in a row. All hands kept the usual Sabbath routine with no collecting party ashore. They had plenty of time to mend shoes and clothing (figure 9.11), and Beck spent part of Christmas day loading shotgun shells in preparation for the next round of collecting. Beck tersely described these two days in his field notes as either "resting" or "nothing doing."

When their Christmas holiday was over, the men went ashore and branched out on longer collecting and exploring excursions. Five of them took the big skiff and pulled around the

FIGURE 9.11 **Field Repairs.** During the expedition, the men had to repair their clothing as well as their boots. They often mention that their clothing was torn by thorny plants and their boots were worn down by the volcanic rocks. *Archives, California Academy of Sciences.*

north end of the island to Adams Cove. The excursion consisted of two groups with different intentions: Beck and Williams planned to go inland for tortoises and to set up camp for three days, while Slevin, Gifford, and King intended to collect near the landing place. Yet getting ashore in the cove proved challenging. They had to go through "quite a heavy surf" before successfully beaching the skiff.

Initially, young Ernest King accompanied Beck and Williams a couple of miles inland with a can of water that was cached for the return trip of the tortoise hunters. King then returned to the coast to join Slevin and Gifford in general "hunting" and collecting in the lowlands. Human history was all around them, and they again intersected the trajectory of tortoise exploitation that extended back many years. They came across the remains of an old tortoise hunters' camp located just back of the beach. It was mute evidence of the rich tortoise fauna that previously existed on James Island.

Whales were another resource that had, in the past, attracted many ships to anchor near this bay. The very land itself bore witness to visitation and exploitation. Near the landing place, the Academy men sighted inscriptions from several old New Bedford whalers, carved in the soft volcanic ash. These nineteenth-century visitors to Buccaneer Cove had left behind clear evidence of their brief landfalls: the names of the captain and mate, the number of days they were out, and the number of barrels of oil obtained. Slevin jotted down this historical information in a small notebook. Unfortunately, it was lost that same day when the skiff was swamped several times while he and Gifford were rowing back out through the surf. To this day, that information remains undocumented.

The tortoise-hunting group of Beck and Williams trekked inland through heavy brush and cactus for several hours before making camp at an elevation of about 1,900 feet. They observed plenty of trees at this elevation, but no green zone. Along the way they stopped at an abandoned tortoise camp and found nothing but bones scattered all around. Perhaps Georg Baur's dire predictions of the imminent demise of the tortoises were not so far-fetched. This type of ghostly boneyard was one of the underlying motivations for the entire expedition. Beck picked up some of the best specimens and set them aside for the return trip to the coast. Williams saw an old piece of dung at the same place, but no fresh signs of tortoise activity. The earlier hunters had left behind more than just the bleached bones of tortoises. Beck found "one blue bottle and pieces of crockery" in the Galápagos equivalent of an archaeological dig.

Rats running around in the bushes kept the men awake at their first campsite. Despite a sleepless night, Beck and Williams ate an early breakfast before pushing on for another mile to the green zone to set up a second camp. After taking a "long tramp around in a northerly direction," Beck discovered the remains of an old tortoise shell. Yet he searched in vain for live tortoises or any sign of them. Beck and Williams managed to follow the good trails through the thick brush made by the numerous donkeys and pigs in the green zone. Still, their main prey eluded them. Of the one dead tortoise Beck found, he wrote, "Tortoise bones along the way probably 10 years old." He found one plastron "in [a] cave where [the] tortoise had fallen in," but the other bones of the skeleton had rotted away.

After Christmas, Hunter continued his bird collecting near the lagoon. In his unique style he jotted down that "an oyster catcher came up to investigate the landing, his curiosity is satisfied, and he is now numbered among the famous." Hunter also continued to pursue the slender pink flamingoes around the lagoon, finding them every bit as delicate as they looked.

"The flamingoes were rather wild," Hunter observed, "but are easily killed even at long range, even if only one or two shot strike their body." Other birds he simply observed from a distance, such as a hawk that "caught a small lizard [and] swallowed it without tearing."

From their base at James Bay, Hunter and Ochsner also set off on a trek inland to camp overnight. They made rather slow progress because of the large amount of water they had to carry and the thick undergrowth, and because they got in each other's way. Judging from Hunter's field notes, the image of these two collectors plodding upward on James Island was almost comical. "The dead branches snapping back in your face and eyes" were partly responsible for the slow progress. They faced other obstacles as well, with plants taking on sinister qualities. "Vines are numerous," Hunter wrote, "and seem inclined to wrap themselves around your neck and feet." By noon they had struggled only to 1,000 feet in elevation. Later that day they made camp at 1,650-foot elevation, putting them just slightly lower in elevation than Beck and Williams, who camped inland farther to the north.

The next morning, Hunter and Ochsner were visited by numerous mockingbirds that came to investigate their cooking fire and general activity. One of the mockingbirds sampled their breakfast of "canned beans, tongue, and hard tack." Hunter made the unique ornithological observation that the birds liked the beans best, but "ate only sparingly" from the unfamiliar food.

Hunter and Ochsner's camp was overrun by rats. Although Hunter set out traps for them, he caught only two during the night. If he was going to be up at all hours, he decided to make use of the time and shot a short-eared owl at 1:00 a.m. From their camp, the men climbed upward, eventually reaching the top of one of the peaks where ferns and grasses grew in abundance. Their barometer read 2,050 feet above sea level. To the north they saw yet another peak that was fully 500–800 feet higher. Hunter became more discerning about which birds to shoot, stating that "the usual common species occur near the top" but that "no rare species were seen."

At night the constant braying of burros that had been introduced to the islands in years past added to the men's sleeplessness. Perhaps out of frustration and loss of sleep, the men shot two burros, a large female and a smaller young male. They only wounded the male with the first shot, and he ran off through the brush. In the next few minutes they heard "the most horrible cry" in the distance from the wounded animal, which was in severe pain. This prompted all the burros in the vicinity to commence braying at once. Through the cacophony, Hunter tracked down the wounded young male, which "had been wounded through the belly and was giving his death cry."

Hunter was persistent in his pursuit of birds and pigs, and tried unsuccessfully to shoot several rails during the three-hour walk back to the coast. Yet they proved too elusive because they "keep under cover very well and it is almost impossible to get them." One afternoon, after Hunter had spent time getting a dozen ducks, he went looking for pigs. "The ducks were as usual very tame," Hunter noted. "I shot twelve out of a flock of fifteen before they got wise and left." Perhaps Hunter acted as an unwitting agent of natural selection. He spent that night on shore trying to get a pig. The pigs, though, turned out to be more elusive than the waterbirds, which were "wild and do not allow very close approach." Hunter was forced to return to the schooner in defeat. "The only thing that I did get," Hunter wrote sheepishly, "were about a million mosquitoes. Between them and watching for a pig I got about half an hour's sleep."

Never one to miss a collecting opportunity, Hunter worked through his nighttime disappointment and frustration to maintain his vigilance for birds. "One flamingo came into the

FIGURE 9.12 **Beck Collecting Flamingoes.** Rollo Beck in the flamingo lagoon at Punta Cormorant on Charles (Floreana) Island. The Academy collectors were intent on bringing back representatives of as many species as possible to the museum in San Francisco. *Archives, California Academy of Sciences.*

lagoon during the night," he recorded. The nervous flock of stilts set up a loud cry when the flamingo landed, which then answered with two or three loud, hoarse honks. "I shot the flamingo and five ducks" (figure 9.12).

Beck and Williams also had mixed success on their camping trip. On their last day inland, Beck went off in search of tortoises, while Williams collected around their campsite. Not finding any live tortoises, the men returned to the coast with only the bleached bones of the long-dead specimens. At the coast, Nelson picked them up in the skiff, as previously planned, and returned them to the schooner for the night. The next day, Beck again set off inland from the landing at James Bay on a two-day solo trip to look for tortoises. He was again unable to locate any tortoises, but he did manage to secure five of the elusive rails that Hunter had failed to get. Beck's technique was to trick the birds into calling out and revealing their location in the brush. He would clap his hands, snap his fingers, or break dead limbs to elicit the rail's call and reveal its location. This, of course, turned out to be the rail's last call. Beck knew just what technique to use to get specific types of birds, and this was another demonstration of his ability as a virtuoso collector.

More frustration was in store for Hunter as he doggedly persisted in his attempt to get a pig. He and Nelson set up camp near the coastal lagoon, but during the night they did not see or hear a pig. However, they reported back to the schooner that they were "nearly eaten alive by mosquitoes." Hunter shot some ducks and a flamingo, salvaging the ill-fated camping trip from being a total failure.

FIGURE 9.13 **Sea Turtles.** In addition to the large and prominent land reptiles, the 1905–1906 expedition members also assembled one of the largest collections of sea turtles from the Galápagos. *Archives, California Academy of Sciences.*

Slevin did not go on any of the overnight upland excursions. He continued to pursue and collect sea turtles, even recording their behavior (figure 9.13). "Had a chance today to see how the turtles move on the beach," he noted. "They raise themselves on all four flippers at once and throw the body forward, coming down with a thump." Their slow movements seemed to be, understandably, quite an exertion when they were out of their element. "But in the water they are extremely fast," Slevin observed, "swimming with the forward flippers and steering with the hind ones." James Bay had abundant sea turtles, and the men saw "a great many copulating." The male turtles were similar in behavior to the one the men had seen in the lagoon on northern Indefatigable Island. When they were in the state of *corpus delicti* with a willing female, the male turtles were "easily approached." The biological drive to copulate overrode the turtles' usual concern for safety. The extent of Slevin's experimental observations and his firsthand experience was revealed when he wrote in his field notes that the males "will still hang to the female if you try and pull them off."

Recuperating from their recent work ashore, and in anticipation of the coming fieldwork, the men mended clothes and repaired shoes on New Year's Eve. Having had multiple failures on the pig and tortoise fronts, the men now gathered together on the schooner for the first time in several days to settle in for the traditional holiday. The only "decorations" for the occasion were a group of balloon fish that swam around the schooner when Beck threw a few bird bodies over the side. Williams scooped up several of the fish with a small net and added them to the collection. As 1905 drew to a close, Theodore Roosevelt occupied the White House. A peaceful settlement had recently been reached in the Russo-Japanese War, and not far away in Central America work was progressing on what would become the Panama Canal. For the Galápagos Islands and for the US government, the canal would be of supreme concern in World War II. But that was still some forty years away, and the First World War, the war to end all wars, was not yet even a remote possibility.

Some of the men on the schooner *Academy* planned to stay up late on New Year's Eve to usher in 1906. They talked among themselves late into the night to find a suitably low-key way to celebrate. Others headed below deck for the comfort of their bunks. All were pleased with their success so far and could only speculate about what the New Year would bring. Loomis would be proud of them.

NOTES

Page 93: *Epigraph*: Darwin (1958, 28).

Page 93: *Epigraph*: Slocum (1954, 294).

Page 93: *The Andes were visible*: RHB, 36, September 19, 1905; LSA, 33.

Page 93: *A sailor explained to a young girl*: Twain (1897, 60).

Page 93: *Neptune came aboard*: RHB, 35, September 18, 1905.

Page 93: *We had no fool ceremonies*: Twain (1897, 60).

Page 94: *Crossing the equator for the first time*: Twain (1897, 50).

Page 94: *Try to shave remainder of crew*: Gifford diary, September 18, 1905.

Page 94: *All hands turned out for breakfast*: Curiously, there are no known photographs of James White, the cook, whom Beck always called Jim in his field notes.

Page 94: *Great abundance of lichens*: Stewart (1912).

Page 94: *Beaten with rifle butts*: "The first bullet had perforated [Cobos's] left lung, causing a massive hemorrhage; the second had passed through his intestine, also causing serious injury. Cobos was mortally wounded" (Latorre 1990).

Page 95: *I have seen the albatross*: Murphy (1947, 82).

Page 95: *Expression to the face*: JSH, 2–3.

Page 96: *Mockers will eat almost anything*: JSH, September 28, 1905.

Page 96: *Odd amalgamation of nesting materials*: Lopez (1998, 53).

Page 96: *Excessive killing by buccaneers and whalers*: Three quotes from VOB, 384.

Page 97: *Approach them slowly; Lay in a few skulls*: RWE, 13.

Page 97: *Laden with specimens*: LSA, 35.

Page 97: *Basis of three papers*: Van Denburgh (1912a, 1912b), Van Denburgh and Slevin (1913).

Page 97: *Weapon for collecting*: RWE, 13.

Page 97: *Coloring under the throat*: RWE, 14.

Page 98: *Lived off the land*: Evans (1985, Williams quote, 207).

Page 98: *A wonder they didn't make them extinct*: JSH, July 24, 1906. Darwin (VOB, 384) observed a young boy "sitting by a well with a switch in his hand, with which he killed the doves and finches as they came to drink. He had already procured a little heap of them for his dinner; and he said that he had constantly been in the habit of waiting by this well for the same purpose."

Page 98: *Cold goat for lunch*: LSA, 36

Page 98: *Beck was a devout Methodist*: Larson (2001, 129); Anonymous (1912).

Page 99: *Variously employed on Sundays*: Nearly any Sunday entry in SLA during the entire expedition.

Page 99: *Following their late breakfast*: LSA, 38.

Page 99: *Usual catch of insects*: Peck (2001, 2006).

Page 101: *An extra heavy swell*: JSH, October 3, 1905.

Page 101: *Others got thoroughly soaked*: JSH, October 3, 1905; LSA, 38; RHB, 40.

Page 101: *Mocker not as inquisitive*: JSH, October 3, 1905.

Page 101: *Usual number of lizards*: RWE, 14.

Page 102: *Barrel Post Office of the Galápagos*: Slevin (1950, 1959).

Page 102: *Post office barrel*: LSA, 39. Slevin (1950).

Page 102: *Hunt for Galápagos specimens*: Larson (2001, 107).

Page 102: *Anti-evolutionist Louis Agassiz*: Louis Agassiz had himself visited the Galápagos in 1871 aboard the US Coast Survey iron steamer *Hassler*. The trajectories of the *Hassler* and the schooner *Academy*, when she was the *Earnest* with the US Coast and Geodetic Survey, had crossed in 1892–1993 when the *Earnest* was towed behind the *Hassler* as a "coal hulk" for carrying coal to Alaska during the Alaska boundary survey. The annual report of the Survey for the fiscal year between July 1, 1897, and June 30, 1898, does not contain any mention of the *Earnest*, except for the following tersely worded sentences: "The steamer *Hassler* and schooner *Earnest*, which formed part of the equipment of the Survey for many years, have been disposed of during the present fiscal year. The *Hassler* having been condemned as unseaworthy and unfit for further use in the hydrographic work, was sold in October, 1897; her purchasers fitted her up for the Alaskan trade, and she was lost on her first trip to Lynn Canal. The *Earnest* was also considered to be no longer available for Coast Survey work, and was transferred absolutely to the commandant of the Puget Sound Naval Station, in March, 1898 (p. 162).

Page 103: *Cats have about exterminated*: LSA, 40.

Page 103: *Back at the beach*: LSA, 39.

Page 103: *Exotic gastronomic treat*: JSH, October 4, 1905

Page 103: *Flamingoes let out a honk*: LSA, 40.

Page 103: *Did not shoot any small birds*: JSH, October 5, 1905.

Page 103: *Beck became exasperated*: RHB, 41, October 5, 1905.

Page 104: *Most amusing manner*: JSH, 14, December 26, 1905.

Page 104: *The greatest curiosity*: LSA, 41.

Page 104: *They are so tame*: JSH, October 9, 1905.

Page 105: *Rainfall and island vegetation*: Stewart (1911, 1915).

Page 105: *Belt of vegetation*: JSH, October 9, 1905.

Page 105: *Corn-starch pudding and ripe oranges*: LSA, 41.

Page 105: *Flamingo tastes somewhat like duck*: LSA, 41.

Page 105: *Cut anybody's throat with pleasure*: LSA, 42.

Page 105: *Judging from their appetites*: LSA, 43.

Page 106: *Fine looking army*: LSA, 43–46.

Page 106: *Cursed by God and Nature*: Larson (2001, 15–34).

Page 106: *Huge red pompon on his hat*: LSA, 43–44.

Page 107: *Beck jumped through hoops*: LSA, 43.

Page 107: *El Niño event of moderate intensity*: Quinn and Neal (1992); see table 32.1 for 1904–1905 El Niño intensity.

Page 107: *A misty rain began falling*: JSH, October 16, 1905.

Page 107: *Namesake of the town*: Alfredo Baquerizo Moreno was vice president of Ecuador under General Leónidas Plaza Gutiérrez in 1901; Baquerizo Moreno later became president of Ecuador twice, in 1916 and 1931. General Plaza was replaced as president by Lizardo García in 1905. García was in turn killed during a coup d'état by Eloy Alfaro in 1906. Alfaro had been defeated by Plaza in the 1901 election.

Page 107: *Naming of Plaza Norte and Sur*: Personal communication, Ricardo Crespo Plaza, November 2001. The Plaza islands were named by Leónidas Plaza in honor of his two daughters, Alégria and María. Leónidas Plaza also had a son, Galo Plaza Lasso, who was president of Ecuador from 1948 to 1952.

Page 108: *Insides on fire*: LSA, 45.

Page 108: *Stewart's own field notes:* West and James (2001).

Page 109: *Natives used to eat them*: RWE, 18–19.

Page 109: *Nowhere is the wind so light*: Melville (1856).

Page 110: *Walking about the island*: LSA, 47.

Page 110: *Barrington goats*: LSA, 47.

Page 110: *Free from their burrows*: VOB, 373.

Page 110: *Pull my tail*: VOB, 373.

Page 111: *Iguanas into burlap bags*: RWE, 19.

Page 111: *Nod their heads vertically*: Darwin (1845, 374).

Page 112: *Aggressive behavior on Barrington*: RWE, 19; JSH, October 23, 1905. Hunter also witnessed aggressive and predatory behavior. He shot and wounded a bird called an "indefatigata," "when almost immediately another bird of the same species flew down beside it and commenced pecking at the wounded bird. The wounded bird flew followed by the other and took refuge under a root where it was viciously attacked by the other. I almost caught the second bird in my hand it was so intent on fighting the wounded." Hunter also saw two hawks "eating a dead iguana."

Page 112: *Cleaned out the entire iguana population*: RWE, 66.

Page 112: *Despite the visits of the natives*: RWE, 67.

Page 112: *Met a fellow named Captain Thomas Levick*: For a photograph of Levick, see Latorre (1999). It is not clear from the literature if this is the same schooner that brought the locals who cleaned out the land iguana colony.

Page 112: *Frankly informed the collectors*: RWE, 66.

Page 112: *Killed off long ago*: RWE, 66. Full quote from Slevin: "Captain Levick of the schooner that runs between the islands informed us that 30 years ago tortoises were found scattered all over Barrington and that he had taken them off of there. He doubted very much whether we would find any as he said they were all killed off long ago."

Page 112: *No signs of tortoise*: LSA, 51.

Page 113: *The agent of extinction*: Latorre (1990, 68, 89).

Page 113: *The meat to my taste is indifferent*: VOB, 362.

Page 113: *Nelson went out fishing*: Even today these fish, locally called *bacalau* (*Mycteroperca olfax* Jenyns, 1840), are highly prized in the islands and are often on the menus of ecotourism boats and in the small restaurants of Puerto Ayora.

Page 114: *The incorrect assumption*: The introduced and nuisance ani bird is assumed to eat ticks, but it does not. The Spanish word *garrapata* means "tick" and refers to the feet of the parasite that grab onto a host.

Page 114: *With the sun riding low on the horizon*: LSA, 48.

Page 114: *Go over the ship*: JSH, 20, October 24, 1905.

Page 114: *Bleached white by the sun*: RHB, 47; RWE, 21.

Page 115: *So he would not travel inland*: RWE, 22.

Page 115: *In tortoise country*: RWE, 21.

Page 116: *Rough time*: JSH, 20, October 25, 1905.

Page 116: *First day of tortoise hunting*: LSA, 49.

Page 116: *Dug a miniature crater in the ground*: LSA, 49.

Page 116: *While the flesh was still pliable*: Most of the 1905–1906 tortoises remain in storage in this posture. The specimens are safely housed in the Department of Herpetology at the Academy's museum in Golden Gate Park in San Francisco.

Page 118: *A little oily and of a peculiar taste*: LSA, 49.

Page 118: *The liver of the Indefatigable tortoise*: LSA, 49.

Page 118: *The female urinated on the nest*: RHB, 47.

Page 118: *Embryos out of four eggs*: RWE, 22.

Page 119: *Blow some eggs*: LSA, 49–50; RWE, 22.

Page 119: *Beck's remarkable toughness*: RHB, 48, October 27, 1905.

Page 119: *Tied her up*: RHB, 48, October 27, 1905.

Page 120: *Our shoulders are getting sore now*: LSA, 50.

Page 121: *Mr. Spreckles's Hawaiian sugar*: Upon leaving San Francisco in June 1905, a tug owned by Mr. Spreckles had towed the schooner *Academy* out through the bridgeless Golden Gate.

Page 121: *Opportunity to wash clothes*: LSA, p51.

Page 122: *Not tasted fruit for some little time*: LSA, 51.

Page 122: *Gave Ochsner's pistol a minute examination*: LSA, 51.

Page 122: *Soft mud and quicksand*: JSH, November 1, 1905.

Page 123: *Very nearly swamped us*: LSA, 52.

Page 123: *Good soaking getting through the surf*: LSA, 52.

Page 124: *A specimen of small ground finch*: JSH, November 2, 1905. Years later, a particular gene of Galápagos tomatoes for salt tolerance would figure prominently in the genetic manipulation of tomatoes.

Page 124: *The potatoes long since used up*: LSA, 52.

Page 125: *Constructed a sturdy tortoise pen on deck*: LSA, 52. Little is known about the specific structure of the schooner *Academy* and her earlier incarnation as the Coast Survey's *Earnest*.

Page 125: *Crossed the channel in good time*: RHB stated that they were "drifting in evening toward Duncan" (50), which is in the opposite direction from Barrington.

Page 125: *We struck on a submerged reef*: LSA, 53.

Page 126: *Hunter saw two pelican nests*: JSH, November 5, 1905.

Page 126: *Named Academy Bay*: James (2012); LSA, 53. Larson (2001, xi) stated that Academy Bay was "named for the California Academy of Sciences in far-off San Francisco." However, the actual namesake of the bay is only a technicality because the schooner was in fact named for the museum when it was rechristened in June 1905, but the bay was specifically named for the schooner by her crew, and that important distinction should be made clear. Sometimes "Academy Bay" is translated into Spanish as "Bahia Académica," which is well intentioned, but incorrect. The Spanish word *académica* is an adjective equivalent to "academic" in English. So if the Spanish is translated in this way, the present bay where the town of Puerto Ayora is located becomes "Academic Bay," which was clearly not the intention of the crew of the schooner *Academy* when they named the bay after their vessel. The Spanish *academía* is equivalent to the English "academy." So in Spanish translation, the name should be either "Bahia Academía" or remain as originally intended and be called "Bahia Academy."

California Academy of Sciences field collectors have had a penchant for naming geographic places during their expeditions. In addition to naming Academy Bay in 1905, they proposed the name "Academy" for the highest volcano on Socorro Island in the Revillagigedo group in 1903. During their 1925 expedition to the Revillagigedo Islands, new geographic names suggested by the Academy and accepted by the Mexican government included Mount Gallegos (highest mountain on Clarion Island, named for the expedition member José M. Gallegos), Mount Evermann (central peak of Socorro Island, named for the Academy director at the time, Barton Warren Evermann), and Ash Heap (highest elevation on San Benedicto Island).

Page 126: *Beck's Sunday reconnaissance*: LSA, 53.

Page 126: *Unwilling Man Friday*: Alexander Selkirk (the Scottish prototype for Daniel Defoe's 1719 novel *Robinson Crusoe*) came to Galápagos in 1709 after his rescue from a self-imposed four-year exile (1704 to February 1709) on Más a Tierra Island (now called Isla Robinson Crusoe) in Chile's Juan Fernández Islands (these are Chile's version of the Galápagos, having been recognized as a UNESCO World Biosphere Reserve in 1977). Upon being rescued by a privateer captain, Woodes Rogers, and his navigator, William Dampier, of the ships *The Duke* and *The Duchess*, Selkirk took part in the sacking of Guayaquil, and the pirates came to the Galápagos for a refit afterward.

Page 126: *As elevation and moisture increased*: LSA, 54.

Page 127: *The night was cool enough*: LSA, 54.

Page 127: *The large male tortoise*: RWE, 25.

Page 127: *Each man was loaded down*: LSA, 54.

Page 127: *Exquisitely envisioned metaphor*: OOS, 489.

Page 128: *Any hour of the day or night*: RHB, 52, November 11, 1905.

Page 128: *They reported back to Slevin*: LSA, 55.

Page 129: *Occasionally Beck also took photographs*: RWE, 25.

Page 129: *Watched the birds fighting*: JSH, November 6, 1905. Regarding the "scrappiness" of this species, Jackson (1994, 188) noted that the mockingbirds often form cooperative breeding groups consisting of adults and offspring that assist their parents in territory maintenance. "At the contested borders of boundaries," he wrote, "confrontations can be exciting to watch. Members of each territorial group will line up opposite each other on either side of the imaginary line defending the border. Amidst much squawking and tail-flicking, birds will rush at their opposite numbers, occasionally fighting."

Page 129: *Ashore in the shallow lagoons*: JSH, November 6, 1905.

Page 129: *Remains of a centipede and the feet of a duck*: JSH, 30–34, November 7–8, 1905.

Page 129: *They kept the hawk for some time*: JSH, November 11, 1905.

Page 130: *Although unnamed back in 1905*: Isla Caamaño was named for José María Plácido Caamaño, who was president of Ecuador from 1884 to 1888. See also: http://www.galapagos.to/TABLE.php, accessed November 4, 2016.

Page 130: *Their immediate goal*: The *Academy* did not visit the Plaza Islands during this expedition. These islands are generally assumed to be named for a former president of Ecuador, the same General Leónidas Plaza Gutiérrez whom the *Academy* crew first met on Chatham Island in October 1905. Leónidas Plaza was the president of Ecuador in 1901 and 1912. His great grandson, Ricardo Crespo Plaza (an attorney in Quito), has informed me (personal communication, November 2001) that the islands were named by Leónidas Plaza in honor of his daughters. Leónidas Plaza also had a son, Galo Plaza Lasso, who was president of Ecuador from 1948 to 1952.

Page 130: *Beck clearly stated*: RHB, 53.

Page 130: *We dropped anchor in 15 fathoms*: LSA, 57; The entry in *RWE*, 26, for November 17, 1905, also states: "Sailed for anchorage near Gordon Rocks and anchored about 5 o'clock in a cove on the NE side of Indefatigable Island."

Page 130: *There were two potential anchorages*: William D. Pitt (personal communication, November 15, 2001). In 1986, a group of paleontologists I was part of retraced Ochsner's steps to find the exact location of the "fossil ledge" that had yielded fossils that are now in the collections of the California Academy of Sciences. The ledge has a reddish-brown color and is approximately one hundred yards long. See also Hickman and Lipps (1990).

Page 130: LSA, 57.

Page 130: *Strong southerly winds*: LSA, 58.

Page 130: *Some good fossil collecting*: LSA, 58.

Page 130: *Beck reported in his notes*: William D. Pitt (personal communication, November 15, 2001). Moving over rough ground at about two miles per hour would have allowed Hunter to reach the fossil ledge 3.7 miles (6 km) away and return to the schooner for its departure at noon on November 20, 1905.

Page 130: *Fossiliferous limestone*: Hickman and Lipps (1990).

Page 131: *While navigator Parker fretted*: LSA, 57.

Page 131: *Extremely desolate*: LSA, 58

Page 131: *Annoyance of the tired walker*: VOB, 373.

Page 131: *Stayed on board all day*: RWE, 26.

Page 132: *Grabbed a second turtle*: LSA, 58.

Page 132: *Sized up the geology*: RWE, 26.

Page 132: *They found fossil marine mollusks*: JSH, November 21, 1905.

Page 132: *Brought back a large collection*: Dall and Ochsner (1928).

Page 132: *On their last day at South Seymour*: This could mean the schooner was anchored in Caleta del Norte, and the lagoon was in Caleta Aeolian.

Page 132: *Nesting time must be rather near*: JSH, November 22, 1905 (incorrectly typed as "Nov. 23" in copy in Archives of the California Academy of Sciences).

Page 132: *The tortoise pickling tubs on deck*: Visitors to Indefatigable Island today arrive by jet airplane at Baltra Island, then cross the Canal de Itabaca in a small water taxi before embarking on a bus ride to Puerto Ayora.

Page 132: *Nearby Daphne Major Island*: From the late 1970s to the present, this tiny, unassuming island has been the location of long-term studies of Darwin's finches as they respond to environmental changes brought by periodic El Niño events. The fascinating story of their changing beak sizes and seed-breaking abilities was told in Jonathan Weiner's 1994 Pulitzer Prize–winning book *The Beak of the Finch*. In 1905, the crew of the *Academy* found the island much the same as the Princeton University biologists Peter and Rosemary Grant did when they began the detailed work of their "Finch Unit" in 1973.

Page 133: *This tiny speck in the archipelago*: LSA, 59.

Page 133: *The best island*: Weiner (1994).

Page 134: *Oars falling out and floating away*: LSA, 60.

Page 134: *A display of seamanship*: JSH, November 23, 1905.

Page 134: *Burned up the potatoes*: JSH, November 23, 1905.

Page 135: *Struck the shark with an oar*: LSA, 61.

Page 135: *Gauntlet of teeth and jaws*: RHB, 57.

Page 135: *After a longer visit to the lagoon*: LSA, 61.

Page 135: *Choppy sea and a stiff head wind*: LSA, 61.

Page 135: *The evening meal*: LSA, 61.

Page 136: *Skinning and pickling the turtles*: LSA, 62.

Page 136: *Loaded down with provisions*: JSH, November 27, 1905.

Page 136: *Water supply too low for comfort*: JSH, November 27, 1905.

Page 136: *Thanksgiving Day*: The expedition reached San Francisco exactly a year from this date; see LSA, 162.

Page 136: *Preparing a suitable feast*: Regarding the doves, Hunter stated in his field notes: "Killed over 200." Beck set the number at 100, and Slevin stated they got "96 doves between us."

Page 136: *Hooked a nine-foot tiger shark*: LSA, 63.

Page 137: *Hoisted the sails*: An anchor buoy allowed the men to secure both the bow with the main anchor and the stern to the buoy with a secondary anchor, thus minimizing the amount of arcing swing the schooner would have if she were anchored only by the bow anchor.

Page 137: *Resident tortoise population*: Anonymous (1897).

Page 137: *Inside the old crater*: JSH, December 2, 1905.

Page 137: *Tied up securely*: Beck reported in his field notes finding sixteen tortoises and tying up ten (RHB, 58).

Page 137: *Chased me for 10 feet*: RHB, 58.

Page 138: *Skin out the tortoises*: RHB, 59.

Page 138: *Comfortable field beds*: LSA, 64.

Page 138: *New role as camp cook*: LSA, 64, 65.

Page 139: *Carry down the day's kill*: LSA, 66.

Page 139: *The problem on Duncan Island*: LSA, 64.

Page 139: *Jotted down in their field notes*: RWE, 33.

Page 139: *The group caught a hawk*: LSA, 65.

Page 139: *Duration of their camp*: LSA, 66.

Page 139: *So full that he could hardly fly*: JSH, December 16, 1905.

Page 140: *Mostly filled with water*: RWE, 32–33.

Page 140: *Retreat from the highlands*: LSA, 65.

Page 140: *Took a wash for the first time in five days*: LSA, 65.

Page 141: *Lost from the field notebook*: RWE, 33.

Page 141: *Greatly reduced numbers*: JSH, December 15, 1905.

Page 142: *Three nameless islands*: These islands were likely the two Islas Guy Fawkes and Isla Venecia, and not to be confused with the more southerly Roca Sin Nombre (Nameless Rock).

Page 142: *Did not appear very promising*: This small island was likely Rocas Beagle, due east of Jervis Island.

Page 142: *Land tortoises below deck in the hold*: RHB, 61, December 16, 1905.

Page 142: *Straightened out as well as possible*: RWE, 33.

Page 142: *Convenient anchorage at Jervis*: The striking red beach of Jervis (Rábida) is a site frequently visited by ecotourism ships today and an official visitor site of the Galápagos National Park Service.

Page 143: *German-born and educated Georg Baur*: When he conceived of his Galápagos expedition, Georg Baur was an assistant to Yale paleontologist Othniel Charles Marsh, whose greatest fame derived from his dinosaur feud with Edward Drinker Cope. Baur's two-month expedition in 1891 was partially funded by paleontologist Henry Fairfield Osborn. See Larson, 2001, pp. 111–116 for a discussion of the genesis of Baur's Galápagos expedition and his notions of island origins.

Page 143: *The head of a slide of cinders*: LSA, 67. Beck gave the weight as 300 pounds (RHB, 62) and Hunter gave the weight as 400 pounds (JSH, December 20, 1905).

Page 144: *So some had to be shot*: RWE, 34.

Page 144: *History-laden James Bay*: James Bay is the name given by nearly all of the pre-ecotourism (or mid-1970s) visitors and expeditions to what is today commonly referred to as Playa Espumilla, where ecotourists now disembark on the long, coffee-colored beach (Heidi Snell, personal communication, December 4, 2001).

Page 144: *Twice the time Darwin had spent ashore*: Two subsequent visits by the *Academy* to James Bay (in July and August 1906) would result in a total of thirty days spent in and around the anchorage at James Bay.

Page 144: *Makeshift grave on the beach*: The events surrounding Cowan's death do not appear to have been known to the men aboard the *Academy*, although in the coming years Slevin would embark on "the search for Cowan's grave," which would continue, without satisfactory resolution, until his death in 1957. See LSA, 58–61.

Page 144: *Exact location of Cowan's gravesite*: John Woram (personal communication).

Page 144: *A fine green zone at the higher elevations*: LSA, 67–68.

Page 144: *Got 20 Graces petrels*: RHB, 63.

Page 144: *But he was disappointed*: Beck likely was referring to the dark-rumped or Hawaiian petrel, which was formerly known as *Astraelata phaeopygia* Salvin, 1876 (it is currently placed in the genus *Pterodroma*) and which Beck usually referred to in his field notes simply as "phaeopygia."

Page 145: JSH, December 22, 1905.

Page 145: *Having demonstrated his resourcefulness*: JSH, December 22, 1905.

Page 146: *Circular platforms*: JSH, December 22, 1905.

Page 146: *Jumped one pig*: JSH, December 23, 1905.

Page 146: *The cook treated the men to a plumduff*: "Plumduff" was a term likely adopted by Slevin in his 1931 *Log of the Schooner "Academy"* from his days in the US Navy. It was a flour pudding, similar to a plum pudding, made with raisins or currants.

Page 146: *Plenty of time to mend shoes and clothing:* RHB, 64, December 25, 1905.

Page 146: *Longer collecting and exploring excursions*: Beck likely insisted that the men stay close to home after they arrived at James Bay in order to observe the double holiday of the Sabbath and Christmas.

Page 147: *Took the big skiff*: Their destination was probably Buccaneer Cove, though it was called Adams Cove by Slevin (RWE, 36) and in Gifford's diary.

Page 147: *Quite a heavy surf*: LSA, 68.

Page 147: *One blue bottle and pieces of crockery*: RHB, 64, December 27, 1905. Beck did not record whether he collected the human artifacts.

Page 147: *Long tramp around in a northerly direction*: LSA, 69.

Page 147: *Numerous donkeys and pigs*: Beck noted that pigs were not common in this area, but those that were there would "go to cactus and eat for water" (RHB, 64, December 26, 1905).

Page 147: *Where tortoise had fallen in*: RHB, 64, December 27, 1905.

Page 147: *Now numbered among the famous*: JSH, December 26, 1905.

Page 148: *But are easily killed even at long range*: JSH, December 26, 1905. Shot refers to individual shot pellets from a shotgun shell hitting the bird.

Page 148: *Inclined to wrap themselves around your neck*: JSH, December 27, 1905.

Page 148: *No rare species*: JSH, December 28, 1905.

Page 148: *Giving his death cry*: JSH, December 28, 1905.

Page 148: *I got about half an hour's sleep*: JSH, December 29, 1905. Ironically, when Hunter returned to the San Francisco Bay Area, his first job was to head the mosquito abatement project in San Mateo and Burlingame, just south of San Francisco. His work was challenging, and he reported to his entomological colleagues about "the importance of the mosquito problem" and how the two saltwater mosquitoes that lived in the marshy areas were "exceedingly troublesome"; *Entomological News* 18, no. 6 (June 1907): 263.

Page 149: *I shot the flamingo and five ducks*: JSH, December 29, 1905.

Page 149: *Nearly eaten alive by mosquitoes*: LSA, 70.

Page 150: *Coming down with a thump*: RWE, 39.

Page 150: *Still hang to the female if you try and pull them off*: RWE, 39.

Page 150: *Beck threw a few bird bodies over the side*: When a bird is stuffed for a museum collection, all of the internal organs, the muscles, and most of the bones (except for the external leg bones and those in the wings) are removed and discarded when the bird skin is temporarily turned inside out (Douglas J. Long, personal communication, November 30, 2001).

10

Rising from the Ashes

⌒——

ON APRIL 18, 1906, a great earthquake and three days of fires devastated San Francisco and severely affected many other cities in the greater San Francisco Bay Area. Had the Galápagos expedition set sail in November 1904 as Loomis had originally planned, the tens of thousands of new Galápagos specimens would have been destroyed along with the thousands of other materials stored in wooden museum cabinets in the Academy's museum building on Market Street. The Academy's building survived the earthquake but perished in the fires that followed. With no building, no income base, and almost no collections, the California Academy of Sciences could have ceased to be. Instead, the bounty of Galápagos specimens that was soon to return to a city destroyed ensured the eventual resurrection and prosperity of the Academy.

San Franciscans who spent the morning of Sunday, January 22, 1905, reading their hometown paper, the *San Francisco Chronicle*, encountered some of the worst advice ever to appear in print. The writer S. G. P. Coryn ridiculed "an old lady resident of San Francisco" who had recently moved from a house on one street to a house on the next street over because the original house was "right over the earthquakes." Setting her up as a geologically naive fool, Coryn proceeded to fool San Franciscans and lull them into a false sense of security. The preceding month in the city had been a time of rattled doors and windows, with numerous small tremors disrupting people's sleep due to the "vibrations . . . of the subterranean Cyclops." Coryn reassured San Franciscans that really bad earthquakes only occurred elsewhere in the world, "in other less favored countries." He included Japan in that category, noting that "sometimes the Pacific Coast has had to suffer for seismic sins not her own." His specific example was the December 23, 1854, Tokai earthquake, which generated a tsunami that reached California.

Coryn did not dismiss earthquakes entirely. He considered San Francisco's earthquake scientists to have been particularly fortunate in having the recent month's abundant supply of "perfectly harmless" material for study. These tremors were "eloquent and instructive messengers from those mysterious depths of the world of which our knowledge is necessarily so meager." Seismologists at the time kept track of data on earthquakes, including the directions from which they came, their intensity, the positions of the heavenly bodies, the hours and seasons at which they occurred, their regularity or periodicity, their influence on the ocean, and their connection to the weather. The author concluded his article, and San Franciscans could thus turn the page of both their newspaper and their concerns, with the falsely reassuring statement, "It is therefore evident that California earthquakes are, as it were, miniatures, perfectly harmless and admirably adapted for scientific observation." One year and three months later, on a Wednesday morning, San Franciscans would learn how devastatingly wrong that prediction was.

At 5:12 a.m. local time on April 18, 1906, the residents of the San Francisco Bay Area were literally thrown from their beds by what has become the most famous earthquake in US history. Immediately dubbed the "Great San Francisco Earthquake," it destroyed 28,000 buildings and 490 square blocks of the city's center (figure 10.1). The devastation is nearly unimaginable to us today. Gone were City Hall, the Hall of Records, theaters and museums, grand hotels, churches and convents, hospitals, schools and universities. A quarter-million people had been rendered homeless.

The writer Jack London traveled to San Francisco from his home in Glen Ellen, some forty miles north in Sonoma County, to witness the devastation of the 1906 earthquake and wrote a firsthand account for *Collier's* magazine. London observed:

The earthquake shook down in San Francisco hundreds of thousands of dollars' worth of walls and chimneys. But the conflagration that followed burned up hundreds of millions of dollars' worth of property. There is no estimating within hundreds of millions

FIGURE 10.1 **Earthquake Damage.** A view of Market Street in San Francisco, looking down the streetcar tracks (still there today) toward the Ferry Building (still standing today), April 1906. The burned-out hulk of the California Academy of Sciences was located on the right-hand side of the street, about halfway down toward the Ferry Building. *Archives, California Academy of Sciences.*

the actual damage wrought. Not in history has a modern imperial city been so completely destroyed. San Francisco is gone. Nothing remains of it but memories and a fringe of dwelling-houses on its outskirts. Its industrial section is wiped out. Its business section is wiped out. Its social and residential section is wiped out. The factories and warehouses, the great stores and newspaper buildings, the hotels and the palaces of the nabobs, are all gone. Remains only the fringe of dwelling-houses on the outskirts of what was once San Francisco.

Within an hour after the earthquake shock the smoke of San Francisco's burning was a lurid tower visible a hundred miles away. And for three days and nights this lurid tower swayed in the sky, reddening the sun, darkening the day, and filling the land with smoke.

London reported further, "All the cunning adjustments of a twentieth century city had been smashed by the earthquake. The streets were humped into ridges and depressions, and piled with the debris of fallen walls. The steel rails were twisted into perpendicular and horizontal angles. The telephone and telegraph systems were disrupted. And the great water-mains had burst. All the shrewd contrivances and safeguards of man had been thrown out of gear by thirty seconds' twitching of the earth-crust."

Among the most notable losses was the California Academy of Sciences, a cherished institution, which had two seven-story buildings at 844 Market Street, between 4th and 5th Streets (figure 10.2). By 1906, its library held fifteen thousand volumes, with a rare accumulation of periodicals, pamphlets, and other scientific publications. Also destroyed were the even more precious natural history collections, including specimens that documented species new to science. The Academy's collections had been built through massive expeditions, including the one to Mexico, as well as more local collecting up and down California and exchanges for duplicates. Loomis in his area of specialty had gathered the largest seabird collection in the world. Although no one realized it at the time, these materials—the most precious things the academy owned—were the most vulnerable to damage because they were stored on high floors and so vulnerable to fire in their wooden cabinets. Inadequate water supplies citywide in San Francisco in 1906 prevented firefighters from doing their job.

The most notable and heroic effort to save anything from the building before it was totally engulfed in flames was made by the botanist Alice Eastwood. The shaking woke the internationally known scientist from her sleep. From her Nob Hill apartment, she looked toward the Academy, where she could see numerous small fires already burning. Realizing the potential magnitude of the disaster, she made her way to the Academy to rescue at least some of the botanical collection. She entered the building with other staff members just two hours after the earthquake hit. She was accompanied by Director Loomis, the librarian Mary Hyde, the insect preparator Carl Fuchs, the reptile curator John Van Denburgh, and Eastwood's friend Robert Porter. The interior was "as still as death" and damaged extensively but remained in relatively good order. A bridge at the sixth floor, suspended above a glass-roofed atrium, connected the Academy's income-producing property with the museum building in the rear (figure 10.3). The bridge had ripped loose and fallen through the skylight over the atrium. An elaborate marble staircase leading from the second to the sixth floor was shattered, although the iron bannisters remained intact.

FIGURE 10.2 **Academy Ruins.** In the background are the ruins of the California Academy of Sciences on Market Street in San Francisco, April 1906. *Bancroft Library, University of California, Berkeley.*

Alice Eastwood and the others carefully worked their way up the crumbling stairs to the fifth and sixth floors, where the scientific records and manuscripts were stored. When working in the field collecting specimens, Eastwood had used torn strips of old undergarments to bundle up the pressed botanical specimens between layers of absorbent paper and cardboard, all held together by a frame of wooden slats. These plant presses, as botanists call them, which

FIGURE 10.3 **Academy Ruins.** The large arched doorway was the passageway from the Academy's income building facing Market Street across stairs and a bridge to the museum building in the rear, April 1906. *Archives, California Academy of Sciences.*

can expand to more than a foot thick when filled with specimens, press the plants flat in preparation for their eventual transfer to permanent sheets of thick paper.

  Eastwood and the others tied as many bundles of botanical type specimens as they could gather—1,497 in all—and quickly placed them flat in plant presses. These were no ordinary specimens. They were the voucher specimens for new species described by many authors over many years. Using a makeshift rope, also strung together with strips of her old undergarments, Eastwood lowered the specimens to the ground floor.

Eastwood saved only a single personal item from the building, her Zeiss lens for studying specimens closely. She would need it in the near future to help rebuild the collection. Loomis and Hyde climbed to the sixth floor and saved the only manuscript copy of Theodore Henry Hittell's history of the Academy, a book I have found invaluable in researching the 1905–1906 expedition. They also saved the Academy's records that I have relied upon. The destroyed library was one of the largest west of the Mississippi River. Carl Fuchs along with Hyde saved 264 type specimens of beetles, Hemiptera (aphids, cicadas, leafhoppers, and their relatives), and wasps, bees, and ants. The complete insect collection comprised 51,000 specimens. In the reptile department, the curator Van Denburgh was able to save only ten type specimens from a total collection of some 8,100 specimens dating back to 1895. In anticipation that day of the arrival of the unstoppable fire marching through San Francisco, Loomis moved on to the Birds and Mammals Department, where he gathered up only two type specimens of the now-extinct Guadalupe storm petrel named by the Academy curator Walter E. Bryant in 1887. With these few specimens removed, books and illustrations tucked under arms, the remaining tens of thousands of specimens of birds, mammals, plants, insects, reptiles, and fossils, and all the rest of the library, were incinerated the same day the earthquake struck. Really, the losses were total (figure 10.4). The best hope for the future of the Academy was the Galápagos expedition already under way and the tens of thousands of specimens stored below decks on the schooner *Academy*.

Loomis desperately needed his Galápagos expedition to succeed. Little did he know how its success would eventually boost the fortunes and continued reputation of his museum—at present he had more immediate problems. He wrote to Rollo Beck three times shortly after the Academy was destroyed, stating his expectations for the expedition in no uncertain terms. Each letter carried exactly the same urgent message, and the last two were dated May 4 and May 18, 1906:

Dear Mr. Beck,

I write a third time so as to make sure that your instructions reach you.

1. You are to carry out your original instructions, received just before sailing from San Francisco, June 28, 1905, without any reference to the earthquake & fire in San Francisco. In my last letter (written several days ago), I informed you of the total destruction by fire of the Academy's building & all the collections. The whole of the business portion of the city and much of the residential portion was destroyed by fire.

2. I ordered the beef for you on April 17, but the next day that part of town was a smoking ruin.

3. On the 17th of April, Mr. Molera [Academy president] cabled the American Consulate at Guayaquil to honor your orders for provisions.

4. On the same day the Crocker-Woolworth bank sent a letter to the Banco del Ecuador, Guayaquil, to establish a credit for provisions for $200. Your signature was enclosed, & instructions were given that your handwriting & signature be accepted as sufficient identification, as you desired to order provisions from Chatham.

FIGURE 10.4 **Interior Damage to the Academy Museum.** The reinforced concrete floors and walls of the Academy's buildings were not significantly damaged in the April 18, 1906, earthquake, but the three days of fires that swept through the city claimed the Academy as a victim. *California Historical Society.*

5. On account of shortness of provisions, if you find it necessary to visit Guayaquil at the end of your stay, do so, and cable from there if you need aid. If you visit Guayaquil, allow no one to go ashore but yourself. [original emphasis]

The work of reorganization of the Academy is being diligently attended to. The fine collections that the Galapagos expedition will bring back will be the foundation of the museum of the greater Academy. Every specimen counts now. The Academy's bird collection now contains two specimens, the types of the Guadalupe Petrel. The more material you secure on the homeward voyage the better. We have a great opportunity, the opportunity to build up a museum based upon recent exploration. Most museums are composed of collections that have been dumped in. Before the fire our work was largely [unclear]; now it is to build new from the foundation. Every collection will be built up according to a definite plan, and the schooner Academy will play an important part in the work.

Give my kindest regards to the members of the expedition. Mr. Molera sends his best regards.

Sincerely yours,

(signed) Leverett Mills Loomis

Without the success of the Galápagos expedition, the Academy as an institution would have been back to where it was in 1853, with no collection, no library, no building, no standing in the community. The efforts of the eight young men aboard the *Academy* provided the nucleus of specimens for a new "cabinet of curiosities," as museum collections were called in previous centuries—and in fact, quite a large one. In time, the building and library would follow (figure 10.5).

The Galápagos specimens lashed down securely in the hold of the schooner would allow the museum to rise, literally and figuratively, out of the ashes of destruction. Joseph Slayton Hunter, one of the eight young collectors, stood on the deck of the *Academy* in the Galápagos after learning of the great conflagration in San Francisco that destroyed the Academy's museum and said, "We are the Academy now." The collectors' safe return to San Francisco was crucial.

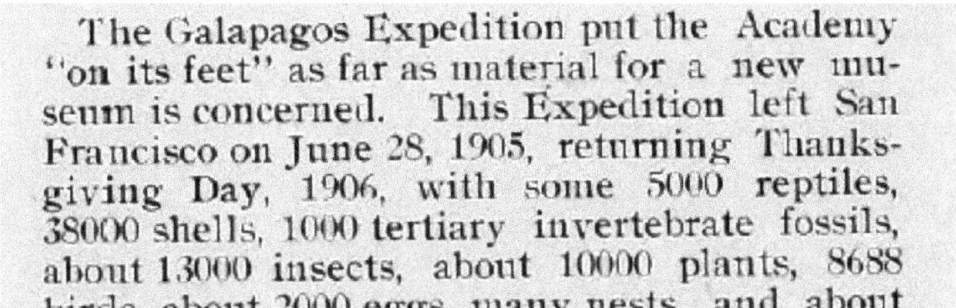

The Galapagos Expedition put the Academy "on its feet" as far as material for a new museum is concerned. This Expedition left San Francisco on June 28, 1905, returning Thanksgiving Day, 1906, with some 5000 reptiles, 38000 shells, 1000 tertiary invertebrate fossils, about 13000 insects, about 10000 plants, 8688 birds, about 2000 eggs, many nests, and about 120 mammals.

FIGURE 10.5. **Snippet from Gifford.** A published report by Edward Winslow Gifford gives a sense of the volume of material collected during the 1905–1906 expedition and the significance of that material for the rebirth of the California Academy of Sciences. Gifford, E.W. 1908. The rehabilitation of the California Academy of Sciences. Source: Public domain, *The Condor* 10: 95–96.

NOTES

Page 160: *A great earthquake*: See Fradkin (2005); Jeffers (2003), Winchester (2006); Klett and Lundgren (2006); London (1906).

Page 160: *Destruction of the Academy museum on Market Street*: See https://www.calacademy .org/blogs/from-the-stacks/1906-academy-building, accessed November 6, 2016; and Leviton, Aldrich, and Elsbernd (2006, 1–34).

Page 160: *Academy's building survived*: Sewell (1907, 76) stated that the "building had cast-iron concrete-filled columns and Ransome reinforced-concrete floor construction. So far as it was possible to ascertain, no damage was done to the reinforced concrete or to the columns by the earthquake. The building was gutted and the floors considerably damaged by the fire, but the columns were not damaged, and on the whole the building stood very well."

Page 161: *Earthquake damage*: Klett and Lundgren (2006).

Page 161: *Earthquake damage and Alice Eastwood*: Svanevik and Burgett (2002).

Page 162: *Cunning adjustments*: London (1906).

Page 162: *Accompanied Eastwood*: M. Bettelheim (2006), http://baynature.org/article/left -to-burn/, accessed May 10, 2016; J. Ronson (2016), https://www.inverse.com/article/14363-the -california-academy-of-sciences-hero-versus-the-1906-san-francisco-earthquake, accessed May 11, 2016.

Page 162: *The most notable and heroic*: For letters from Alice Eastwood and Leverett Mills Loomis concerning the earthquake destruction, see: Nelson, E.W. 1906. The California Academy of Sciences. Science N.S. Vol. 23, No. 595 (May 25, 1906), pp. 824–826.

Page 165: *Alice Eastwood undergarments*: Story via Peter Raven, director of the Missouri Botanical Garden, October 13, 2015, at the California Academy of Sciences Fellows dinner, confirmed in person by me with Dr. Raven over dinner; story also heard by all in attendance at Dr. Raven's presentation that evening.

Nature never wears a mean appearance. Neither does the wisest man
extort her secret, and lose his curiosity by finding out all her perfection.

RALPH WALDO EMERSON (1803–1882)

In my simplicity I remember wondering why every gentleman
did not become an ornithologist.

CHARLES ROBERT DARWIN (1809–1882)

How I did enjoy shooting, but I think that I must have been half-consciously
ashamed of my zeal, for I tried to persuade myself that shooting
was almost an intellectual employment.

CHARLES ROBERT DARWIN (1809–1882)

11

Ship's Log, 1906

FOR ALL THE men aboard the schooner *Academy*, the year 1906 opened at James Bay under cloudy skies and a strong southerly wind. The extent of their New Year celebration was really nothing more than staying up late and blowing the ship's foghorn when midnight arrived. Having finished their first three months in the islands, and after a total of 188 days since leaving California, the men still had nine more months to go before sailing back north to San Francisco. They held a subdued New Year celebration because they realized it would still be many months before they saw their families and friends back in California.

The foghorn blasts, which lasted only seconds, signaled the intention of the men to stay in the islands for many more months, a signal to all the animals onshore that collecting on this island was not yet complete.

The Academy men spent New Year's Day resting and doing light work aboard ship, with "nothing doing" being the focus after a late breakfast. They had their second double holiday in a week, because their usual Sabbath break corresponded with New Year's Eve and they continued it into New Year's Day, 1906. The cook, White, made pumpkin pie for supper to start the New Year right.

Preparations were made during the day, however, for an overnight excursion the next day. The men had collected extensively along the shore and planned an inland trip for additional collecting. Their biggest disappointment so far on James Island was the lack of tortoises. They had three more days on the island to rectify the situation.

Only four men remained on the schooner (Beck, Slevin, White, and Parker) while the other seven landed after breakfast on January 2, 1906. The shore party carried water and provisions for three days and set their sights for the top of the island. They made slow progress because of their heavy packs and the thick underbrush. After a lunch stop at the 1,000-foot elevation, they pushed on, only to run into thicker vegetation and the same vines that had previously bedeviled Hunter and Ochsner with their inclination to "wrap themselves around your neck and feet."

Late in the afternoon, the shore party found burro trails and followed them. Hunter collected a variety of birds during the day and made some astute behavioral observations, one in particular concerned one species of Darwin's finches. "One *Camarhynchus psittacula* [the large tree finch] was noticed working on a small composite twig for nearly five minutes, industriously tearing away at the bark," Hunter observed. "Whether he was eating the bark I cannot say, but it appeared as if he were."

The party pitched camp at the edge of the green zone at 2,100 feet, within striking distance of what they hoped would be good tortoise territory. Scattered showers passed over during the night as the men huddled and slept fitfully under their temporary shelter. Although they were unsure if they would find tortoises, the night certainly confirmed that mosquitoes were abundant. Under such field conditions, getting an early start up the mountain was more liberation than burden. If ever there was a night when they prayed for sunrise, this would have been it.

A CONTRAST IN TAMENESS: THE RAIL AND THE HAWK

No two Galápagos organisms demonstrate the broad spectrum of fearlessness experienced by the Academy collectors to a greater extent than two dramatically different birds: the rail and the hawk. Joe Hunter actively pursued both of these birds on James Island, and his field notes contain remarkable vignettes of his experiences and frustrations. The Galápagos rail was timid and elusive, while the Galápagos hawk was tame and approachable to a fault.

JANUARY 3, 1906

On January 3, 1906, while on the inland trek to the top of James Island, Hunter started out early to hunt for the elusive Galápagos rail. Frustratingly, he could hear the birds in the grass and tangled ferns "clattering away" in groups of two or three, but they were "extremely cautious, never showing themselves outside of the brush." He could easily elicit a response from them, the same way Beck had done, by clapping his hands and then standing still to hear "a loud cackle that can be heard 100 yards through the brush." The rails moved very quickly, unseen in the brush, in four-inch-high "runways" that formed a sort of rabbit warren beneath every bush. Hunter knew the birds were in there, but he just could not catch a glimpse of them.

As he slunk down to their level, Hunter had only moderate success: "The only way to secure specimens seems to be when one is located in a patch of fern, to lay down where a clear sight can be had below the vegetation for a few feet ahead and then watch for the bird. After watching carefully for some time without any noise, you may be rewarded by seeing the little pink-eyed fellow cautiously look from behind some clump of vegetation. At the least motion he will be away, and if your gun loaded with a very light charge is not aimed directly at him, you will not stand any show of getting that rail. Often the bird will not be more than a foot or so from your face and you dare not shoot him. I saw several birds, but only succeeded in getting one good one; another was blown to pieces."

The contrasting experience with hawks would come later, following another significant discovery about one of Darwin's finches. But the men's trek to the top was not over yet.

While down on his belly hunting for the cryptic rails, Hunter experienced the classic foggy and rainy *garúa* conditions that prevail during the Galápagos cool season but can occur at any time of year at the higher elevations. "A heavy fog and drizzling rain made things about as disagreeable as they could be," lamented a soggy and frustrated Hunter. The fog was so thick they even had to halt their progress for a while. But about ten in the morning, Hunter and the others were delivered from their soggy misery when "the fog lifted and a glorious view was opened to us."

With their spirits lifted by the sweeping panorama, the men collected for several more hours as they worked their way steadily up to the top of the island. Just four of the seven men pushed on to the very top, which they measured with their aneroid barometer to be 2,750 feet above sea level. As they stood at the summit looking down to the ocean, Ochsner, Williams, Hunter, and Gifford could clearly see the masts of the schooner, and their panoramic view extended to the north as far as Bindloe (Marchena) and Abingdon (Pinta) Islands. The lunch they ate sitting at the top of the island must have been one of the most pleasant and memorable of the entire expedition. That evening, looking up the mountain from the schooner, Slevin wrote, "Tonight the mountain is fine and clear and plainly visible from the schooner."

Hunter continued his pursuit of rails in and around their mountain camp. One night he "hunted rails until supper time but with no success," and another time he "went out looking for rail and shot one all to pieces." He found that the rails were "not at all as inquisitive as they were yesterday and would rarely answer the clapping of the hands." Although lying on his belly, he caught mere glimpses of them running in the low runways through the ferns. The birds would not, he lamented, "wait for me to get the gun to bear on them." That was the extent of his luck. As Hunter retreated down the mountain, the rails scurried furtively in the low bushes, mocking his attempts to turn them into museum specimens.

The story of approachability was quite different with the tame Galápagos hawk. In contrast to his disappointing efforts with the timid rail, Hunter had no trouble getting close to and collecting specimens of this sharp-eyed, sharp-clawed bird of prey. As with the rails, Hunter included in his field notes insightful observations about the behavior of the hawk: "There were about 12 or 15 hawks around camp. Four were eating the body of a hawk that had been killed by one of the fellows the day before. They were taking their turn tearing out the flesh and entrails, not fighting among themselves at all but acting much like well-trained children, each waiting until the other had his bite." No sooner had he finished writing these insightful notes than he quickly made the transition from observer to collector and "these 4 were shot." Their table manners did not assuage his urge to collect these convenient specimens. With four more specimens to stuff back on the schooner, Hunter observed that "inside of 15 minutes, five others were down tearing away, [while] two others were waiting for room to get in to eat." The tameness of the hawks meant they were infinitely easier to collect, and for Hunter the "catch per unit effort" was seductive. "The hawks are very tame," Hunter observed. "Almost any of them allowing themselves to be hit with a stone or stick and then to be handled. Very soon they become accustomed to being handled and seldom offer to claw or bite."

These two extremes of tameness illustrate the spectrum of behaviors that island isolation can produce. Hunter did not address why one bird species would lose all of its protective wariness, while another species would retain all of its ancestral fearfulness. This is the type of evolutionary question that Hunter and his compatriots on the schooner *Academy* did not address in their field notes or, disappointingly, in their published work. They just took it on face value that not all birds in the islands were equally tame and easy to catch.

A TOOL-USING FINCH

One of the most significant behavioral observations of the expedition was made while the men were camped inland on James Island. This historic observation was perhaps the first documentation of the remarkable phenomenon of tool use in the woodpecker finch, *Camarhynchus pallidus*. In his field notes, Hunter described King watching the bird use a twig to secure insect grubs, and this remarkable behavioral adaptation is now fully incorporated into the broader picture of the evolution of Darwin's finches. "King reports seeing a very interesting antic on the part of a pallada [= pallidus]. It was pecking woodpecker-like on a dead branch an inch and a half in diameter, and broke through to a borer hole. It tried to reach to the end of this hole with its beak, but was unable to do so. It then flew to a branch nearby and broke off a small twig and with it flew back to the hole. Arranging the twig in its beak so that the end pointed parallel with the beak, it then ran the twig into the hole for a feeler for grubs. In this particular case there was not [a grub], and the bird dropped the stick."

This simple behavioral observation, another first for the Academy expedition, made high on the slopes of James Island, is now standard fare in natural history documentaries whenever Darwin's finches are featured. The men recorded it first, and Hunter attributed the observation to the field assistant, King.

Hunter followed this with another observation that was of equal relevance to the history of Darwin's finches and to the human dynamics of the expedition itself. "Beck reports having seen this same performance gone through with two years ago," Hunter quipped, "[but] he is such a liar that I do not know whether to believe him or not." Either Hunter was commenting on Beck as a good-natured joker, or else this was a more serious indication of personal conflict on the schooner. In either case, personal conflict between two of the other men would soon erupt into violence.

LIVING OFF THE LAND

For the time being, though, personal relations were calm in early January 1906. On the way back down to the coast, the importance of pigs on James Island became abundantly clear. Not only were pig trails everywhere, but also many old carcasses were scattered from the extreme top of the island all the way down to the coastline. Nelson shot four pigs in one small area, with hardly any effort. Another pig Nelson had wounded in another place exhibited some fierce behavior to the men, but "a shot from his revolver stopped it." Of the pigs Nelson shot, which included a big boar and three smaller sows, "the younger [ones were] not . . . such bad eating," as the men found out when back on the schooner (see figure 11.1).

Along with pigs, burros were also abundant. The men shot one old male that measured more than five feet from ear to rump and had nine-inch-long ears. Hunter observed in his field notes, "The males seem to be quite hostile at times, one made a rush for the camp when King brayed, and [the burro] was apparently very ready for a fight." The burros used the same

FIGURE II.I **Nelson the Provider.** On multiple occasions during the expedition, the mate of the schooner *Academy*, Frederick T. Nelson, went out foraging for food for the men. One of his favorite targets was wild pig. *Archives, California Academy of Sciences.*

well-worn tracks as the pigs, also wandering up to the very top of the island. "The life of a jack on this island must be happiness supreme," Hunter mused. "Nothing to do but eat, no dogs or any other animal to disturb them as the ones on Charles have." Having "a very satisfactory feed in the luxuriant grass that is green the year through" reinforced this image of Burro Eden.

Amid slaughtered pigs and the cacophony of aggressively braying burros, one of Hunter's more poetic observations was that "doves were seen yesterday drinking out of the orchids." Here were a peaceful bird and a delicate flower in evolution's outdoor laboratory. The alien pigs and burros added an air of discourse and disrupted the otherwise placid island environment where all the native species had evolved in unison, if not always in harmony. Stewart and King returned to the schooner a day earlier than the other men, collecting plants from the summit to the coast. Once back on board, Stewart spent his time pressing plants, while King returned to shore for an unsuccessful pig hunt.

With their spirits uplifted by the sweeping panorama from the top of James Island, the remaining men collected for several hours before heading back down to the coast. They had to admit total defeat in the tortoise-hunting department, settling for the decade-old bones that Beck had found elsewhere on the island. Slevin rowed to shore in the skiff to pick them up. Even on this mundane shore trip, Slevin wasted no opportunity to collect. He reported that while the boat was being loaded on the beach, he "caught two turtles and brought them on board" (figure II.2). As with Hunter and the hawks, whenever specimens were relatively easy to collect, the men did not pass up the opportunity.

FIGURE 11.2 **Sea Turtles on Deck.** Rollo Beck and his colleagues on an earlier expedition collected sea turtles as did the 1905–1906 expedition. *Archives, California Academy of Sciences.*

The prevalence of pigs and burros on this island clearly had had a detrimental effect on the tortoise population. The men were starting to see a trend that vindicated Georg Baur's dire warnings. Wherever humans, donkeys, rats, and pigs were abundant, tortoises were scarce. The abandoned tortoise hunters' camp and the introduced mammals were indications of the one-two punch that could send the tortoise population down for the eight count, or even knock it out completely.

BECALMED AND ADRIFT

The collectors' extended visit to James Island was concluded early on the morning of January 5, 1906. After an early breakfast, they raised anchor and set sail for a place that now, thanks to them, had a name. They set out for Academy Bay on Indefatigable Island, but they didn't get very far with "very light southerly winds during the morning and light unsteady winds during the afternoon." The unexpected events of the next seven days were unwelcome, but they allowed the men to take "advantage of this time at sea to pack up the collections and straighten up in general."

Over the course of the next week (January 5–11, 1906), the men and the schooner were becalmed in the center of the archipelago. During this frustrating time, they had only occasional control of the schooner's progress. When not completely abandoned by the wind, they had no better than a fresh breeze to sail by. Even this would inevitably die down after a few hours, returning them to the mercy of the equatorial currents running between the islands.

On their first day of drifting in the center of the archipelago, they prepared for what could have turned quickly into a bad situation and made the most of the isolation imposed on them by the equatorial calm. Getting too close to any of the islands put them at risk of going aground, so they became hypervigilant, especially at night. As an insurance policy against ending up on the rocks, Nelson broke out the hawser and kedge. He kept this anchor and stout line close at hand in case they needed to "pick up bottom somewhere" in the event that they drifted too close to an island. All hands were "busy standing watches" around the clock.

Twice they ended up drifting as far south as the Crossman (Cuatro Hermanos) Islands, just off the southeast coast of Albemarle (Isabela) Island. They also drifted about as far north as Conway Bay on Indefatigable (Santa Cruz) Island. Most of this movement was random and all of it frustrating. Their intended destination, Academy Bay, was tantalizingly close several times as they drifted up to the island and then back away from it. The little active sailing and headway they accomplished toward their goal was reversed as they drifted back and forth in the waters south of Duncan Island.

Despite not being able to get the schooner up into its namesake bay, the men stayed busy. General maintenance included hoisting the large skiff on deck for overhauling and repairs. Numerous sea turtles taken in and around James Bay were skinned, and birds were stuffed at a steady rate. There was a sense of urgency to get all of the birds skinned and stuffed before decay set in.

The sea turtle meat and fat thrown overboard into the calm water by Slevin and King served as a magnet for seabirds. Jumping at the chance to capitalize on the fortuitous chum that was floating around the schooner, Beck and Williams put off in a skiff to do what Beck did so well with a shotgun and a bag of cartridges. Hundreds of birds circled around the schooner and eagerly swooped down to pick up the tasty morsels. Two whales swam by and spouted, adding to the bountiful image of the archipelago.

Within a short time and a short distance from the schooner, Beck shot several dozen seabirds. These included white-vented (Elliot's) storm petrels, wedge-rumped (Galápagos) storm petrels, band-rumped (Madeiran) storm petrels, Hawaiian (dark-rumped) petrels, and Audubon's shearwaters. Back in San Francisco, Loomis had no idea of the bounty that was accumulating in his taxonomic specialty. He would have shearwaters and petrels to keep him busy for years to come.

The turtle chum attracted flocks of seabirds, and then the bird bodies that were thrown overboard attracted sharks, which circled the schooner as it continued to drift during the week. So despite the calm water and good weather, swimming was out of the question.

Finally, on January 11, 1906, after the crew had played cat and mouse with the wind and currents for a week, "a fair sailing breeze" brought the schooner up to the coast of Indefatigable Island, only to die yet again. This time, though, they would not let the island recede on the horizon. They grabbed hold of the island "during a calm spell" with their kedge anchor. Finding themselves "in twenty-five fathoms of water about a quarter of a mile off shore" meant they had to row ashore. They were still a few miles west of Academy Bay, but it felt good to get off the schooner.

Beck, Hunter, and Slevin rowed the skiff over to a small, unnamed lagoon for a few hours of collecting on Indefatigable. They landed at what is most likely today called Bahia Tortuga, because Hunter described the location as being "on the almost extreme southerly end of the island." Once they made it ashore, they found several small lagoons just behind the beach with a dozen ducks and one flamingo. Hunter dispassionately described what happened next when he wrote, "Beck shot the ducks and flamingo." One of the current Galápagos bird books considers the greater flamingo to be "one of the shyest of Galapagos birds—should never be made to fly or otherwise disturbed." Another Galápagos bird book states, "The flamingo population is small and vulnerable, and birds must be left alone." The conservation sentiment of today's birders and ecotourists was not part of Beck's and the other Academy collectors' mindset. A guidebook author, Mike Jackson, wrote of Galápagos flamingos, "A small group of flamingos feeding quietly and moving gracefully through a brackish lagoon is a most peaceful sight." The same peaceful sight likely greeted Beck, but was answered not with the click of a camera shutter, but with the sharp report of a shotgun.

Although the men did not find any live tortoises during their brief visit, Hunter found the remains of two dead specimens that had been there for not more than a year or two. The two small tortoises were found a few feet back from the beach, turned on their backs with the belly-side bones, or plastron, detached, and machete cuts were visible on the shell. Finding these "old signs of tortoises" in this condition was clear evidence that the tortoises had been killed by people, and Hunter attributed their deaths somewhat vaguely to "the natives." One of the upside-down tortoises had dried dung in the shell, indicating there had been undigested food in the intestines when the animal was killed. Paradoxically, no skulls were found.

Among the discoveries the men made at this coastal lagoon were some of the largest marine iguanas they had yet seen, some more than four feet in length. These impressive specimens were "quite plentiful along the rocks at this part of the coast," and Beck brought seven of them back to the schooner. Beck also ended up thrashing around in shallow water in a wrestling match with two sea turtles. He noticed them feeding in the bay near shore in two feet of water. This prompted Beck to take off his clothes and grab hold of one, "but he got away, then I got another and finally forced him into shallow water where I got meat and skull." This was not a bad performance for someone who couldn't swim.

BACK AT ACADEMY BAY

The three men rowed back out to the schooner with an impressive collection resulting from only a few hours of work near the landing place. Confident that they would be able to reach a new anchorage for the evening, the crew hoisted anchor at noon and "beat up against a light wind towards Academy Bay." Although the distance was only a few miles, they spent the next seven hours rounding the southern tip of Indefatigable Island and coaxing every possible knot out of the schooner. At 7:00 p.m. they finally dropped anchor in three fathoms of water at what Beck called their "old place in the bay." There must have been a sense of homecoming. Of all the bays they anchored in during their year in the islands, Academy Bay was the spiritual homeport of the schooner *Academy*.

The collectors were welcomed to Academy Bay that evening not by friends or family, but by a multitude of mosquitoes that kept them awake most of the night.

Once the collectors had finally reached Academy Bay on January 11, 1906, their week of frustration and drifting was over, but their misery was not. That first evening in the bay, Slevin reported, "the mosquitoes gave us a warm welcome and kept us awake most of the night." The descendants of those very same mosquitoes still pester visitors in and around Academy Bay. So do the mosquitoes in the town of Puerto Ayora, which today bustles with tourist and commercial activity along the shore of Academy Bay. But these coastal mosquitoes were not the only ones on the island to cause the Academy men misery. Many more were waiting for them in the moist highlands.

Not one building existed in Puerto Ayora during the visit of the schooner *Academy*, and not one soul was to be seen along the shore. The only evidence of human visitors to the newly named Academy Bay was a series of footprints in the sand, the origin of which puzzled the collectors. Only later would they learn that a group of shipwrecked sailors were desperately looking for someone to rescue them as they moved from island to island, but the paths of the schooner and these men would never cross.

Despite the fact that no one lived at Academy Bay in January 1906, the crew of the *Academy* made use of the bay much the way vessels do today, to get fresh water and resupply their vessel. On their first full day, the men filled their "water breakers" at the water hole near the coast and engaged in general collecting around the shore.

After working so hard just to get to Academy Bay, the men planned to stay put for the next eleven days, from January 11 to 22, 1906. During this second of three visits to Academy Bay, they would fan out in all directions for general collecting. The men were busy making day trips to the highlands and setting off on individual overnight excursions to the rainy highlands. After their disappointment with tortoises on James Island, they had no trouble finding tortoises within a day's walk from Academy Bay. Beck and Slevin were the first to venture away from the shore, and they had immediate success. They found two good-sized tortoises and tied them up to be skinned on "the morrow" (figure 11.3).

The tame Galápagos hawks that congregated around the watering hole at Academy Bay were quickly and easily added to the ship's collection, and in this case by non-ornithologists. Ochsner and Parker killed nine of them by knocking them down with sticks, and Hunter noted, "Hawks still continue to eat their dead relations." One hawk he shot had "the remains of a number of centipedes and what was left of a mocker" in its stomach.

Hunter continued his astute observations on birds. He went ashore and walked the coast of Academy Bay in a big arc looking for birds of any type. The most common bird he found that day was the cactus finch, *Geospiza scandens*. The birds were actively singing, and the adults were "in worn plumage, much more so than they were when we were here before." Visiting an island multiple times as the men did allowed them to make these sorts of comparisons before, during, or after the breeding season. Adults and juveniles of the cactus finch were of equal interest to Hunter, and specimens of both were collected. "Two young were shot," Hunter wrote in his field notes. "They could not have been more than a week or ten days from the nest." Although he could not identify the parents of these two birds, the juveniles were "in the same bush and near an old nest and I think that they were from the same parents." They were evidently still under the care of their parents, as the "beak and feet were soft."

FIGURE 11.3 **Skinning Tortoises.** The process of tortoise skinning often involves cutting through the shell with a handsaw. Here the men in an earlier Rollo Beck expedition work at the task. *Archives, California Academy of Sciences.*

One poor bird that Hunter put out of its misery was a large-billed flycatcher, *Myiarchus magnirostris*. The one he shot was "in very poor plumage, probably due to the great number of lice that were on him." Hunter found two or three lice at the base of each head feather, as well as on the neck. Others of the same species were in good plumage and not afflicted with lice.

SIDE TRIP TO VILAMIL

But not all the food they needed or wanted could be had at Academy Bay. To get other food items the mate, Frederick Nelson, left the schooner after breakfast on January 12, 1906, in the small sailboat he had rigged up and set sail for Vilamil across the channel where they had just been becalmed. For one man sailing alone across fifty miles (eighty kilometers) of open ocean, this turned out to be an adventurous but largely uneventful sojourn.

Nelson's objective in Vilamil was to get a load of the sweet potato–like *otoys* (which they called *atoyas)* by trading the hundred pounds of flour that he loaded into the small sailboat. After sailing all day on the twelfth, Nelson reached the horseshoe-shaped Brattle Island (Isla Tortuga) toward evening and anchored in the lee of the island for the night. Sleeping in the little boat was probably as uncomfortable as it was difficult, and he got an early start the next morning, setting sail again at 2:00 a.m. for Vilamil. The winds must have been weak and variable, because Nelson spent the next twelve hours covering the eight miles from Brattle to Vilamil, which he reached on the afternoon of January 13.

Nelson spent the next three days at Vilamil, essentially on "shore leave" away from his shipmates. What he did during those days is not recorded, but he had been there previously and he knew several people in the coastal town and at the little settlement higher up the volcanic slope. He did not depart with his load of *atoyas* until the afternoon of January 15, 1906. The plan was to retrace his steps back to Academy Bay, spending the first night again in the lee of Brattle. After another night in the small boat, he did not get under way until 7:00 a.m., a slow start compared with his 2:00 a.m. departure on the way over to Vilamil. Perhaps he was tired after three days of partying in Vilamil, or perhaps the fickle winds dictated his course of action.

For Nelson, the sail back to Academy Bay was not particularly fast, although a fresh southerly wind carried him steadily along. After fifteen hours of sailing, he had reached only the southwest coast of Indefatigable Island, so he anchored again for the night. Trying to get up into Academy Bay at night would have been extremely foolhardy, and Nelson was experienced enough not to try. On the morning of January 17, he got under way at 6:00 a.m. with the schooner as his destination. This last leg of his solo excursion, despite being close to Academy Bay, took eight hours, and he finally reached the schooner in the early afternoon on the seventeenth.

Nelson's much-anticipated cargo was three hundred pounds of *atoyas*, which Slevin described as "a very necessary addition to our stores, as they are the only fresh vegetable we can procure." He also got a small bunch of bananas, several "mushmelons," and a couple of watermelons.

Hunter continued his quest for rails at Academy Bay, with less effort and more success than on James Island. He continued to comment on and compare their tameness or timidity on the two islands, evidently attempting to generalize about these birds.

In the mangroves lining Academy Bay, he heard four of five rails in about an hour of careful watching. "Two were shot," reported Hunter. They were "not as inquisitive as the James rail, but will answer rather well if a sudden noise is made." Compared with his experience high on the slopes of James Island, where he had to get down on his belly in the wet grass at Academy Bay, they were "much easier to get a shot [at], as they came out into the open mangrove trees quite often."

Hunter continued his quest for rails when he joined Williams, Ochsner, and Gifford on the four-day trip to the green zone for general collecting. They enthusiastically left the schooner on January 15, returning tired and weary to the coast on January 18.

From their first camp at an elevation of about six hundred feet, Hunter could hear rails "cackling in the brush on all sides," and he shot two that evening. Hunter's recent experience with the Galápagos rail on two islands led him to generalize about their behavior and degree of tameness. "The rail on this island [Indefatigable] are much more easily obtained than those on James. Individuals are at times very cautious, running away at the first move. Others will calmly come up to investigate when a noise is made and will slowly move through the brush, not at all like the nervous little fellows on James Island that will not allow you to wink without running away." Gifford agreed with Hunter's observations: "They are very fearless and inquisitive, quite the opposite of the James Island birds."

Hunter observed that the Galápagos rails flew weakly, if at all, and he never saw the birds "more than a few inches from the ground." He speculated that they made "no use of their wings, unless to assist in running." This reduced flight ability made the rail appear to Hunter an unlikely colonist of the remote Galápagos, but its strong-flying relatives had successfully colonized islands throughout the Pacific.

Although rails are more likely to be seen walking than flying, some members of this chicken-like group have good powers of flight and have colonized some of the world's most remote islands. Hunter observed the general trend toward flightlessness in rails on remote islands when he inspected the flight muscles of the specimens he shot. "The breast muscles [are] very much reduced," Hunter observed from his dissections, "and would not give very good service in flying."

After working so hard, and largely in vain, to secure rails on James Island, Hunter was pleased with his success on Indefatigable Island. He described his innovative collecting technique as follows: "The best way to get the rail here is when one is located by his cackling in a place not too thickly covered with brush, to lay flat so that the surrounding ground for a few feet around can be seen. After a few minutes, the little fellow will be seen slowly picking his way through the tangle. Often one will come within a couple of feet. It is necessary to shoot them with very light loaded shell. In all, about thirty were taken."

Gifford made two insightful observations about rails during the inland trek on Indefatigable Island. In addition to concluding that the rails were much tamer on Indefatigable, he was able to discern after listening to their calls that the notes of the birds on Indefatigable were "somewhat different" from those of the rails on James Island. Gifford's notice of inter-island differences in behavior and vocalization is an indication that he was thinking along Darwinian lines. Inadvertently, one of the Galápagos birds most important to Darwin provided comic relief for the men on the inland trek. Gifford noted that the Galápagos mockingbird was very common all the way up the island and was "very amusing in actions" and "inquisitive." That inquisitiveness nearly cost one bird its life when it "jumped into live coals and jumped out very much surprised."

RAIN, BED BUGS, AND MOSQUITOES

Up in the green zone of Indefatigable, Hunter, Williams, Ochsner, and Gifford found much to marvel at, but their forward progress was slower than anticipated. They eventually got as high as 1,100 feet above sea level, but encountered difficulties day and night. In order to stay dry during the night, they cut some massive Canna leaves to form a crude shelter, which was only partially effective.

The moisture at that elevation proved troublesome. "With the help of some kerosene," Slevin reported secondhand, "they started a fire to cook supper." "Cooking is somewhat difficult," Slevin continued, "as the forest is kept continually wet by the clouds that cover the top of the island most of the time." The well-known Galápagos *garúa* was keeping the highlands green and the progress of the men to a minimum.

Although ambitious collectors, the four men were unprepared for fieldwork in what they knew from their previous experience would be a rain-soaked part of the island. After a dinner of beans, coffee, and canned fruit, they turned in early and hoped for a good night's sleep, which would prove as elusive as the rails Hunter was trying to shoot. Their difficulty in maintaining a fire was mirrored by their difficulty in staying dry through the night. Before daybreak, the rain started to fall on the huddled men, and despite their leaf shelter, "all hands got thoroughly drenched."

After such a wet and disagreeable night, the field party found it impossible to make any headway against the jungle and started back down the trail they had just cut. Once they reached "dry country," they halted for a much-needed meal of coffee and flapjacks, because "cooking in the wet zone is next to impossible." Although at this lower elevation they were out of reach of the misty rainfall, they were still fair game for the mosquitoes. On their last night of three nights away from the schooner, the men "were nearly eaten alive by mosquitoes." They managed to salvage the trip by engaging in general collecting all the way back to the schooner. When they arrived at the schooner, Beck commented that they were "sleepy and tired."

The misery of the men who remained behind on the schooner echoed that of the four men journeying into the mosquito-infested highlands. On the night of January 13, 1906, they "burned a smudge fire to get some relief from the mosquitoes." Having had moderate success with insect deterrence, the next night they upped the ante considerably. Bedbugs had tormented the men for many weeks in their births below deck. Something had to be done and now was the time. War was declared.

It was a Sunday and all hands were observing the traditional Sabbath respite. Hunter and Beck put together a concoction of "six ounces of potassium cyanide and other dope" to fumigate the schooner's cabin. The results were less than perfect. They still needed a silver bullet for their tiny nighttime tormentors. The toxic fumes from their makeshift concoction killed most of the flies, cockroaches, and mosquitoes, "but no bedbugs." Beck was thoroughly unsatisfied with the results when he "found lots [of bedbugs] in my bed," and he proceeded to "kill a half dozen in the morning as they climbed up through the mosquito net." In the struggle for existence below decks on the *Academy*, the bedbugs were proving themselves fit for survival.

Bed-bug fighting aside, Sunday was a typical day of rest for the collectors. Their activities were not particularly strenuous, but they were productive nonetheless. After a late breakfast, Slevin, Williams, and King went ashore to wash clothes at the water hole. Beck engaged in his characteristic Sunday activity suitable for a good Methodist: taking photographs. Beck rarely took his camera ashore during the week, choosing instead to concentrate on the heavy work of collecting. He took photographs almost exclusively on Sundays, as did Gifford.

On this particular Sunday, Beck and the cook, Jim White, went to a nearby lagoon for photographs. They also picked up a dozen "crawfish," which White made into a delicious salad for supper. The other men stayed on board repairing shoes (Hunter's specialty) and preparing for the upcoming four-day trip to the highlands.

Tortoise hunting continued smoothly between January 13 and 21, 1906. The daily forays were led by Beck, who stayed on the schooner each night during this eleven-day visit to Academy Bay. The party hiked about three and a half miles inland to find good tortoise country. One female that Beck and Slevin skinned was "very fat and the liver large and light-colored. Just the opposite to those on Duncan, probably on account of the food, these getting more green food while the Duncan tortoises get the grass and lichen also." Slevin attributed the differences in liver color between the tortoises on the two islands not to genetic factors (which would suggest species-level differences and would be an evolutionary explanation) but to the variation in foods eaten, the temporary effect of environment.

Slevin reported that another big male "had more fat than any male taken so far. But as a rule they don't have near as much as the females." The females also possessed "much larger livers than the males." Yet another tortoise "was found eating cactus and stomach was full and bladder full of water. Lots of cactus spines were stuck in the throat."

One unusual male, collected on January 20, 1906, escaped and traveled two miles from where the men had tied it to a tree. Marveling at this tortoise's rapid locomotion, Slevin wrote, "He had traveled about 2 miles in an afternoon." Oddly, though, this tortoise was "fairly fat for a male and had a stub tail." Slevin speculated that "some accident probably happened to him when he was small."

A rare bit of personal commentary appears in Beck's field notes at the end of the collectors' stay at Academy Bay. Although subtle and indirect, the remarks seem to indicate some personal friction between Beck and Stewart. Beck stated in his field notes on January 15, 1906, "Stewart sick from drinking cup of water ashore," and wrote the next day, January 16, 1906, "Stewart 'didn't feel like going' but helped fill [water] casks, a load of water each day." Beck rarely slowed down for either illness or fatigue, and he expected everyone around him to be equally resilient.

But two days later, on January 18, Stewart appears to have recovered when he accompanied Beck, Slevin, and King on a trip inland to find a medium-sized male tortoise that had been tied up by Beck the day before. On the following day, Beck went "in with boys, all but Gifford, for tortoise." Some sort of drama might have been brewing, because the very next day, January 20, Beck wrote, "Ashore with boys & mate, except Stewart who asked Captain to take him ashore while rest getting shoes on."

Whatever drama was simmering between Stewart and any of the other men, it boiled over on the day the schooner departed from Academy Bay, January 22, 1906. It was a Monday morning, and the men were variously engaged in preparing to leave. Four of the men, Beck, King, Williams, and Slevin, went ashore to the water hole to wash clothes. Slevin noted that "the rest of the party" remained aboard and was busy scraping barnacles and other marine growth off the schooner's sides.

The exact events that occurred on the schooner were not reported by any of the men, and Beck wrote, "Stewart and Ochsner had a scrap, honors about even. Optic discolored on one; no official knowledge; occurred while we were washing clothes [at the water hole]." The surprising aspect of this incident is not that it occurred, but that it didn't occur more often. Tempers flared and the bout of fisticuffs left one of them with a black eye. A crowded vessel with eleven men living under harsh conditions is a tinderbox for occasional violent and unfortunate outbreaks.

JANUARY 22–23, 1906

Despite the testosterone-driven shenanigans between Stewart and Ochsner at Academy Bay, Beck and Slevin had time to go "shooting birds on the boat" before they hoisted all boats on January 22, 1906. The crew weighed anchor at midday and "beat out of Academy Bay against a light breeze and set course for Chatham Island." They hoped the wind would carry them quickly over the forty nautical miles to their destination, but they barely covered half the distance before the wind gave out. The trip to Chatham Island would give the men an opportunity to reunite with the friends they had previously made and to check for any mail that may have arrived from San Francisco. And where there were people, there were deals to be made.

By 8:00 p.m. the schooner was only eight miles to the northeast of Barrington Island, sailing in light breezes. The wind picked up later in the evening of January 22 and into the early hours of the next morning. Sunrise on January 23 found the *Academy* "up under the lee side of Chatham Island," with the wind blocked by the mass of the island. Throughout the day the crew continued to work the vessel "close in to shore to get the benefit of the current." The sailor-scientists continued to learn the subtleties of Galápagos navigation. Transiting from island to island within the archipelago was tricky and sometimes risky with variable winds and no auxiliary engine, as most sailboats have today. They learned to enlist the aid of ocean currents as an alternative source of power.

The winds continued their fickle puffs, and the collectors maneuvered the *Academy* within a mile or two of Kicker Rock for a frustrating twenty-four hours. They were, in essence, becalmed just to the northeast of Wreck Bay, along the western coast of Chatham Island. But with a freshened breeze and short tacks, they were finally able to coax the schooner down the coast and up into Wreck Bay late in the afternoon on January 24, 1906. While waiting for the wind to carry them the last short distance to their destination, they caught forty gallons of rain in a canvas awning for drinking water because "the Indefatigable water was none too good." Overnight, while anchored in the harbor the next day, they picked up two hundred more gallons from the awning.

The men would spend the next five days in and around Wreck Bay, which they also called Puerto Chico, or Little Port. Although the heavy January rain provided them with decent drinking water, it ruined field collecting. It also foreshadowed darker events. Chatham Island was the island recently ruled by Manuel J. Cobos. They used favorable winds and currents to navigate back to the main port and town on Chatham Island, but they also navigated into the dark undercurrent of life on this island. Although Cobos was dead, evidence of lingering resentment confronted the scientists.

JANUARY 25–30, 1906

Putting into the port town on Chatham meant both business and pleasure for the expedition members, the second of six visits by the schooner *Academy* to Chatham Island. While anchored in the harbor, the men would renew old friendships with local residents, collect more specimens, barter and trade for supplies, get mail, and pay a debt.

Three months had gone by since their last visit to Chatham in mid-October 1905. One of the first orders of business after dropping anchor in five fathoms of water and after having supper was to check for mail. Ochsner and Gifford went ashore to ask up at the plantation if any mail was waiting. The plantation owner informed them it would be brought down to the schooner in the morning. In addition to providing them an opportunity to go ashore and stretch their legs, the mail excursion afforded time apart for Ochsner and Stewart, in case there were still hard feelings about the scrap they had recently had.

The next morning was so rainy that almost no collecting could be done. Williams and Slevin spent the morning collecting along the road to the hacienda, but they found that "the road was a river" and the island "green from the shore up." Those who ventured up the road toward the settlement at El Progreso slogged back down to the coast in time to be there when the mail came down at noon. The dual excuses of bad weather and news from home kept the men aboard the schooner for the rest of the day reading letters and entertaining company. That evening, the governor of the island came aboard for supper, but the details of this social visit were not recorded in anyone's field notes.

Despite the bad weather, several of the men did their best to collect. Acting on a published report by Georg Baur in 1891, Slevin was hoping to find specimens of the gecko called *Gonatodes*, but lamented after several hours in the rain, "Have not had the good fortune to run across" any specimens. The geckos might have been scarce on the island because that particular group appears to have been introduced in historical time to the Galápagos from the Ecuadorian mainland. The specimens found by Baur were real, and the species still occurs there today. Here is a clear example of how the published work of a previous expedition influenced the collecting activities and expectations of the 1905–1906 expedition. Although Slevin does not explicitly state that he was acting on Baur's published report, Baur's was the only previous expedition to report these geckos on Chatham, and Baur had done his collecting only fifteen years earlier (in 1891). So Slevin was running around in the driving rain looking for a critter that doesn't occur there naturally (and was possibly locally extinct there anyway) and feeling disappointed by his lack of "good fortune." Georg Baur was still pulling the strings, even from his grave. Over the next several days, Slevin continually looked for, but never found, the elusive *Gonatodes* gecko. Despite drier weather and finding other species of gecko, he wrote that he "failed to find" it and that "I have not been able to get the *Gonatodes* as far as I can see so far," and a couple of days later he lamented again that he was "unable to find any of the *Gonatodes* yet." He didn't give up looking until they left the island.

Other collectors had equally spotty success. Hunter noticed that birds along the shore were much more common than they had been during their last visit. Several of Darwin's finches were in breeding plumage and some revealed eggs internally, ready to lay, when dissected after being shot. But for Hunter "the rain spoiled the collecting" and the birds "were in poor condition for specimens."

Beck walked up to the sugar mill at El Progreso and found it shut down for lack of sugarcane to process. This was probably a result of the murder of Cobos just one year earlier and the resulting disintegration of the various enterprises on Chatham.

The next day (January 26, 1906) the weather dried up a bit and the men fanned out to collect. Hunter went up the road to El Progreso and turned off onto a dirt track about two-thirds of the way up. Birds were common in the green vegetation, and his specimen of the woodpecker finch was the first ever taken from that island; he noted its call as "a pleasing mellow whistle."

Hunter frequently heard the call of the dark-billed cuckoo but he seldom saw one. Beck had better luck. He found an adult cuckoo incubating an egg in a nest "in a tree in a big rock pile," which then "flew off into a tree and stayed there until shot."

Hunter had an encounter with two adult mockingbirds protecting their nest, which contained three young birds, just starting to grow feathers. Both parents came around to see what was happening while Hunter was at the nest. "The female was the most courageous.

She making repeated attacks on my hat and hand. She made no cry, except when I made a grab for her. Then she would give vent to several sharp shill screams. The male was not at all demonstrative, calmly watching the affair. He was first seen with a smooth caterpillar about an inch long in his beak and later with a grasshopper." These protective parents, defending their nest and bringing food to their young, met a predictable end. Hunter sealed the fate of the family of five: "Both the young birds and the parents were collected."

Nests made good collecting areas for Hunter, as he could collect both the adult birds and their young, or the eggs if they were unhatched. But things didn't always go smoothly. Hunter found a nest of the Galápagos flycatcher in a stand of cereus plants on Chatham Island. Climbing up about seven feet off the ground, Hunter gave the branch holding the nest a slight tap to see if the bird was home, and the whole branch fell to the ground, splitting open the nest and cracking the eggs. But Hunter did not leave empty-handed. When the female parent lit on a branch a few feet away, she was "killed with a stone." Another nest belonging to one of the ground finches was found with eggs, but because the parents were not nearby, the nest was "left for a later day." Hunter was patient and let the specimens come to him.

In addition to collecting, Hunter made numerous observations about the general condition of the birds. One of these observations concerned a disease affecting the feet of birds. He noted that the last time they collected on Chatham, in October 1905, "nearly all of the birds showed signs of having the peculiar disease of the feet that at times is common among the birds of the U.S." He was pleased to see that the disease (likely Marek's disease) had abated, with the occasional bird missing a toe or two, some without toenails, and rarely, a bird missing an entire foot.

PINEAPPLE DIPLOMACY

Although the heavy rain slowed the collecting of biological specimens, it did nothing to impede the social interactions between the schooner's men and the residents of Chatham. The evening after the governor came aboard the schooner for dinner, Beck and Williams stayed onshore to have supper with the plantation manager. His name is not recorded, but he would have taken over the plantation after the murder of Manuel J. Cobos—a dark chapter in Galápagos history.

The *Academy* crew still owed money for an anchor they had bought from the plantation manager back in September 1905. Beck and Williams went ashore on January 28 to hand over payment, which included 150 bars of soap, 500 pounds of flour, and $27.50 in gold coin. Beck also spent $15 in gold coin for 100 pounds of local coffee "to take the place of the 'coffee' in the ship's stores."

Additional bartering was conducted on the schooner and on the harbor dock throughout the day. Beck noted that several "natives" came on board to trade ponchos and bananas for socks and pants. Bananas and pineapples appeared to be the common currency among the locals, and the schooner ended up with fourteen bunches of bananas hung from the stern davits and at least a dozen pineapples. The Academy men also found guavas and lemons that they collected and brought on board.

Ochsner lived up to his nickname of "Doc" by providing medical treatment to the governor, who was suffering from an unspecified leg ailment. Beck wrote cryptically, "Doc fix

Governor's leg." Later in the trip, Ochsner would tend to other, more serious wounds near the El Progreso plantation.

The men's habit of keeping the Sabbath took an evangelical turn on January 28, 1906. Slevin noted, "A Barbados negro was given a hymn book, which he seemed to prize highly." Perhaps the men on the schooner had an extra copy, or perhaps they weren't using their copy much.

The devotion of the collectors to the tasks of the expedition manifested themselves in different ways. Gifford writes of a day in January 1906 while the schooner was anchored in Puerto Chico that he spent "on board all day but on deck little." He was so focused on skinning birds that he put his head above deck for only a few minutes all day, just long enough to see a frigatebird and a blue-footed booby fly overhead. On that same day, other men hiked and explored, just to see what was out there. An all-day trip to the top of Chatham Island by Slevin and Williams revealed "a remarkable absence of bird life and no reptiles at all." They found the top of Chatham covered in "beautiful green pastures and hedges of agave," as well as tree ferns that gave the rolling plateau a verdant, tropical appearance. The numerous cattle roaming the higher elevations had plenty of food and water from the several small lakes scattered about. Despite the poor collecting for larger animals, they found a few land snails clinging to the moss, which they brought back and added to Ochsner's collection of shells. They would become "shells of contention" years after the expedition.

All hands rose early on January 30, 1906, and made the vessel ready for sea. Beck and Williams made their last trip ashore to buy more coffee and some barrel hoops to repair the wooden tubs used for soaking tortoises. After the small boat returned from shore, the anchor line was "hove short" and the crew was about to get under way when a sloop was sighted. The very desirable prospect of getting more mail from home was reason enough to wait to see if it had just returned from the mainland. The men were disappointed when they hailed the sloop and learned that she had just come from Indefatigable (Santa Cruz) and South Seymour (Baltra) Islands. On board the sloop they could see a bunch of live goats, which they speculated had been taken off South Seymour Island and were being transferred elsewhere.

At the last minute, as the men "broke out the anchor" and the schooner was leaving the harbor at Wreck Bay, a local man was seen "in a canoe paddling towards us for all he was worth, so the schooner was held close to the wind until he came alongside. He had two pineapples for us, presented with the compliments of the Governor." The men on the schooner seemed to be aware of the harsh punishments meted out by those in power on the island. As an indication, Slevin wrote that the man in the canoe "no doubt was fully aware what would happen if he failed to deliver the pineapples; hence the burst of speed." Perhaps little had changed after the death of Cobos.

With their pineapples and bananas stowed, the schooner and her crew navigated out of Wreck Bay and headed south toward Hood Island—and right into the remnants of a world-class natural disaster.

JANUARY 31, 1906

One of the world's largest recorded earthquakes occurred off Ecuador on the morning of January 31, 1906—and the men on the schooner *Academy* probably didn't feel a thing. They departed from the harbor at Puerto Chico at midmorning on January 30 and set sail almost due south for Hood Island. They sailed the schooner "full and by," making their best speed into the wind by keeping the sails full of wind, as they navigated across the thirty-mile channel separating the two islands. The sky clouded over in the afternoon and they had light southeast winds. Several of the men noted in their field notes that there was already a "considerable swell running" while they were under way.

By 3:00 in the afternoon on January 30, the schooner passed a shallow area called McGowan's Reef, having made it that far on one tack from Puerto Chico. During the night they made short tacks back and forth for safety, not intending to make substantial forward progress but still heading in the general direction of Hood Island. By 6:00 a.m. on the thirty-first they were eight miles from Hood Island. This was a good position from which to navigate safely to their destination, and with steady wind they made short tacks close to shore to get themselves into Gardner Bay by early afternoon. They anchored halfway between Gardner and Hood, where Hunter noted, "We get the full force of the swell." Slevin also noted there was "quite a ground swell and schooner rolls considerably." Whatever effect the earthquake had had on the ocean, the heavy seas obscured it.

When the earthquake had hit at about 9:30 on the morning of January 31, it shook the history books. On the open-ended Richter scale, it had a magnitude of 8.8 or 8.9—making it even today one of the four or five largest earthquakes of the twentieth century. The schooner at that hour was likely very close to Hood Island, and the men were fully involved in sailing close to shore in a fresh breeze with heavy seas running. The earthquake occurred on the sea floor in the Pacific Ocean offshore from Point Tortuga in Esmeraldas Province in Ecuador, some 460 miles from Hood. Even if the direct effects of the earthquake were minimal, it generated a tsunami that struck the coasts of Ecuador and Colombia, killing at least a thousand people. Wave heights in Ecuador were estimated at more than thirty feet. With all this devastation occurring on the morning of January 31, the schooner *Academy* was not all that far from the epicenter but ended up completely safe.

How could such a large earthquake and such a devastating tsunami occur so close to the Galápagos and yet be unnoticed by the men on the *Academy*? Part of the answer lies in the fact that they were under way in a heavy swell that obscured the tsunami. With substantial swells all around them, the tsunami likely passed right under the schooner and through the Galápagos on its way across the Pacific.

Once the schooner reached Gardner Bay on January 31, and having missed the tsunami, the men dropped anchor in three fathoms of water. A strong current was running in the channel between Hood Island and little Gardner Island just offshore. Hunter was apparently not pleased with their anchorage halfway between Gardner and Hood, because it was a place "where we get the full force of the swell." Their first order of business, before several days of collecting on Hood and around Gardner Bay, was to clear up the decks and spread out a canvas awning for shade. A boat called the *Matilde* came in from Indefatigable Island, bringing goats and fish.

Beck and Nelson went ashore immediately to hunt for goats. Within a short time, Nelson got two goats and Beck one, which would be plenty of meat for a few days during the hot season with no refrigeration. Gifford also went ashore for birds and made an observation that is consistent with a tsunami hitting the islands. Just behind the bright white sand beach that is a common "wet landing" for present-day Galápagos ecotours, Gifford noted that "two small lagoons have appeared behind the beach, probably from an exceptionally high tide." Perhaps it was a high tide, or a tidal wave, another name for a tsunami.

However, Gifford's experience upon setting foot ashore at Gardner Bay was very different from that of today's visitors, who must adhere to a strict hands-off policy. When the inquisitive Hood mockingbird approached Gifford, as these birds readily do today, Gifford, the collector of evolution, had this experience: "I had hardly stepped ashore before one [mockingbird] came trotting up to inspect me. I gave him a warm reception. His testes were quite enlarged."

Even the simple activities of the expedition concealed hidden dangers. Several of the men could not swim, yet all of them remained safe during the entire trip. But there were several close calls. At 5:00 in the afternoon on that first day at Gardner Bay, Slevin and Williams rowed ashore to pick up the three men in the shore party. On the way back to the schooner, the five men "shipped a couple of heavy seas getting off the beach, but no damage was done." Had the heavy seas swamped and sunk their skiff in deeper water, the outcome could have been tragic. But fortunately, none of the men had to test their swimming skills, and for dinner that night they had "Chatham Island pineapples and Hood Island goats"—as exotic as it was satisfying.

How they slept that night with a heavy sea running and with the schooner incessantly rolling back and forth was not mentioned by any of the men. But they likely wished for calmer waters to ensure that their exotic dinners remained in their stomachs.

I HAVE SEEN THE ALBATROSS

For the first full week in February 1906, the men collected productively on Hood Island and nearby islets. This was their fourth visit to Hood; they would return twice more. One of the largest and most ambitious trips on this visit involved five men taking the big skiff down the rocky coast to the large albatross colony on the southeastern tip, at a place called Punta Cevallos today. The group consisted of Beck, King, Gifford, Williams, and Slevin. On the way, perched in the rocky sea cliffs along the coast, the men saw hundreds of swallow-tailed gulls. There seemed to be more of these birds in the rocks and ledges than they had seen on a previous trip along this coast, confirming the importance of repeated visits to document seasonal variation. Gifford reported that the swallow-tailed gulls did not build a nest but instead simply laid their eggs on the bare rock, usually under an overhanging rock ledge. On the other hand, Hunter reported that the nests were built up an inch or so with small pebbles, and the eggs laid directly on top.

In the process of collecting swallow-tailed gull specimens, Gifford learned a few things about their behavior. He found that the older birds have "two or three cries, one is made by opening the mouth very wide." On the other hand, he found that the young ones "also squawk and snap viciously at one when grabbed." Gifford did in fact grab some and learned that "when grabbed both young and old often vomit squids." Two young ones that he had tied up were "killed I think by a hawk as one was near when I returned." When they were not tied up, Gifford noted that "they run about very lively, even when having practically nothing but down on them." Sadly, he also noted, "When a wing is broken the birds like most others bite savagely at the wounded member."

Gifford also learned of the bravery of red-billed tropicbirds as they resisted his intrusion into their nests along the coast of Hood Island. "One bird which I disturbed," Gifford wrote, "was sitting on its egg with wings drooping and every feature showing rage at my intruding. Both young and old threw up fish and squids."

A bird in the air is worth two in a burrow, and Gifford wrote that there were "quite a number" of Audubon's shearwaters flying about the cliffs. "They sail very gracefully over rocks and water, often wheeling very sharply." His observations paralleled those of John James Audubon, the premier wildlife artist of the first half of the nineteenth century. Audubon wrote about seeing what would become his namesake birds in 1826 in the Gulf of Mexico, "They skim very low over the sea," and "they flap their wings six or seven times in succession, and then sail for three or four seconds with great ease, having their tail much spread, and their long wings extended at right angles with the body." As beautiful as the birds in flight seemed to both Audubon and Gifford, they were also valuable as museum specimens. Audubon sent a specimen back to the Academy of Natural Sciences of Philadelphia for naming, and Gifford intended that his specimens end up at the Academy on the other coast. Significantly, they belonged to the group of birds that the California Academy of Sciences director, Leverett Mills Loomis, considered his specialty, the widely ranging seabirds called the tubenoses. Gifford extracted two or three pairs of these delicate shearwaters from their burrows among the rocks and added them to the growing collection.

When the skiff reached the albatross colony at Punta Cevallos, four of the men went ashore, and Slevin pulled back out to a safe distance and anchored "to await the return of the hunters." The rocky shore made it impossible to beach the skiff, so the shore party would work for several hours while Slevin bobbed offshore. The surprising discovery was that all the albatrosses had left the colony, and the only signs of them were "a few rotten eggs and feathers." The albatrosses had departed from their breeding ground to begin the nomadic part of their lives. Although Hunter was not with the hunting party that day, he wrote that the albatrosses had "for the time left the island," and he knew there was more to the story. "Where the birds go is somewhat of a mystery," Hunter mused. "They have been seen only once to be reported by scientists in the southern hemisphere and that is where they go." The colony had been packed when the men visited in September 1905, and they had returned to the schooner "loaded down with albatrosses." But for now, these mysterious birds from these mysterious islands had flown the coop.

Years later, the extensive fieldwork of scientists such as Dave Anderson of Wake Forest University would reveal where the albatrosses had gone. Using satellites and radio transmitters in the 1990s, Anderson and his research group discovered that in 1906 the albatrosses of Punta Cevallos did in fact fly south as Hunter speculated.

The deserted nests and scattered, broken eggs of the albatrosses contrasted with the well-adorned nests of the noddy terns observed by Beck. He found the nests of these brown seabirds to be composed of "wet seaweed from the ocean, lined with crabs legs and backs, sometimes a little shell or two, or a few bird quills."

Armed with the knowledge that February was not the month to collect albatrosses, the men left Punta Cevallos and "pulled back to the schooner in a fairly heavy sea." Back at Gardner Bay they picked up the other shore party from opposite the anchorage and returned to the schooner for dinner. They were rewarded with goat meat for supper and banana pies made from their Chatham Island bananas. They much preferred this fresh food over "some of the menus we have had," Slevin remarked.

While the collectors were ashore, the cook, White, prepared tasty meals to satisfy their hearty appetites, and the mate, Nelson, stayed busy oiling down the decks and scraping seaweed off the sides of the schooner. Nelson's other diversion was to hunt goats and other game, and he was quite successful. Whenever they needed fresh meat, Nelson could usually deliver.

Collecting onshore one day was usually followed by a day of preserving the specimens. Despite the disappointment at the albatross colony, the next day Slevin recorded, "Ornithologists busy on board skinning seabirds taken at the albatross colony yesterday." In the tropical heat, any specimen that wasn't gutted and preserved within twenty-four hours would begin to rot and smell. In fact, Nelson was kept busy with daily goat hunts because "it is necessary to get fresh meat each day, as in this climate it will not keep any length of time."

The success of their general specimen collecting on Hood was dampened by the fact that they had not found any tortoises. Slevin wrote, "We are always in hopes of picking up a tortoise and try to visit the greener parts of the island." But instead of tortoises, Slevin and his assistant, King, collected mainly lava lizards and marine iguanas. King reported to Slevin that the lava lizards on Hood Island were exceedingly tame, sometimes running up and picking up the crumbs from one's lunch. Examining the stomachs of six males and one female, Slevin learned that they normally feed on the leaves of a juicy green shrub that grows along the shore, when not nibbling people food. Over on little Gardner Island, Slevin found the lava lizards opportunistically "feeding on maggots on a dead seal and some eating flies," but concluded that they ate leaves more than anything else.

A typical Sabbath was spent on February 4, 1906, with some of the men staying aboard all day and others doing light work ashore. Beck went the farthest afield, taking his camera all the way down to the abandoned albatross colony at Punta Cevallos. Gifford stayed below skinning birds and reported that "I was on deck very little," popping his head up only long enough to see frigatebirds, boobies, and pelicans flying over the schooner. Williams unsuccessfully went shark fishing over the side of the schooner. The leisurely pace of the day allowed Beck time to write more than the usual few sentences in his field notes. In fact, that day Beck revealed his talent for observation and comparison, especially of nesting birds he saw on his half-day trip down the coast. All the species of seabirds in the cliffs down the coast came under Beck's careful inspection.

Beck also had a tendency to anthropomorphize when he observed bird behavior, and he did so abundantly on his photography trip to Punta Cevallos. In the midst of a frigatebird colony, Beck saw a mockingbird that "picked at wing of frigate to get him off egg, seemingly, but soon quit. Very inquisitive." Other mockingbirds were attempting to pick holes in tern eggs when the adults flew away or to pick holes in gull eggs. Beck assumed the mockingbirds were conscious of their sneaky, predatory behavior when he wrote that they "watched always the frigates flying over, seemed like he knew [he] had no business there." Beck, the devout Methodist, might have been reading too much into the life-and-death struggles of these birds, but moralizing about proper behavior would have made an appropriate Sunday sermon right there on the sea cliffs, had anyone been so inclined to speak or to listen.

The next day, Beck, Williams, King, Ochsner, and Slevin rowed down the coast again to Punta Cevallos. It was a productive collecting location and worth several trips. This time

Williams stayed with the boat, allowing Slevin to go ashore, where he added to the reptile collection with two sacks full of marine iguanas and a small green sea turtle. The iguanas on Hood were characteristically "very brightly colored in green, red, and black." Although Slevin attributed the bright coloration of the iguanas to the onset of the mating season, he saw no real signs of mating. He marveled that the iguanas could be seen laying close to the rocks with their long claws and could "hang on resisting the wash of heavy surfs that break over." Darwin had noted, "Their limbs and strong claws are admirably adapted for crawling over the rugged and fissured masses of lava, which everywhere form the coast." Slevin's observations about marine iguanas clinging to the rocks, feeding on seaweed, and the timing of their reproductive season paralleled Darwin's observations some seventy years earlier.

Slevin was disappointed by not seeing the fabled feeding behavior of the marine iguana, diving down in shallow water to eat green seaweed. Slevin lamented, "I saw none feeding or swimming today; in fact, you seldom see them in the water." Charles Darwin had cut open the stomachs of several marine iguanas during his visit in 1835 and "found them largely distended with minced sea-weed" that grew on rocks below the tidal zone.

Unlike Darwin and all those who had undertaken scientific expeditions before them, the men of the 1905–1906 expedition had an important advantage. They could return to the same site multiple times, and in different seasons, to document changes in abundance and behavior. Hunter observed on Hood Island that warbler finches were much more common near Gardner Bay than they were back in October 1905 and that "50 could be taken in half a day without difficulty." Beck's comparisons with their previous visit in October 1905 revealed that masked boobies had replaced blue-footed boobies in the same nesting places along the coast of Hood Island and that young shearwaters were nesting in the same places he had seen them in four months earlier. Gifford's observations spanned several months, as well as the past few days, with red-billed tropicbirds nesting in the rocky sea cliffs on Hood Island. "One with a fresh egg," Gifford wrote, "I took from a hole where last time [October 1905] gulls were nesting. Many were occupying nests, whose former owners I had killed less than a week ago."

Gifford kept up his steady pace of collecting and observing bird behavior at Gardner Bay. For Gifford, any bird worth observing was also worth collecting, as he demonstrated in this passage:

> One nest I found occupied by a female and four fresh eggs—greenish-blue, covered thickly with light brown spots. The nest was built at the junction of two bushes, about four feet above the ground. The foundation was of twigs. The inner part consisted of fine grass, goat's hair, and cotton and was about two inches deep—shallow compared to two or three mocker nests I have seen here. The bird did not get off until I was quite close, and then went only a few feet. Bird is No. 2058. [The four eggs are] Set no. 615.

King also had a good eye for bird behavior, whether it was on the ground or in the sky. Although King did not leave any field notes, occasionally he would relate his observations to the other men, and they would record the information. One day on Hood Island, King saw two frigatebirds taking turns chasing a swallow-tailed gull that was returning with food in its stomach. Frigatebirds exhibit a behavior of forcing other birds to disgorge their stomach, the contents of which the frigates catch before it hits the water. Even the Hawaiian people knew of this behavior and called frigatebirds in their archipelago 'iwa (pronounced ee-vah), which means "thief." The aggressive frigatebirds seen by King injured the leg of the gull by "grabbing it and swinging the bird, which would not disgorge." The disgorgement of food by their victims is the energy payoff frigatebirds get from being kleptoparasites, harassing other birds to force them to regurgitate their food.

Gifford must have been thinking about Darwin's phrase "the struggle for existence" after observing the shortage of nesting sites for seabirds in the rocky cliffs of Hood Island. The cramped conditions and the constant replacement of one species by another indicated to Gifford that nesting space was at a premium. Observing the complicated nesting situations of a dozen seabird species on Hood led Gifford to speculate, "There must be a constant struggle for nests judging from the examples given above." That Gifford was speculating about the interactions within and between species, and doing more than just collecting specimens, shows that despite his young age he was functioning in an evolutionary context.

But before nesting, there must be mating. Gifford was able to provide some insights into a particularly acrobatic version of mating in frigatebirds when he observed, "Yesterday I saw two *Fregata aquila* copulating. The female was sitting on the nest, and the male on top of her. He was not holding himself on with his beak as a rooster does but simply balanced himself while performing the feat." Nature could indeed be "red in tooth and claw," but it could also be a seductive balancing act on a rocky cliff overlooking the tropical Pacific Ocean.

After a week on Hood Island, the next goal was to sail north to Chatham Island again, not to revisit the town at Wreck Bay, but to explore and collect on the eastern side of the island. On the morning of February 7, 1906, all hands turned out at 5:00 a.m., and after coffee they weighed anchor and set sail for Freshwater Bay on the southeast side of Chatham.

On the short sail from Hood to Chatham, a note of bitterness about Parker's navigation and sailing abilities reappeared in Rollo Beck's field notes. Beck was the Galápagos expert and Parker was a newcomer to this equatorial archipelago. Over the previous seven years, Beck had sailed these waters in four schooners, and he possessed more knowledge about the fickle winds and currents than any living collector of his time. If you wanted someone to get you around the islands competently and securely, then Rollo Beck was your man. On this expedition, Beck allowed Parker to set the course but never failed to note Parker's mistakes and shortcomings, as did the other sailor-scientists. The resentment continued to build and would get worse.

On the way from Hood to Chatham, Beck and Gifford stayed below all day skinning birds. Parker and the men manning the sails faced light winds in the morning and a freshening breeze toward noon. By midafternoon they tacked the schooner a mile southeast of Whale Rock but found they were not in a favorable position to make Freshwater Bay. This shortcoming was Parker's fault, as far as Beck was concerned. Beck thought it was inexcusable that on one tack Parker lost half a mile, forcing them to "turn around and make Bassa Point anchorage" instead. Parker was guilty of "bad navigation due to not holding vessel up to windward more" and not showing better control of the lines that control the sails, or in Beck's staccato shorthand, "sheets too free." Small mistakes such as this could be forgiven and compensated for, but bigger mistakes by Parker could and would jeopardize the safety and success of the expedition.

Resigned to their failure to make Freshwater Bay, the men settled for Bassa Point, just south of Stephens Bay on the southwestern shore of Chatham. Dropping anchor at 7:00 p.m. in five fathoms of water, the men spent that night under a full moon in a smooth sea.

FEBRUARY 8, 1906

The next morning, making the most of their alternative situation, all hands went ashore collecting. This was fortuitous because they had not previously visited this portion of Chatham. An excursion up to the 1,000-foot level by Williams and Slevin revealed very rough country and nearly impenetrable brush in places. The calm seas of the night before gave way to strong southerly winds with squalls and rain. On the way back down the mountain, the afternoon rain hit Williams and Slevin, and continued until they reached the landing place, soaking them to the skin. Nelson rowed one of the skiffs ashore to pick them up, and in the choppy sea returning to the schooner, they shipped two or three waves into the skiff while getting off the beach "and got an additional soaking," this time with salt water—certainly a wet landing in today's terminology.

Beck hiked to the top of a nearby hill on February 8 and could see ranch houses a few miles off in the distance. Along the coast there were a few small ponds, the largest perhaps as big as an acre, with a group of blue-winged teal that flew off when Beck approached within seventy-five yards. But the tamer Galápagos teal remained in the pond, and Beck "waded in and shot 1 with pistol." That day Beck also saw "orchids by 100s in some trees," the same observation that moved Barry Lopez to write in 1998, "You extend your fingers here to the damp, soft rims of orchids, blooming white on the flanks of dark volcanoes."

While the collectors were ashore, Nelson spent the day "cleaning off seaweed from the waterline." They were picking up wispy bits of marine flora in the islands, in addition to the more obvious terrestrial flora collected by the botanist, Stewart. They also had hitchhikers in their freshwater supply: abundant mosquito larvae. They often found the mosquitoes "thick and troublesome" both aboard the schooner and ashore.

FEBRUARY 12, 1906

Going ashore always held a certain amount of uncertainty in terms of what specimens the collectors might find, as well as the rough topographic conditions and the often rather pesky vegetation. On February 12 (Charles Darwin's birthday), Slevin and Beck set off on a tortoise hunt, with King helping partway, shouldering all the water they could carry and three days' provisions. Finding tortoises for the museum came at the cost of traveling through rough country much overgrown with plants incompatible with human skin and clothing: thorn bushes and cacti, two of the hallmarks of the Galápagos. Early afternoon rain soaked the men thoroughly as they kept walking inland. Around nightfall they found a good-sized tree growing up through a lava tube. Starting at the upper branches at their ground level, they used the tree as a ladder to descend into the lava tunnel. There they gathered firewood and built a fire to cook dinner and to dry their clothes. Well in from the mouth of the "cave," they found perfectly dry ground to spend the night and made the cave their headquarters for the next two and a half days.

After a breakfast of hardtack and coffee, Slevin and Beck explored the lava tube for fifty or sixty yards and found the skeletons of seventeen tortoises. Lava tubes are common features in the geologic genesis of the Galápagos, Hawaii, and other volcanic islands. In the Galápagos, lava tubes are a dangerous death trap, into which tortoises have fallen and starved to death on several islands. In this case, the tortoises likely stumbled into the lava tube while wandering around the lava flow looking for water. Or perhaps the collapsed skylight opening was obscured by vegetation and they blundered in and tumbled down, never to escape. All were doomed to die of eventual starvation and dehydration, perhaps up to a year after a headlong plunge into their stygian graveyard. Some of the tortoises were found inverted, lying on their back, and others were in a perfectly natural position, dead in their tracks.

From this lava tube, Slevin and Beck took two of the best-preserved shells and all the best skulls and bones, packed in moss for transport back to the schooner.

They used the lava tube for another full day, but they were unsuccessful in their tortoise hunting. A day of walking through heavy undergrowth yielded no sign of living tortoises, so they returned to the cave for another night's sleep.

FEBRUARY 15–17, 1906

On February 15, 1906, the men hoisted all boats aboard and hove short, only to wait an hour for light wind; they touched bottom on a ledge as they sailed out of their anchorage, but no damage was done. Hunter sarcastically described the incident as "Parker blames the striking on the ledge as treachery on the part of the mate. About as much reason for it as there would be burning coal in a gas engine. It was lack of ability on the part of Parker that we struck."

Two days later (February 17) they were uncomfortably near Kicker Rock (Leon Dormido) and were nearly swept into it by a strong current. But a freshened wind carried them away as they tacked sails to head offshore. On some days they progressed little, making short tacks in the light winds, often not being able to make progress against the strong currents.

FEBRUARY 21–28, 1906

During the expedition, the men on the schooner had a series of social interactions with the people living in the islands, who were certainly connected to events outside the archipelago. On February 21, 1906, after finally dropping anchor in Wreck Bay after several days of struggling with light winds, Hunter went up to the hacienda to see if there was any mail and commented, "Much formality was gone through in the giving of the letters."

Hunter learned at the hacienda that a revolution had taken place in Ecuador since their last visit to Chatham Island and a new governor had been installed. This was the revolt in which Eloy Alfaro (1842–1912) deposed the elected president, Lizardo García (1844–1927). Alfaro remained in office until 1911, was exiled to Panama, and was eventually assassinated back in Quito in 1912.

Up at the hacienda, they worked with "a negro," who acted as their interpreter and guided them to a small lake south of the plantation. They found a stream that flowed out of the lake toward Freshwater Bay, but noted a scarcity of bird life and reptiles. In the afternoon, some sailors from the trading schooner *Manuel J. Cobos* came aboard the schooner *Academy* and brought some bananas as a gift. Back ashore the next day, Gifford and Slevin bought some Guayaquil chocolate at the store, and they traded hardtack with "the natives" for several bunches of bananas. Their thermometer hit 80°F, the highest temperature they had recorded so far in the islands. Williams collected moths ashore in the evening with a light and secured many excellent specimens. That night Beck had a pleasant dinner and talk with Captain Thomas Levick.

Beck and Williams went ashore one morning—Beck to settle their accounts with the plantation manager and Williams to get food for some caterpillars he was raising. Later that morning as they hove short and were about to break the anchor out of its grip on the bottom, the governor came by to say goodbye and a boatload of sailors from the trading schooner came alongside with bananas. When Beck asked them if they wanted some whiskey, there was a loud and immediate response of "Sí, sí, señor!!" Not having enough real whisky on board to share with all the sailors, Ochsner and Slevin came up with an ingenious solution. "Hurrying down below, [we] got some clean alcohol out of a barrel, diluting it to about 75 per cent. A hurried trip to the galley was then made, and some juice bailed out of a dish of stewed prunes gave our home brew the desired coloring. After shaking well, we poured the concoction into a whiskey bottle. In less time than it takes to tell it, we had a drink that seemed to hit just the right spot, as the sailors drank it down in great gulps." Slevin concluded his retelling of this interaction stating, "The Governor made some queer facial contortions and seemed to think all was not well, as he declined a second drink, a thing never heard of in these latitudes. However, he was too polite to say anything and thanked us profusely. Bidding our guests goodbye, we got under way, and clearing the entrance to Wreck Bay, shaped our course for Charles Island."

Sailing around tiny Onslow Island, the crew let go anchor in Cormorant Bay for more collecting ashore in the fine lagoon "with great flocks of ducks and flamingos." Beck and Ochsner got twenty-four flamingoes and some ducks from the lagoon; then the schooner departed for Black Beach Roads.

MARCH 2, 1906

Leaving Charles Island for Albemarle Island on March 2, 1906, and passing Brattle Island along the way, the men arrived at Turtle Cove with the brigantine *Nellie* at anchor and her captain bearing an astonishing revelation. In recent days, volcanic disturbances had occurred on the island. Magma forced upward had caused deformation and uplift of the land, resulting in the shape of the bay being altered and made much shallower. Nearby lagoons tilted and the water drained out over time, stranding fish in the shallow water and providing a feast for yellow crowned and plumbeous herons, gulls, gallinules, and frigatebirds, along with stilts, curlew, and semipalmated plovers.

MARCH 17–19, 1906

The dangers the men faced during the expedition were quite real. March 17, 1906, opened with calms and light airs, and the schooner sailed up to Iguana Cove on Albemarle Island midmorning. After measuring a safe depth in the bay and installing a buoy to anchor to, the men brought the schooner up to the buoy and let go the anchor in nine and a half fathoms with a rocky bottom. There was heavy surf breaking over the point, and they understood the intrinsic danger. The shoreline was most forbidding and ended in steep cliffs, with heavy rollers breaking against them. Beck, Williams, King, and Slevin took the small skiff and pulled about two miles down the coast, with the remainder of the shore party landing at the cove. The group in the skiff located a small boulder beach, landed with no great difficulty, and proceeded inland after tortoises. They found three large males a very short distance from the beach, so they drove them down to the landing place by prodding them with sticks. As it was getting late, Beck decided to load all three tortoises into the skiff and pull back to the schooner. The tortoises were so heavy, though, that they could not launch the skiff. So Beck decided they would load one and launch the skiff; then they would be able to "parbuckle" the other two, using a loop of rope like a sling to raise them up. The men got the skiff off the rocks with one tortoise inside, but found it was not so easy to parbuckle the other two, as they were extremely heavy and the skiff bobbed and dipped. They got the second one partway in by listing over the skiff, but its front leg got caught on the edge, or thwart, of the skiff, and they could get it neither in nor out. During this maneuver, the skiff turned broadside to the swell and an extra-heavy roller capsized the skiff, throwing out both tortoises, the oars, and the remaining contents into the surf. Fortunately, King, who could not swim, was left on the beach and watched the tortoises and some of the oars go floating out to sea. Beck could not swim either, but hung on to the stern of the skiff. Both Williams and Slevin felt perfectly at home in the water and finally got the skiff righted. Beck crawled in over the stern and grabbed one oar and paddled for the shore. They landed on a rough ledge of boulders, getting somewhat cut up on the sharp edges, but were otherwise none the worse for wear. Beck grabbed the line at the bow of the skiff and started walking along the ledge of rocks, when an especially heavy swell came in and crashed the skiff down on a sharp point of rock, smashing it to pieces. All that was left on the line was the ringbolt. With darkness approaching, they decided to walk down the coast back to the schooner. Beck had lost his shoes and had to go in stocking feet. Slevin had left his pants, shoes, and hat on the beach, so he was much better off, having only lost a shirt. Slevin recounted, "As we walked through the brush in the dark, I felt as if there was not a cactus or thorn bush on all Albemarle Island that I missed running into. However, I wouldn't have traded places with Beck for anything." The other men on schooner became concerned when the skiff did not return after dark and mounted a search party in the other skiff. As the shipwrecked men walked slowly down the shore, they saw a light and heard a whistle, which were coming from Nelson, Ochsner, and Hunter. Landing on the rough shore was a bad idea, so they headed back to Iguana Cove to be picked up. "After what seemed like a journey that would never end, we reached the cove at 9:15 p.m., tired out and very hungry," recounted Slevin. They had a much-belated supper on the schooner, bringing to an end one of the most exciting, and dangerous, days of the expedition.

The next day was Sunday, March 18, and Williams noticed one of the tortoises drifting past the anchorage "bobbing about like a cork." Its long neck protruded far out of the water, giving it a grotesque appearance. The men put out a boat to the "rescue," tied a line around one leg, and towed it back to the schooner. Several pieces of the skiff also floated by, but they saw nothing worth picking up. Later, another tortoise was sighted and they went out to rescue it as well, but found it had had a rough passage, with the shell broken and punctured by being battered against the rocks. On March 19, Beck and Slevin went back down the coast to where they lost the skiff and picked up some wreckage and the third tortoise that they left behind. They also found two more tortoises, tying them up until they could come back the next day (figure 11.4). The tortoises were too big and heavy to pack out alive, so they lowered them over the cliffs with ropes and let them drift out to sea, where they picked them up and towed them back to the schooner. Ironically, two of them died of injuries received while being lowered over the cliffs into the water. There were lessons to be learned and generally the men were doing whatever was necessary to secure specimens. In light of the fate of the Academy museum in San Francisco, their efforts were greatly appreciated.

ADRIFT AND AGROUND

The expedition members were assailed by two unwelcome and recurring complications in the course of their otherwise systematic and meticulous specimen collecting. On the one hand, they were bedeviled by not enough wind, sometimes for weeks at a time, as they drifted aimlessly and helplessly many miles from the archipelago. And on the other, they were occasionally challenged, suddenly and unexpectedly, by a lack of water under the keel when they ran aground. One of these recurring events involved unforeseen and uncontrollable acts of

FIGURE 11.4 **Tortoises Ready for Transport.** After the men transported the tortoises to the shore they would row them back to the schooner for further processing. *Archives, California Academy of Sciences.*

nature; the other involved avoidable errors in human judgment. The lapses in judgment and navigational acumen were criticized and blame quickly placed on the "guilty" party. Things got so bad toward the end of the expedition that one of the men "wanted to fight with knives." The acts of nature merely had to be tolerated as the windless days went by and the sails listlessly flapped as the schooner rolled, pitched, and yawed in the gentle swells near the equator. Being adrift and running aground endangered the expedition's success and the men aboard the *Academy*. Having no engine aboard the schooner contributed to the seriousness each time they were adrift and made running aground a potential nightmare. An engine could have enabled the crew to return to a safe anchorage and await a rising wind. But they were 100 percent sail-powered. If the keel touched bottom near shore as they hopped between islands, an engine could have been thrown hard in reverse and minor groundings quickly avoided. But the sailor-scientists could get themselves free of the bottom when grounded only with the time-consuming routine of rowing out into deeper water, dropping a kedge anchor that would bite into the bottom and provide a firm purchase, and winching the vessel off the rocks with the line attached to the anchor. By that time, swells could readily push them farther ashore and set them harder aground. When vessels are truly hard aground, almost nothing can get them free. An engine would have been a form of insurance for avoiding some situations and finding a solution to other predicaments.

At times the navigational charts were inaccurate, sometimes dangerously so. At Bassa Point on Chatham, the men anchored near the sloop *Josephine Cobos*, which was undergoing hull cleaning and minor rudder repairs, and measured the water depth at low tide at only six or seven feet, when the chart showed fifteen feet. After that discovery, they needed to again use the kedge anchor "to hove the schooner a little further into the cove," as she was touching bottom at low tide. This maneuver was followed by more kedging, as the schooner touched on the reef forming the outside of the cove. The next day they had to set the kedge anchor astern to hold the schooner in position, as she was again touching the bottom at low tide.

Even an activity as seemingly simple as leaving one island to sail to another could be fraught with difficulty and danger. Sometimes the crew had to wait with all hands ready for an hour or more for the wind to pick up before they could hoist sail. Preparing to leave, they would "hove short," or pull themselves right up to the anchor with the anchor chain pointing straight down, having brought aboard all the anchor chain and line that normally stretches out underwater in a downward-curving arc from the vessel at the surface to where the anchor rests on the bottom. The curving arc and the total amount of anchor chain and line is called the "rode" and has to be a greater distance in deeper water. Each time they hove short, all the rode was hauled aboard and only the minimum of chain and line remained in the water. With wind and with at least one sail raised, they could weigh anchor and get under way.

The health of the specimens on the schooner posed a problem several times during the expedition. The schooner was infested by bugs, and the collectors had to take drastic action time and again. On March 26, 1906, Nelson fumigated the vessel, because bedbugs, dermestid beetles (that could potentially destroy specimens by eating the flesh, fresh or dried), flies, and cockroaches "were getting too thick for comfort." Nelson released what he thought was a lethal dose of sulfur and closed up all the hatches, but the next morning they found plenty of cockroaches still alive. Again on May 26, 1906, the same menagerie of insect pests was back causing grief among the men. After dinner they lit pans of sulfur on fire, battened down the hatches, and closed and caulked shut the companion way ladder doors. That night all hands slept on deck. The next morning at 8:00, they opened up the hatches and found that not all of the sulfur had burned out and that, although there were many dead cockroaches and flies scattered about, "the bedbugs were as good as new." On July 1, 1906, Williams put a live mockingbird in the fore hold, and it made itself at home by eating some of the dermestid beetles infesting the vessel. Nelson again fumigated the cabin, and the next morning the men found "quite an assortment of dead flies, cockroaches, and dermestids. and Nelson fumigated again on September 8, 1906.

To anyone visiting the Galápagos today on an ecotourism trip, or even scientists going ashore, every landing can elicit a certain amount of trepidation, bordering on anxiety. Wet landings require hopping out of an inflatable Zodiac, or panga, into ankle-deep to knee-deep water and wading ashore. Dry landings require extreme skill by the boat driver manning the outboard engine, getting the panga just close enough to a rocky outcrop of basalt lava or, worse, to an algae-slickened concrete platform. Assistance exiting the panga is thoughtfully provided by crewmembers or naturalist guides. F. X. Williams had his own landing experience when he went ashore for some night collecting of insects, a favorite technique of entomologists. He beached his skiff as the tide was going out and it was left high and dry, so he could not get it off alone. To summon help, he built a fire and Slevin and his young assistant, King, went in to get the skiff back into the water.

The men on the schooner *Academy* learned of the San Francisco tragedy less than two weeks after the earthquake occurred, via San Francisco newspapers carried from the mainland to the islands in an Ecuadorian gunboat, the *Cotopaxi*. They initially learned on April 30 of the tragic events in San Francisco that occurred on April 18.

All the men were concerned about the safety of their families. Rollo Beck, the leader of the expedition, was asked by the crewmembers to sail to the mainland of Ecuador so that they could seek news of their families and friends back in San Francisco. But, like Jason resisting the conflicting demands of the Argonauts, Beck forced the schooner to stay in the islands and made the men focus on collecting. Any deviation from the mission, he believed, would jeopardize the entire expedition. A few months later when visiting the settlement at Wreck Bay on July 3, 1906, the men learned from a load of letters and newspapers for the first time some of the details of the earthquake and fires. Letters from home told them their families were safe.

Williams summarized the feelings of all the expedition members when he wrote from Chatham Island on July 8, 1906, to the museum director, Loomis:

> I am very sorry the Academy suffered such a great loss, many others are in the same boat, I hope our schooner will be back safe & well laden with fine material for the new collection.

The number of times the schooner either touched bottom or went aground is frightening, given that each event could have resulted in total disaster. The eight young men felt that Parker, in his role as navigator and due to his lack of skill, was responsible for the groundings. They squarely placed the blame on Parker when near the end of the expedition they staged a "mutiny." They wrote and signed as a group (except the mate, Nelson, who had maritime legal reasons not to sign) a letter to the "Master of the Vessel Rollo Beck" that said, "Mr. J. J. Parker, navigator of the schooner Academy, having proved himself, in our belief, incompetent, the last and most serious exhibition of incompetence having occurred to-day (August 9, 1906)." They insisted that Beck divest Parker of his navigation duties::"We consider that we and our employers, the California Academy of Sciences, have altogether too much to lose to allow the navigation to rest in Mr. Parker's hands."

Parker responded on August 10, 1906, to the crew's demand by writing to Beck, "I believe from the past experience on board I think best to agree with the request. I shall always be ready to obey any order you or your choise [sic] to take my place may give with the exception of common ordinary labor."

On September 5, 1906, all of the crew, including Nelson, wrote Beck again, stating, "We the undersigned members of the crew of the American Schooner "Academy" demand to be discharged from said schooner unless J. J. Parker is returned to San Francisco by some other vessel as we object to his presence on board." Parker left the vessel at the town of Vilamil on Albemarle (Isabela) Island and refused any help from Beck to return to San Francisco. Williams wrote home as early as October 16, 1905, stating, "Our Captain 'King Dodo' is the most cold-footed individual for 'keeping off the rocks' you ever saw." Williams subsequently annotated the typed letter with a footnote: "[Parker] once mistook a nearly submerged manta ray for a rock—not an unreasonable mistake." Parker was not perfect, and his errors became increasingly egregious and unforgivable.

NOTES

Page 169: *Epigraph*: Ralph Waldo Emerson, *Nature*, ch. 1 (1836, revised and reprinted 1849; A facsimile of the first edition with an introduction by Jaroslav Pelikan [Boston, Beacon Press, 1985]).

Page 169: *Epigraph*: Darwin (1959, 45).

Page 169: *Epigraph*: Ibid., 55.

Page 169: *Wrap themselves around your neck and feet*: JSH, December 27, 1905.

Page 170: *Whether he was eating the bark*: JSH, January 2, 1905.

Page 170: *Another was blown to pieces*: JSH, January 3, 1906; in this passage I have inserted five commas for clarity.

Page 171: *Glorious view*: JSH, January 3, 1906.

Page 171: *Measured with their aneroid barometer*: A current map of the Galápagos Islands shows the top of James Island to be 920 meters above sea level, or 67.5 meters higher than the value obtained by the Academy collectors.

Page 171: *Tonight the mountain is fine and clear:* LSA, 71.

Page 171: *Although lying on his belly*: JSH, January 3–4, 1906.

Page 171: *Four were eating the body of a hawk*: JSH, January 4, 1906.

Page 171: *The hawks are very tame*: JSH, January 3–4, 1906.

Page 172: *A very interesting antic*: Due to the rules of taxonomic nomenclature and Latin grammar dealing with gender and number, the name *pallada* (Sclater and Salvin 1870, 327) was changed to *pallidus* when the species was transferred to the genus *Camarhyncus*.

For studies of tool use see Tebbich et al. (2001, 2002).

Page 172: *It was pecking woodpecker-like*: JSH, January 4, 1906.

Page 172: *Attributed to King*: In some circles, this observation is attributed to Edward Winslow Gifford.

Page 172: *A shot from his revolver stopped it*: JSH, January 4, 1906.

Page 173: *Very satisfactory feed in the luxuriant grass*: JSH, January 4, 1906.

Page 173: *Doves drinking out of the orchids*: JSH, January 4, 1906.

Page 173: *An unsuccessful pig hunt*: RHB, January 3–4, 1906.

Page 173: *Caught two turtles*: LSA, 71.

Page 174: *Very light southerly winds*: LSA, 71.

Page 174: *Advantage of this time at sea*: LSA, 72.

Page 175: *Busy standing watches*: LSA, 71–72.

Page 175: *Their intended destination*: As an example of the men's frustration, Hunter did not even make any entries in his field notes between January 5 and 11, 1906. He likely refrained from writing down what he did during this interval because he felt that the time was completely wasted. He indicated his frustration by finally writing on January 11, "Have spent all of the time since leaving James Bay trying to get around the south end of Indefatigable to anchor in Academy Bay" (JSH, January 11, 1906).

Page 175: *Grabbed hold of the island*: JSH, January 11, 1906.

Page 176: *Almost extreme southerly end of the island*: JSH, January 11, 1906.

Page 176: *Beck shot the ducks and flamingo*: JSH, January 11, 1906.

Page 176: *One of the shyest of Galapagos birds*: Harris (1982, 85).

Page 176: *The flamingo population is small*: Heinzel and Hall (2000, 117).

Page 176: *A most peaceful sight*: Jackson (1994, 169).

Page 176: *Quite plentiful along the rocks*: RWE, 40.

Page 176: *For someone who couldn't swim*: RHB, 69, January 11, 1906.

Page 176: *Hoisted anchor at noon*: LSA, 73.

Page 176: *Dropped anchor in three fathoms*: RHB, 70, January 11, 1906.

Page 177: *Kept us awake most of the night*: LSA, 73.

Page 177: *Filled their "water breakers"*: A water breaker was a "small cask for fresh water carried in ship's boats." See http://www.usmm.org/terms.html#anchor238834, accessed December 29, 2016.

Page 177: *Hawks still continue to eat their dead relations*: JSH, January 13, 1906.

Page 177: *Remains of a number of centipedes*: JSH, January 12, 1906.

Page 177: JSH, January 12, 1906.

Page 177: *Beak and feet were soft*: JSH, January 12, 1906.

Page 179: *Nelson's much-anticipated cargo*: LSA, 74.

Page 179: *A small bunch of bananas*: Beck stated that Nelson had struck a good bargain for the *atoyas*. The locals at Vilamil charged $2.50 gold for 100 pounds of *atoyas*, so the bill for 300 pounds came to $7.50. Nelson traded the 100 pounds of flour for $5.00 worth of *atoyas*, and Beck concluded, "We owe them $2.50 gold" (RHB, 71, January 17, 1906).

Page 179: *The green zone*: Beck reported on the food the four men carried on this four-day trip: "28 cans of pork and beans, 15 cans of salmon, 3 cans of corned beef, two bags of buckwheat flour, and a can of butter" (RHB, 71, January 14, 1906). They were probably glad to see Nelson and the 300 pounds of *atoyas* he acquired on his sailing excursion to Albemarle when they got back to the coast.

Page 179: *Not allow you to wink:* JSH, January 15, 1906.

Page 179: *Very fearless and inquisitive:* Gifford bird notes, January 15, 16, 17, 18, 1906.

Page 180: *Made no use of their wings:* "Once established on an island, however, rails tend to become flightless or nearly so and thus highly vulnerable to introduced ground predators (rats, mongooses, etc.). As a result, many Pacific island rails are endangered or extinct" (Pratt, Bruner, and Berrett 1987, 124; Ripley 1977).

Page 180: *Breast muscles much reduced:* JSH, January 15, 1906.

Page 180: *Innovative collecting technique:* JSH, January 15, 1906. The number thirty is confirmed by RHB, 71, January 17, 1906.

Page 180: *Insightful observations about rails:* Gifford bird notes, January 15–18, 1906.

Page 180: *Up in the green zone:* The 1,100-foot elevation that they reached is not quite halfway to the top of Albemarle Island; the highest point, Mount Crocker (named several years later by a subsequent California Academy of Sciences expedition) is 870 meters (2,854 feet) high.

Page 180: *Started a fire to cook supper:* LSA, 74.

Page 181: *Managed to salvage the trip:* RHB, 72, January 18, 1906.

Page 181: *Potassium cyanide and other dope:* Three quotes from RHB, 70, January 14, 1906.

Page 181: *Typical day of rest for the collectors:* LSA, 73.

Page 181: *Tortoise hunting went smoothly:* RWE, 41.

Page 182: *Much larger livers than the males:* RWE, 42.

Page 182: *Bladder full of water:* REW, 42.

Page 182: *Rapid locomotion:* RWE, 42. The fast-moving male had escaped from them on the seventeenth, but King quickly found it and tied it up again; the party did not return to skin it until the twentieth.

Page 182: *Personal friction:* Because Stewart's field notes and botanical notes (perhaps kept together in one journal) are thought to have been lost, we do not have his account of any events during the expedition. See West and James (2001).

Page 182: *Didn't feel like going:* Beck's comments on Stewart: RHB, 71, January 15 and 16, 1906.

Page 182: *In with boys:* RHB, 72, January 19; Slevin (LSA, 75), wrote that everyone except Gifford and Nelson went ashore, with Nelson busy watering the ship and Gifford reporting in his diary that he stayed aboard the schooner skinning birds.

Page 182: *Some sort of drama:* RHB, 73, January 20, 1906.

Page 182: *Busy scraping barnacles:* JSH, January 22, 1906; Hunter also reported that he "scrubbed side of schooner" before the crew "pulled up anchor and headed for Chatham during the P.M." Gifford also reported, "Scrub vessel in morning. Leave for Chatham at noon" (Gifford diary, January 22, 1906). The "rest of the party" mentioned by Slevin would have included the other four collectors, Stewart, Gifford, Hunter, and Ochsner, as well as Nelson, Parker, and possibly White.

Page 182: *Stewart and Ochsner had a scrap:* RHB, 73, January 22, 1906. Beck did not punctuate his handwritten field notes, and the punctuation here has been added. My assumption is that the "darkened optic" means one of the two men, unspecified, ended up with a black eye.

Page 182: *Testosterone-driven shenanigans:* It is interesting that the two men who got into multi-year intellectual fights after the schooner returned to San Francisco (Stewart and Ochsner) were the same two who engaged in fisticuffs in the islands.

Page 183: *Weighed anchor at midday:* LSA, 75.

Page 183: *Sailing in light breezes:* LSA, 76.

Page 183: *Winds continued their fickle puffs*: Kicker Rock is also known by its Spanish name of León Dormido, or Sleeping Lion. Both English and Spanish names refer to the shape of the rock, which is actually two rocks, both rising some five hundred feet out of the water. A small opening between the two rocks can be navigated by a small boat, but the crew of the *Academy* would likely not have attempted this in a rowboat.

Page 183: *Indefatigable water was none too good:* RHB, 50, January 24, 1906.

Page 184: *The road was a river*: Gifford bird notes, 65, January 25, 1906; heavy rains suggest an El Niño year.

Page 184: *Species still occurs there today*:

"Before the turn of the century, Dr. George [*sic*] Baur collected four geckos on San Cristóbal Island, which belonged to a then-unknown species. It was described by Garman in 1892 and given the name of *Gonatodes collaris* (Van Denburgh, 1912b). Later expeditions were unable to find this species. The California Academy of Sciences Expedition of 1905–06 made extensive collections of geckos from San Cristóbal—148 specimens of *Phyllodactylus leei*, an endemic species, and 21 of *P. tuberculosus*, a common mainland species that had become established on the island. No *Gonatodes* was found (Van Denburgh, 1912b).

"It was discovered much later that *G. collaris* is a mainland species. It is, of course, possible that the Baur specimens had been collected in Guayaquil and became mixed up with his collections from San Cristóbal. However, there is another possibility—Baur had collected a newly introduced gecko, either catching the lot or taking so many that they could not become established.

"The next collection of *Gonatodes* that I know of consisted of three specimens I caught in the general area of the landing at Puerto Baquerizo Moreno, San Cristóbal, in 1961 or 1962. These I gave to Dr. André Brosset, then Director of the CDRS [Charles Darwin Research Station]. Unfortunately, Dr. Brosset left the Islands before he received information about their identity. Later, I saw several geckos that appeared to belong to the same species, or a closely related one, in the same area. Dr. Marinus S. Hoogmoed believes my specimens could be *Gonatodes caudiscutatus* (Hoogmoed, pers. comm., 1991). This particular species has been reported from both the towns of Puerto Baquerizo Moreno and El Progreso.

"Another gecko (*Lepidodactylus lugubris*), of pantropical distribution, has been reported from Santa Cruz (Hoogmoed, 1989). However, Dr. Hoogmoed did not meet with this species while he was doing a preliminary study of another gecko (*Phyllodactylus reissi*), also a recent introduction, which had been observed for the first time in 1975, near the Puerto Ayora landing. This last species gives good reason to fear for the survival of the gecko native to the island (*P. galapagensis*), since the latter disappears from areas invaded by its larger relative, the introduced *P. reissi* (Hoogmoed, 1989).

"There is still much to learn about these small invaders and how they affect the island species with which they must compete. Unfortunately, their small size is their greatest advantage. It makes it easy for them to hide in cargo from the mainland and to remain unnoticed for a long time. Once successfully established, they may turn out to be impossible to eradicate. The little red fire ant is an example familiar to every Galápagos resident and can serve to raise local awareness of the insidious invaders." http://www.darwinfoundation.org/articles/n5900049810.html, accessed July 10, 2003.

Page 184: *Even from his grave*: Baur died June 28, 1898; see Hay (1898).

Page 184: *The rain spoiled the collecting*: JSH, January 26, 1906.

Page 184: *Beck had better luck*: RHB, 50, January 26, 1906.

Page 185: *Hunter sealed the fate of the family of five*: JSH, January 26, 1906.

Page 185: *Still owed money for an anchor*: RHB, January 29, 1906.

Page 186: *Doc fix Governor's leg*: RHB, January 29, 1906.

Page 186: *Given a hymn book*: LSA, 77.

Page 186: *Shells of contention*: LSA, 77.

Page 186: *Failed to deliver the pineapples*: LSA, 77–78.

Page 186: *Large earthquake*:

"Some of the world's largest recorded earthquakes—on January 31, 1906, off the coast of Colombia and Ecuador, and on March 2, 1933, off the east coast of Honshu, Japan—had magnitudes of 8.9 on this scale, which is open ended."

http://pubs.usgs.gov/gip/earthq3/magnitude.html, accessed January 14, 2003. "The largest earthquakes known measured an estimated 8.9 on the Richter Scale. There were three—Lisbon, Portugal 200 years ago, killed 60,000 and affected 6 million square miles; Colombia and Ecuador, January 31, 1906 and Honshu, Japan, March 2, 1933." http://www.canadianrockhound .com/junior/geology_quake.html, accessed January 14, 2003.

"The earthquake generated a tsunami that struck the coasts of Ecuador and Colombia, killing between 500 and 1500 people. The height of the waves in Ecuador is estimated to have been 5 metres. The tsunami was observed all along the coast of Central America and as far away as San Francisco and San Diego, U.S.A and Japan." http://www.asc-india.org/gq/ecuador.htm, accessed January 14, 2003.

Page 187: *The full force of the swell*: JSH, January 31, 1906.

Page 187: *A strong current was running*: JSH, January 31, 1906.

Page 188: *Exceptionally high tide*: Gifford bird notes, 42, January 31, 1906.

Page 188: *I gave him a warm reception*: Gifford bird notes, 42, January 31, 1906.

Page 188: *Shipped a couple of heavy seas:* LSA, 78.

Page 188: *When a wing is broken:* Gifford bird notes, February 1, 1906.

Page 189: *Wheeling very sharply, skim very low*: Gifford bird notes, February 2, 1906.

Page 189: *They skim very low over the sea*: http://www.audubon.org/bird/BoA/F43_G4c.html, accessed July 28, 2003. Original of Audubon in Sonoma State University Library in Sibley Rare Book Collection.

Page 189: *A few rotten eggs and feathers*: Gifford bird notes, February 1, 1906.

Page 189: *Wet seaweed from the ocean*: RHB, February 1, 1906.

Page 190: *Menus we have had*: LSA, 79.

Page 190: *Get fresh meat each day*: LSA, 79.

Page 190: *Picking up the crumbs from their lunch*: RWE, 43.

Page 190: *Beck tendency to anthropomorphize*: RHB, February 4, 2016.

Page 191: RWE, 44.

Page 191: *Darwin had noted*: VOB, 371.

Page 191: *Seldom see them in the water*: RWE, February 5, 1906.

Page 191: *Cut open the stomachs of several marine iguanas*: VOB, 371.

Page 191: *Any bird worth observing*: Gifford bird notes, February 5, 1906.

Page 192: *Grabbing it and swinging the bird*: Gifford bird notes, February 6, 1906.

Page 192: *Constant struggle for nests*: Gifford bird notes, February 6, 1906.

Page 192: *Yesterday I saw two Fregata aquila copulating*: Gifford bird notes, February 7, 1906.

Page 192: *Sheets too free*: RHB, February 7, 1906.

Page 193: *Waded in and shot 1 with pistol*: RHB, February 8, 1906.

Page 193: *Blooming white on the flanks of dark volcanoes*: Lopez (1998, 53); RHB, February 8, 1906.

Page 194: *Coal in a gas engine*: JSH, February 15, 1906.

Page 194: *Much formality gone through*: JSH, February 21, 1906.

Page 195: *If they wanted some whiskey:* LSA, 84–85.

Page 195: *Drank it down in great gulps*: LSA, 85.

Page 195: *Thanked us profusely*: LSA, 85.

Page 195: *Cormorant Bay*: LSA, 85. Cormorant Bay is most likely named for HMS *Cormorant*, which visited Galápagos in ca. 1886, and not for the flightless cormorant that does not occur at the bay. See http://www.galapagos.to/TEXTS/CORMORANT.HTM, accessed July 24, 2015. For information on HMS *Cormorant* see http://www.galapagos.to/SHIPS/INDEX.php, accessed July 24, 2015.

Page 196: *Dangers the men faced*: LSA, 91–93.

Page 198: *Too thick for comfort*: LSA, 95.

Page 199: *Bed bugs good as new*: LSA, 112.

Page 199: *Could not get skiff off alone*: LSA, 81.

Page 199: Unpublished letter, F. X. Williams Collection, Archives of the California Academy of Sciences.

Page 200: *They squarely placed the blame*: Unpublished group letter, Archives of the California Academy of Sciences.

Page 200: *Ready to obey any order*: Unpublished letter, Archives of the California Academy of Sciences.

Page 200: *We object to his presence on board*: Unpublished letter, Archives of the California Academy of Sciences.

12

That Sinking Feeling

⌒──

TODAY, TO ANYONE visiting the Galápagos Islands or reading about them from a distance, two scientific ideas are rightly presented with great certainty and extensive supporting evidence. First, the islands are of *recent volcanic origin*, plumbed into the Earth's molten mantle, built up slowly from orange-hot magma and lava emitted from a volcanic hot spot on the Pacific Ocean floor, the same ongoing mechanism that forms all the Hawaiian Islands. And, second, *evolution by natural selection*, the gradual process described by Charles Darwin in 1859, is the mechanism of permanent genetic alteration of organisms over generations and change as a result of permanent genetic modification from their ancestors over time. These are textbook ideas today in geology and biology. But in 1905, competing ideas in geology and evolution vied for attention from scientists, including the eight young men on the schooner *Academy* and their scientific supervisors back in San Francisco.

One of the strangest and least-known chapters in Galápagos science began in 1889, when the Yale University paleontologist Georg Baur made up his mind one day about the history of the islands. He concluded that iconic Galápagos tortoises and the geology underfoot were intertwined in a novel way.

Baur arrived at his momentous and unwavering Galápagos conclusion on January 9, 1889, two and a half years before setting foot in the Galápagos on his own expedition. He made his a priori decision, by happenstance, after unpacking at the Yale museum a large land tortoise from the Miocene fossil beds of Nebraska. The tortoise fossil closely resembled the giant Galápagos tortoises, the namesake organisms of the archipelago. For Baur, this resemblance was a revelation that required an explanation, and not just the standard explanation. It struck Baur and stuck with him until the end. He immediately wondered, "How did these large *land*-animals come to the *islands*?" Baur found it inexplicably incongruous that land animals could somehow occur on islands far out to sea, far distant from a continent. He insisted on an explanation, and it had to be his explanation, not the explanation of Darwin or anyone else. Later that very night he wrote in his diary the words that laid out the research program he doggedly pursued for the remaining nine years of his life, until he died at age thirty-nine in June 1898: "What is the origin of the Galápagos fauna? It is not introduced, but left there: the Galápagos originated through subsidence of a larger area of land; they do not represent oceanic islands, as generally believed, but are continental islands." It did not matter to Baur that few other scientists agreed. Baur would have been well served by heeding the 1891 advice of Sir Arthur Conan Doyle: "It is a capital mistake to theorize before one has data. Insensibly one begins to twist facts to suit theories, instead of theories to suit facts."

Undeterred even by harsh published criticism from his contemporaries, Baur spent the rest of his short life desperately trying to prove he was right and everyone else was wrong (figure 12.1). Like someone staking a gold claim on the side of a mountain, Baur was staking his claim on an idea and a geological mechanism. Baur was set for the battle of his life. He would protect his golden idea against all interlopers.

The alternative geological explanation favored by Darwin and nearly everyone else before and after Darwin was that the islands were built slowly upward from persistent volcanic activity originating on the sea floor. Baur and others referred to this version of island geology as the *elevation theory*. For Darwin, and for geologists today, the Galápagos were *oceanic islands*, never connected to the mainland as a landmass or by land bridges. Oceanic islands are sterile when first formed, due to the intense heat of the formative lava. No organisms exist in this primal stage. Darwin knew organisms must arrive from afar to the pristine islands on wind, water, or wings. These are the many contrivances of long-distance dispersal, and Darwin was firmly in favor of dispersal. For Baur, however, the Galápagos were *continental islands*, formerly connected to the South American continent or more distantly to Central America. And the islands were fully populated from day one by the very same plants and animals that had always been on the now-subsided landmass. In the Galápagos, no two ideas could be more diametrically opposed. Were the tortoises stranded or transported? Did the islands rise up or sink down?

Simply put, Baur believed the individual Galápagos islands of the archipelago were the summits of mountains or volcanoes of a former landmass that had dropped down below the waves. It was like the lost continent of Atlantis, except the peaks of the mountains still poked out of the ocean—or like a sailboat with tall masts that sinks in shallow water while the masts still stick out of the water.

Sinking or subsidence was just enough to make islands out of mountaintops, but not too much to cover everything. Baur invoked an element of "Goldilocks," not too much, not too little, in his vision of a subsiding landmass. What was left as an archipelago was "just right." Baur's islands are ghost islands, remnants of their former continental greatness. For Baur, all Galápagos organisms, particularly the giant tortoises, were left stranded in this new landscape. Creatures found themselves wandering around on newly subdivided islands that only recently were interconnected mountains on a larger landmass. With no means to get from the mainland to the islands or to move between islands in Baur's view, the tortoises were frozen in time. Baur had no need for Darwinian divergence or change over time on the islands from South American mainland ancestors, or even from tortoise ancestors millions of years ago in Nebraska. Did the islands rise from the sea or sink down into it?

Baur gave a lot of thought to his ideas. He staked his claim for the subsidence origin on a line of reasoning he called "harmony" of species on the now-submerged islands. He envisioned the flora and fauna to be "more or less harmonic,—that is to say, the islands will be like satellites of the continent from which they developed, and the whole group comparable to a planetary system." Baur allowed that *continental islands* would have both the original inhabitants, marooned on the islands since before they sank, and a few newly arrived inhabitants, perhaps orphans of the storm. But true *oceanic islands* would have only the newly arrived plants and animals, populating islands that started off pristine and sterile. Baur allowed for a low level of dispersal to the islands and between islands, but not for the tortoises, and he even conceded, "As is well known, the whole group is volcanic."

Baur examined the flora and fauna, declared them in "absolute harmony," without adequately defining "harmony." As a further violation of the scientific method, he failed to show how harmony could be distinguished from disharmony. Thus, he presented an idea that could not be refuted, that had to be true by his logic. He concluded, "The Galapagos are continental islands, originated through subsidence; they all formed at a past period one large island, and this island itself was at a still former period in connection with the American continent." He clearly knew and clearly stated that he was in direct opposition to "all authors who have worked on this group of islands," including Charles Darwin, Alfred Russel Wallace, the co-discoverer of natural selection, and Darwin's closest friend, the botanist Joseph Dalton Hooker. Baur staked out the battle lines when he wrote, "All declare that these islands are of recent volcanic origin, that they have emerged out of the sea through volcanic activity, and that they have become peopled from the continent successively." Somehow Baur could not reconcile the obvious volcanic origin with his desire to promote a non-Darwinian agenda. Baur's only known ally during his lifetime was Henri Milne-Edwards, who also maintained that the Galápagos represented the remains of a former continent, now subsided.

Baur's entire subsidence idea is a backhanded slap at Darwin. Baur fixated on the foundation of his new, lifelong passion, the *subsidence* theory. For Baur, promulgating an alternative, nonvolcanic origin for the islands was a way to usurp some of the glamour associated with Darwin and the Galápagos. If the Galápagos were merely submerged mountaintops as Baur maintained, there was really nothing much, after all, for Mr. Darwin to explain. Darwin needed remote, isolated, oceanic islands that were previously, and originally, devoid of life due to their hot, violent origin. Baur could not comprehend that down in the bowels of the Galápagos there was a rumbling of magma. And rumbling magma held the answer to the origin of the islands. Baur's hope was that Darwin's groundbreaking mechanism of natural selection as the origin of Galápagos species would sink, like a captain going down with his ship, into oblivion as Baur's own sinking continent receded slowly into Davey Jones's locker. If Baur could sink a continent, perhaps he could also sink Darwin.

At the same time that Baur was formulating and publishing his ideas about Galápagos geology, and before he actually went to the islands, he was already being vigorously denounced by Alexander Agassiz of the Museum of Comparative Zoology at Harvard University. Agassiz began his critique in simple and clear terms by stating, "Dr. Baur has expressed views on the origin of the flora and fauna of the Galapagos entirely at variance with what is known of their geological structure." Agassiz continued his surgical dissection of Baur, stating, "Dr. Baur also mentions the case of the Sandwich [Hawaiian] Islands as having originated by subsidence. No more unfortunate suggestion could have been made regarding their origin. All we know of their geology seems to show that the different islands have been gradually built up around a central nucleus by successive eruptions, much in the same way that the Galapagos were." Baur was bucking a groundswell of magmatic intensity. Even "the biggest liar in the

Pacific," the American sealer Benjamin Morrell (1795–1839), could discern the ancient history of the islands when he wrote, "There is no doubt that these islands are all of volcanic origin," and "Each island in the Galapagos group is of [volcanic origin], as I am fully convinced by a careful inspection." Morrell knew from his visits in 1824 and 1825 that the volcanic origin was "a remarkable fact, and worthy of the notice of scientific men." And scientific men did take notice, but not Baur.

Agassiz also wrote in 1892 after extensive dredging and depth sounding on board the US Fish Commission steamer *Albatross* around the Galápagos and in the eastern Pacific, "We may therefore look upon the Galapagos Islands as a group of volcanic islands, gradually built up by successive flows of lava upon a huge mound, itself perhaps raised by the same agencies from the floor of the ocean." Baur and his ideas seemed destined to suffer the fate of a castaway sailor, marooned on a desert island with no rescue in sight. Baur responded readily to Agassiz in the journal *Science* in the same year, stating, "When Professor Agassiz or any one else is able to explain this [harmony of distribution] by the elevation theory, I shall be the first one to adopt it. But until this has been done, I believe in subsidence." Baur's belief in subsidence was an unerring belief, from which no evidence would sway him.

GEORGE BAUR

FIGURE 12.1 **Georg Baur, 1859–1898.** Baur's published ideas about the geological origin of the Galápagos Islands and the evolutionary relationships of the organisms on the islands, particularly the giant tortoises, were arguably as important as those of Darwin in shaping the collecting agenda of the 1905–1906 expedition. Source: Public domain, from Wheeler (1898).

Late in 1898, Stanford University sent two collectors to the Galápagos, and they inevitably formed an opinion of Baur's subsidence mechanism of island formation. Robert E. Snodgrass and Edmund Heller had special collecting privileges that allowed them to collect on all the islands. They traveled and collected widely in the archipelago on the schooner *Julia E. Whalen* from December 1898 to the end of June 1899, a model expedition for the Academy's in 1905–1906. Heller wrote in 1903, "All the islands are obviously of volcanic origin, yet Dr. Baur invokes a continental origin or connection" to explain the geology and flora and fauna. He pointed out that the coast of South America "has been rising through a long period of geological time," and yet Baur envisioned the Galápagos sinking down, just the opposite of what was clear to Heller. Baur's mechanism was unnecessary and at variance with the geological and biological evidence on other island groups that were populated by "strays" dispersing over long distances and then differentiating into many species. His frustration with Baur was palpable when Heller wrote, "We do not see the necessity of making the Galapagos an exception" and "Dr. Baur [is] alone in adopting the subsidence theory."

GALÁPAGOS LAND

Back at the California Academy of Sciences and writing in 1913, John Van Denburgh and Joseph Slevin put together their evolutionary history of the lava lizards in the genus *Tropidurus*, which in the plural they called *Tropiduri*. The worked on more than two thousand specimens from twenty-four islands, which was "an enormous task." One of the main differences they found between species had to do with size and number of scales. They counted scales around the middle of the body, along the back, and along the belly of fifteen hundred specimens. Those data were charted island by island to show the "differentiation" or speciation that occurred in the islands. "This has involved the counting of more than a quarter of a million scales."

From previous collections and the 1905–1906 collection, they knew that the lava lizards occurred on nearly every island of the archipelago. How did they get there and how did they differentiate into unique species? Van Denburgh and Slevin had a ready answer: "These *Tropiduri* must have reached these islands either by land or by water," which are the two obvious choices. This kind of long-distance dispersal requires that individuals were carried to the islands by some physical means, such as floating driftwood. A more attractive alternative for Van Denburgh and Slevin was that "they were already on each island at the time when it became separated from a larger land-mass." Here was Baur's subsidence hypothesis rearing its head and being applied to the results of the 1905–1906 expedition. They sided with "those who believe that these islands all formerly were connected, and formed part of a single large island which, sometime, must have been connected with continental America." They could not even entertain the explanation of long-distance dispersal from continent to island or from island to island, because "the intervening water" would prove too great a barrier to dispersal. Dispersal was too accidental or occasional for the wide distribution of these lizards in the islands. Van Denburgh and Slevin addressed the modern concept of gene flow when they wrote, "Were such means of dispersal sufficient to bring about the wide distribution of these reptiles, we must believe that the interchange of lizards between the islands [= gene flow] would result either in preventing differentiation on the various islands [= speciation], or in the transportation of differentiated races from island to island [= gene flow and hybridization]." Their evolutionary views and expectations did not match what they knew to be true. If dispersal was occurring, they expected to find "either one kind of lizard on all the islands," which is not the case, or they expected to find "all kinds of lizards" on each island, also not the case. But because each island has its own peculiar kind of *Tropidurus*, and no island has more than one kind, "the evidence all points to complete isolation during a long period of time," which is exactly opposite of the modern view. They continued, "We, therefore, adopt the other theory: that there formerly was a single large island inhabited by one species of *Tropidurus*; that through partial and gradual submersion this island became divided into the many islands of the present archipelago; that each island after its separation was occupied by those animals which inhabited it before; and that the present fauna of each island is directly descended from its original inhabitants." This is what they meant by "the lizards were already there" when the islands were created. They did, however, allow for the possibility that the separation of the islands on the sinking landmass could have occurred at different times rather than simultaneously, but that is just too convenient. By having the islands form at different times, one sinking here and another sinking later, there is time for the morphological divergence actually seen between the lizards of different islands. This divergence and the visible differences are what give rise to the different scientific names applied to the various species. They managed to get Darwinian evolution layered on top of Baur's sunken landmass.

Van Denburgh and Slevin concluded, "Thus we find that the evidence gathered from a study of the *Tropiduri* points to the gradual depression and partial submersion of a former Galapagos Land, resulting in its division into many smaller islands and islets. The story agrees in almost every detail with that which we have previously gathered from an investigation of the snakes and geckos."

For Galápagos snakes collected during the 1905–1906 expedition, Van Denburgh likewise sided with Baur. He recognized that the snakes have very close relatives throughout the West Indies and South America and must have had a common origin. But what were the details of that common ancestry? One option was that the West Indian and Galapagos species were derived from the snakes of South America. An alternative was that all of the species must be descendants of species that occupied a great central landmass in a former geological time, and the landmass had now sunk below the level of the sea. This signalled a very clear allegiance with Baur, but Van Denburgh felt that not enough evidence existed to choose one alternative over the other. As Baur proposed, after the landmass sank, it left species in Central America, northern South America, the Antilles, and the Galápagos. But for Van Denburgh, both alternatives implied a former land connection and a continental origin for the Galápagos snake fauna. Darwin advocated long-distance dispersal to the Galápagos by the three Ws—wind, water, and wings—with species drifting on rafts of vegetation, flying, blown in on the wind, or swimming. Van Denburgh distanced himself from Darwinian dispersal by stating, "I cannot bring myself to share the opinion of those who believe that the fauna of the Galapagos has reached these islands by the more or less accidental agency of the winds and ocean currents." Van Denburgh went on to state that if he had correctly figured out the story of the snakes, there was "nothing in the least suggestive of an unconnected group of volcanic islands thrust independently above the surface of the ocean, to become the home of such animals as might reach them through more or less accidental or occasional agencies of dispersal. Instead of telling of the elevation of new islands, the evidence points to the gradual depression and partial submersion of a more extensive land-mass which must have had more or less direct connection with continental [South] America."

Prior to the 1905–1906 expedition, two species of snakes were described from the islands, and with the new material available Van Denburgh named four more. The type specimen of one new species, *Dromiscus occidentalis*, was collected by Joseph Slevin on Narborough Island on April 18, 1906.

Baur's extensive and repetitive Galápagos writing was arguably more important than Darwin's as a motivating factor for the 1905–1906 expedition. Its historical impact was far from insignificant, because several scientists such as Van Denburgh and Slevin bought into the subsidence idea. Baur had a remarkable level of influence for a scientist few have heard of today. Even geologists now working in the islands barely know Baur's name. Yet Baur guided what was done daily during the Academy expedition. His views influenced for many years, and possibly do so even today, how museum specimens and species were viewed following the Academy expedition.

Writing in 1953, the Academy's honorary curator of entomology, Edwin Van Dyke, examined the 1905–1906 specimens of beetles, the Coleoptera, which were Darwin's favorite group of organisms in his university years. He realized that "the fauna of the Galapagos Islands, because of its long isolation, has acquired certain peculiar features which throw light upon many of the problems of nature." This is an evolutionary statement by Van Dyke, echoing Darwin's words in *The Origin of Species*: "When on board H.M.S 'Beagle,' as naturalist, I was much struck with certain facts in the distribution of the organic beings inhabiting South America, and the geological relations of the present to the past inhabitants of that continent. These facts, as will be seen in the latter chapters of this volume, seemed to throw some light on the origin of species—that mystery of mysteries, as it has been called by one of our greatest philosophers."

Van Dyke realized the depauperate (species-poor) and endemic (unique) nature of the beetle fauna, which was "limited in extent and most of its species and certain of its genera are confined to the Archipelago." The magnitude of evolutionary change seen in their anatomy was evidenced by the "great deal of divergence shown in the species"; "most species are restricted to certain islands or even to special areas on the islands." Realizing this was "not a haphazard distribution," and "there is a relationship between the variations of the species and their distribution." All this sounds consistent with a Darwinian view or the view of Alfred Russel Wallace, the co-discoverer of natural selection. Van Dyke considered the changes to be "modifications which have taken place . . . by the gradual isolation of the different species such as could be produced by the breaking up of larger islands into smaller ones." Here he starts to diverge from Darwin and side with Baur. Van Dyke makes his geological views quite clear, noting that the beetle fauna is related to that of the western portion of South America, with no extraneous element. He sides with Baur when he writes, "It would appear as if a certain portion of western South America was isolated by the *subsidence* of the intervening area between the Islands and the mainland. This *land mass* was later broken up into small islands, probably by *subsidence* subject to extensive volcanic action and desiccation." The volcanic activity was obvious to everyone, from Baur to Van Dyke. The important evolutionary mechanism for Van Dyke was that "these factors in time reduced the fauna [few species compared to the mainland] and accelerated the various divergent modifications [endemic or unique species]." Van Dyke acknowledged that "because of the extensive volcanic action on the Islands and the deep ocean bed between them and the mainland [also noted by Agassiz], they have been listed as *oceanic islands* by Alfred Russel Wallace and others. *I do not believe this*. I believe that the Islands are *continental* and that an extensive *subsidence* of the coastal area of western South America coincided with the elevation of the northern Andes which the geologists believe has occurred in later geological times, has accounted for their isolation." Van Dyke was clearly in the Baur camp in 1953.

To his credit, Baur could be considered the father of Galápagos conservation for his early, urgent, and accurate calls to action to save species. Baur recognized the importance of Galápagos organisms for addressing evolutionary questions and the great need for conservation. Writing in April 1891, the same year the new museum of the California Academy of Sciences opened its doors on Market Street in San Francisco, he wrote, "There is no place on the whole earth which affords better opportunities for such work than the Galapagos. Here we have the original natural conditions, hardly influenced by man. If all the variations of the forms [species] on this group of islands, or only the variations of a few genera, are studied, and the conditions of each variation are examined, then we may perhaps be able to express a more definite opinion on the causes of variation itself." For Baur, the visible variation in genera and species was the raw material that could either cause or prevent evolution from moving in a Darwinian direction. In the late nineteenth and early twentieth centuries, writing about "variation" was code for writing about evolution. Baur's obituary in 1898 stated, "It was, perhaps, fitting that one born in the year of the publication of the 'Origin of Species' should gain inspiration from the spot where the idea of the 'Origin' was conceived. At the very outset, however, he announced a far-reaching opinion utterly at variance with Darwin's view of the origin of the Galapagos."

Baur's most fervent, even shrill, call to action motivated the organizers of the 1905–1906 expedition. Baur emphatically wrote further, "Such work ought to be done *before it is too late*. I repeat, before it is too late! Or it may happen that the natural history of the Galapagos will be lost, as it has unfortunately been lost in so many islands; for instance, of St. Helena and the Mascarenes, lost forever, irreparably!" Even today, conservationists could follow Baur's suggestions.

None of the Academy curators back in San Francisco and none of the field collectors on the schooner *Academy* were explicitly attempting to show whether Darwin was correct or not. As we phrase it today, none of them were trying to support or refute Darwin's hypothesis of the mechanism of natural selection in 1905–1906. They were certainly aware of Darwin and his published work, but the geological and dispersal ideas they were testing were those of Baur.

Not only did Georg Baur disagree with Charles Darwin about Galápagos geology and dispersal of organisms, he also disagreed with Darwin on evolution in general. Baur was a fervent and steadfast neo-Lamarckian, as was the American paleontologist Edward Drinker Cope and others who rejected natural selection and adhered to the notion of organisms acquiring characteristics during their lifetime and passing them to their offspring. This was called the eclipse of Darwin from about the time Darwin died in 1882 until the 1920s, when it was itself eclipsed by population genetics and the modern evolutionary synthesis. As implausible as Baur's ideas seem today, at the end of the nineteenth century some scientists took him seriously and expended great effort to support him. One of those efforts was made, in part, by the 1905–1906 expedition. At the time, it was a plausible, mainstream alternative to Darwin.

Baur also took issue with Darwin and geology. Darwin wrote in his 1839 summary of the *Beagle* voyage, his *Journal of Researches*, based on soundings made by the ship's crew, that "the profound depth of the ocean between the islands, and their apparently recent (in a geological sense) volcanic origin, render it highly unlikely that they were ever united." For Darwin the origin of the islands was clearly the opposite of what Baur later concluded. And Darwin, like Baur, based his conclusion on the animal life, writing, "And this, probably, is a far more important consideration than any other, with respect to the geographical distribution of their inhabitants." Baur took those same geographical distributions and turned them on their head to support his subsidence idea fifty years later.

The trajectory of Baur's life intersects two significant dates in Charles Darwin's life. Baur was born in 1859, the year Darwin published *On the Origin of Species*, and he completed his PhD in 1882, the year Darwin died. Born and educated in Germany, Baur became a laboratory assistant in vertebrate paleontology under Professor Othniel Charles Marsh (1831–1899) at Yale University, infamous for his role in the highly competitive dinosaur "bone wars" in the late 1890s with Edward Drinker Cope (1840–1897) of the Academy of Natural Sciences in Philadelphia. Despite these temporal affinities and overlaps with Darwin, Baur was decidedly anti-Darwin on several levels. His Galápagos perspective flew in the face of mainstream science, and yet his impact cannot be dismissed out of hand. Baur was by all accounts a brilliant and energetic researcher, but he failed to be an effective university teacher due to his pedantic style and heavy German accent.

Once his mind was made up, Baur largely abandoned his productive career in anatomy and paleontology, and devoted the remainder of his life to confirming his eccentric ideas. Baur prided himself on having figured out almost everything he needed to know about the geological origin of the Galápagos before gathering even a shred of evidence. To his credit, Baur spent two and a half months in the islands during June–August 1891, but mainly to reaffirm his own ideas and predictions, thus falling short of the scientific ideal. From then until the end of his life, he staunchly argued and extensively wrote about Galápagos geology, often repeating himself and quoting himself at length in his own publications. Baur was one of the most theory-driven researchers to visit the Galápagos. A main component of his flawed argument about geology was based on how the organisms, particularly the giant tortoises, were related to each other and how those supposed relationships could only support his version of geological history. Baur's attempt at developing a sound hypothesis constitutes a textbook example of how *not* to do science. He made up his mind on a whim that day at Yale and would not allow any line of evidence to sway him. Rather than form a hypothesis to be supported or refuted with data and observations by himself and other researchers, Baur wanted to form a new "law" of Galápagos geology. Unfortunately for Baur, very few scientists would subscribe to his novel idea—very few, that is, except those from the California Academy of Sciences.

With two radically competing schools of thought to choose from, how did the California Academy of Sciences vote? The Academy's herpetologist, John Van Denburgh, who, like Baur, made up his mind before ever visiting the Galápagos, wrote in 1907: "Early in 1905 the California Academy of Sciences decided to send an expedition to the Galapagos Islands.

The general purpose was to explore this group more thoroughly than the opportunities of previous investigators had permitted, and to secure large collections of the plants, mollusks, insects, birds, mammals, and reptiles in the hope of throwing more definite light upon *the origin of the archipelago*." Van Denburgh did visit the islands but published widely on the material from the 1905–1906 expedition, collected by men who served as his proxies.

Van Denburgh did *not* say that he sought to shed light on the origin of species or to support or refute the ideas of Darwin, who made the islands famous, or even the views of Lamarck, but rather to examine "the origin of the archipelago." That *origin* is the geological *origin*, and Van Denburgh supported Baur. Van Denburgh continued, "Particularly, it was determined to study the geology of the islands, to make a very careful search for fossils, and to spare no effort to secure specimens or remains of those races of the gigantic land tortoises which long had been thought extinct." Like Baur, Van Denburgh considered tortoises and geology to be closely associated, and went on to become an expert on all Galápagos reptiles. Both believed that studying one topic would yield insights about the other. Van Denburgh wanted to show that even if an island's race or variety or species of tortoise became extinct, its fossil remains would vindicate Baur.

Van Denburgh aligned himself and the Academy with Baur when he continued, "Study of the published results of previous expeditions had convinced me not only that these islands must all, at some former period, have been parts of a single land-mass, becoming later, by partial submersion, separated into the various islands." This amounted to a concise and accurate rephrasing of Baur's subsidence hypothesis and leaves no doubt about the Academy expedition's agenda for collecting tortoises. Van Denburgh combined recent local uplift with the overall trend of subsidence by writing, "Albemarle [Isabela] Island, which possesses several races of tortoises and on which [Stanford University researchers in 1898–1899] Heller and Snodgrass found evidence of an elevation amounting to several hundred feet had much more recently been formed by the union of several smaller islands corresponding, probably, to its five great volcanoes." So Van Denburgh combined Baur's generally subsiding large landmass with Heller and Snodgrass's active volcanoes altering the landscape and increasing the sizes of islands. Sometimes nearby islands could merge, like lumps of cookie dough placed too close together on a cookie sheet in the oven. Each island formed separately from a source of lava, a hot spot, on the sea floor. The same thing happened in the Hawaiian Islands: Maui and Oahu are each made of two volcanoes, and the big island of Hawaii is made of five volcanoes. The volcanoes merged together, their lavas intermingled over time, and the two became one island. Albemarle Island is the best example of the merged-cookie idea in Galápagos. It has five substantial volcanoes that were likely separate islands in the past. Each erupted enough lava to meet and merge, resulting in the characteristic backward-L shape of Albemarle with an uncanny resemblance to a seahorse. Van Denburgh's was a hybrid idea, but it was fundamentally incorrect.

Van Denburgh's desire to vindicate Baur was clearly part of the eight young men's fieldwork agenda. "Accordingly, the members of the expedition were instructed to collect on Albemarle exactly as though it still were five islands." Baur's thinking was at the theoretical heart of the expedition. Van Denburgh laid out the pragmatic framework for their daily collecting. On the largest island of Albemarle, they fully expected to find unique tortoise species stranded on each of the five major volcanoes. Finding those species would nicely vindicate Baur. The hand of Baur rested on the shoulders of the men, guiding their fieldwork, as they moved about the islands for a year and a day.

In 1912, a half dozen years after the Academy expedition returned, Van Denburgh firmly rejected the ideas of Darwinian dispersal and de novo volcanic origin:, "If we read this story of the [Galápagos] snakes correctly, there is nothing in the least suggestive of an unconnected group of volcanic islands thrust independently above the surface of the ocean, to become the home of such animals as might reach them through more or less accidental or occasional agencies of dispersal." Van Denburgh supported subsidence of the "Baur landmass": "Instead of telling of the elevation of new islands, the evidence points to the gradual depression and partial submersion of a more extensive land-mass which must have had direct or indirect connection with continental America."

The Academy expedition's botanist saw things slightly differently. Alban Stewart did not agree completely with Baur or Van Denburgh on geology or the distribution of organisms. The plant life of the islands argued against a connection to the continent: "There appears to be little evidence to show that there has ever been a land connection between the islands and the mainland." But Stewart hedged and did not rule out Baur's subsided landmass: "Yet there is no very strong evidence opposed to the view that the islands may have been connected with each other, at some not distant geological period, either as one large island or as two or three smaller ones." Stewart saw the "harmonic zoological relationships" that Baur predicted would preferentially indicate "such a condition [subsidence], than if each island had been formed separately." As with any subject as complex as the distribution of organisms and how they came to be the way they are, Stewart also had contrary evidence: "While all of the above facts seem to point to a general subsidence of the islands, there are a few evidences of elevation." He went on to discuss several sedimentary deposits of marine invertebrates, mainly corals, mollusks, and echinoderms, that indicated to the expedition's geologist and paleontologist, Washington Henry Ochsner, that local uplift had occurred in places like Tagus Cove on Albemarle Island, as well as in the sea cliffs on either side of the Itabaca Channel on Indefatigable and South Seymour Islands. Despite these areas of local uplift, Stewart believed that "there is no evidence of a general elevation . . . and it is not improbable that during the period of general subsidence there might have been times in which it ceased and during which local elevation took place." With this contradictory evidence, the movement of the islands could be likened to a yo-yo.

The expedition's geologist, Washington Henry Ochsner, was firmly in favor of subsidence, mentioning the word seventy-six times in his unpublished field notes. Ochsner provided a previously unpublished summary of his geological findings to Stewart, quoted in Stewart's plant monograph of 1911. Ochsner firmly stated, "I am much in favor of [Baur's] theory of subsidence. With additional thought and study given that matter I feel that the testimony of my collected facts and observations [on geology and paleontology] will go to prove this theory nearly a fact."

In his published work on the plants of Cocos Island, Stewart was more definitive in his opposition to Baur when he wrote, "In a paper written some years ago by Dr. George [sic] Baur, an attempt was made to establish a former land-connection between the Galapagos Islands and the American continent, the connection presumably having been somewhere in the Mexican region. The improbability of such a connection has already been shown, and it seems that the great difference in the floras of Cocos and the Galapagos islands strongly opposes Dr. Baur's view."

Francis Xavier Williams, the expedition's entomologist, stated in 1926 with respect to the bees and wasps of the islands, "The scanty list of bees and wasps shows a series of insects which have, I believe, reached their destination by the flotsam and jetsam method, and by flight."

Baur's theoretical framework also appeared in Williams's 1911 paper on the hawk moths of the islands. Williams could accept the sunken-landmass idea, but he insisted that the insects and other organisms reached the islands not as marooned remnants in the Baur scheme, but through Darwinian long-distance dispersal. He wrote, "I have considered the Galapagos as oceanic as regards their natural history; whether they issued in the first place from the bed of the ocean, or whether they were of continental origin, provided they were once completely submerged, or all living organisms thereon otherwise totally destroyed simultaneously by volcanic activity, as the flora and fauna would still be of oceanic character, i.e., transported across water to the islands, a condition the writer believes has happened." For Williams, the islands had to be devoid of life at some time in the past, then populated from the mainland across the intervening ocean. This view was a hybrid of Baur and Darwin. Williams conceded in a footnote, "There is good evidence that the Galapagos Archipelago was once one large island which by subsidence has formed the many smaller islands. This view makes it easier to us to explain the existence on all or most of the islands of closely allied species or varieties." Conveniently for Baur, the anti-Darwinist, this view required neither Darwinian long-distance dispersal nor Darwinian evolution.

Baur and Van Denburgh expected to find species on certain islands to verify the subsidence idea. Baur named two species: one from Sierra Negra Volcano on Albemarle, *Chelonoidis guentheri* Baur, 1889 and a species from Floreana Island, *Chelonoidis galapagoensis* Baur, 1889. And Van Denburgh named four species, all in the year following the return to San Francisco of the Academy's expedition. By naming new species where they "ought" to be on what Baur viewed as formerly separate islands, he was lending taxonomic credence to his unfounded geological ideas. This again points to the function of taxonomy in telling a story about a place, whether it is a geological story or an evolutionary story. Ironically, Baur was vehemently anti-Darwin, but by putting species on each "subsided island" he was allowing a scenario that included subsequent evolutionary divergence on those former islands, exactly what Darwin predicted. This kind of reasoning became a self-fulfilling prophesy for Baur and Van Denburgh. New species should have an objective reality: they are new because they have never been seen or studied or described by science. For Baur and Van Denburgh, species could be new because they filled in missing pieces of a geological puzzle. For Baur and his adherents, the present-day distribution of tortoise species mapped out the size and extent of previous larger islands, now submerged except for the peaks of the tallest volcanoes.

Species named by Van Denburgh in 1907, partly in response to Baur's notion that each island "needed" its own tortoise species, because in his view those species already existed long ago in the past before the hypothetical landmass subsided, included the following:

Chatham Island tortoise, *C. n. chathamensis* 1907
James Island tortoise, *C. n. darwini* 1907
Hood Island tortoise, *C. n. hoodensis* 1907
Narborough Island tortoise, *C. n. phantastica* 1907

In the final analysis, not requiring the benefit of hindsight, Darwin got it right and Baur got it wrong. Baur was wrong about the geological origin of the islands, even in light of geological knowledge in his day. But he was absolutely right about the need for conservation, and we can thank him for that. Eventually, Baur's version of the geological origin of the Galápagos Islands withered away in the punishing heat of scientific scrutiny and subsided into obscurity.

The expedition's entomologist, Francis Xavier Williams, summarized in 1911 the confusion about the tension existing between competing hypotheses for the geological origin of the archipelago: whether the islands are volcanoes formed from the bottom of the ocean upward by innumerable eruptions of lava from volcanic vents on the sea floor or whether the islands are the remaining peaks of a submerged continental landmass. The issue was not resolved by geologists, biologists, or paleontologists for many years after the expedition, despite Darwin's having written in *The Voyage of the Beagle* that the islands "are all formed by volcanic rock." But Darwin did not take a stand on whether the islands were made of *uplifted* volcanic rock (the modern view) or of *subsided* volcanic rock, as preferred by Williams and also by Ochsner. Given what we know today about plate tectonics and volcanic island-building hot spots, it would be easy to harshly judge their geological interpretations.

Charles Darwin's evolutionary legacy is embedded in the way organisms change genetically over time from their ancestors, generation by generation. Darwin's ideas themselves also changed over time, even after his death, especially in the time period known as the evolutionary new synthesis. The enduring usefulness of museum specimens resides in their function as an archival record of the flora and fauna of an area. New generations of researchers can interpret specimens on the basis of new knowledge and information.

NOTES

Page 206: *Momentous and unwavering*: Baur went to his grave adhering to his unconventional geological ideas.

Page 207: *Continental islands*: Baur (1891, 307).

Page 207: *Theories to suit facts*: Doyle (1891, 7).

Page 208: *Continental islands vs. oceanic islands*: Baur (1891, 310; 1895, 68).

Page 208: *Recent volcanic origin*: Baur (1891, 312; 1895, 69).

Page 208: The son of Henri Milne-Edwards (1800–1885), Alphonse Milne-Edwards (1835–1900) named the blue-footed booby, *Sula nebouxii*, in 1882.

Page 208: *Geological structure*: Agassiz (1892, 70–71).

Page 208: *Successive eruptions*: Agassiz (1892, 71).

Page 209: *Biggest liar in the Pacific*: Mills (2003).

Page 209: *The notice of scientific men*: Morrell (1832).

Page 209: *The floor of the ocean*: Agassiz (1892).

Page 209: *I believe in subsidence*: Baur (1892).

Page 209: *No evidence would sway him*: Baur (1895).

Page 210: *Stanford expedition*: Heller (1903).

Page 210: *From previous collections*: Van Denburg and Slevin (1913).

Page 211: *Single large island*: Van Denburgh and Slevin (1913, 145); Van Denburgh (1912b, 417).

Page 211: *Partial submersion*: Van Denburgh and Slevin (1913, 175).

Page 212: *Galapagos snakes:* Van Denburgh (1912a, 331–332).

Page 212: *Writing in 1953*: Van Dyke (1953); see also Peck (2006) for a treatment of Darwin's darkling beetles.

Page 212: *Mystery of mysteries*: OOS, 5.

Page 212: *Subsidence of the intervening area*: Van Dyke (1953, 3); emphasis added.

Page 213: *I do not believe this*: Van Dyke (1953, 3); emphasis added.

Page 213: *Causes of variation itself*: Baur (1891).

Page 213: *Origin of the Galapagos*: Wheeler (1898).

Page 213: *Before it is too late*: Baur (1891); emphasis in original.

Page 214: *They were ever united*: (Darwin, 1839, 398).

Page 214: *Geographical distribution of their inhabitants*: VOB, 383.

Page 214: *Born and educated in Germany*: Schuchert and LeVene (1940, 302–308).

Page 214: *Confirming his eccentric ideas*: Baur died of general paresis, or paralytic dementia, a neuropsychiatric disorder affecting the brain, caused by late-stage syphilis. A brother also died of this affliction (Wheeler, 1898, 17).

Page 215: *Origin of the archipelago*: Van Denburgh (1907, 1); emphasis added.

Page 215: *Long thought extinct*: Van Denburgh (1907, 1).

Page 215: *Single land-mass*: Van Denburgh (1907, 1–2).

Page 215: *Five great volcanoes*: Van Denburgh (1907, 2).

Page 215: *Five islands*: Van Denburgh (1907, 2).

Page 216: *Connection with continental America*: Van Denburgh (1912a, 331–332).

Page 216: *Each island had been formed separately*: Stewart (1911, 233–235).

Page 216: *No evidence of general elevation*: Stewart (1911, 238).

Page 216: *Plant monograph of 1911*: Stewart (1911, 7–288).

Page 216: *Prove this theory nearly a fact*: Ochsner quoted in Stewart (1911, 238).

Page 217: *Strongly opposes Dr. Baur's view*: Stewart (1912b, 382).

Page 217: *The flotsam and jetsam method*: Williams (1926).

Page 217: *Transported across water to the islands*: Williams (1911).

Page 218: *All formed by volcanic rock*: VOB, 357.

Page 218: *Evolutionary New Synthesis*: Smocovitis (1996).

His methods are the practical application of the
theories of Darwin.

DAVID STARR JORDAN (1909)

13

Darwin's Islands Meet the Plant Wizard of Santa Rosa

THE TWO GREATEST minds known for their work on natural selection and artificial selection crossed paths, at least in spirit, in the Galápagos Islands during the historic voyage of the schooner *Academy* in 1905–1906. Had Charles Darwin (1809–1882) and Luther Burbank (1849–1926) actually met in person in the islands, they could have discussed plant propagation and selective breeding and natural selection, interests they shared and that formed the legacies of the two men. But instead of meeting face-to-face, they were brought together symbolically, long after Darwin's death, by the field collectors sent from the California Academy of Sciences.

The group that went inland found walking through the thorny vegetation difficult at times. The immense tree cacti of the genus *Opuntia* presented a stark contrast to the normally low-growing form of this plant. Clumps of these cacti, with their unusual reddish brown trunks devoid of spines, made a favorite hiding place for flocks of goats. Resting in the shade, the goats became easy targets and quickly entered the food chain.

A few months before the expedition left San Francisco, the renowned plant breeder Luther Burbank wrote to the Academy director, Leverett Mills Loomis, and requested seeds from the Galápagos *Opuntia* cactus that might be collected from any future expeditions. Burbank wanted to use these seeds, which Rollo Beck collected, in his ongoing research on developing spineless cactus for use as animal food. Burbank's fame rests partly on his having developed spineless cactus on his property and experimental greenhouses in Sonoma County, California. Burbank's record keeping was chaotic, and whether he used the Galápagos seeds in his genetic crossbreeding is lost to history.

Burbank pioneered experimental plant breeding in the late nineteenth and early twentieth centuries. During his lifetime he created such recognizable varieties as the russet Burbank potato (the world's predominant potato in food processing; think of McDonald's french fries), the Shasta daisy, the freestone peach, the Santa Rosa plum, the July Elberta peach, and the white blackberry, among many others. In addition to varieties of everything from apples, strawberries, and cherries to quinces, chestnuts, and pears, Burbank developed thirty-five kinds of fruiting cacti.

Burbank's goal with cactus in the genus *Opuntia*, the common prickly pear, was to maximize the amount of forage available to cattle while minimizing thorns. Burbank bred the plants from seed, then made crosses between varieties and finally backcrosses between generations, all common techniques in the horticulturalist's bag of genetic tricks. The resulting plants that Burbank deemed most favorable were allowed to live and reproduce. Burbank played his role as an effective agent of artificial selection in the glass greenhouses and experimental plots located on his properties in Sonoma County. What Burbank did artificially in these greenhouses is exactly what Darwin saw as the role of nature in natural selection. Some individuals live and some individuals die, and Burbank made those artificial decisions, uprooting and discarding any seedlings not fitting his desired outcome. The evolutionist and president of Stanford University, David Starr Jordan, wrote in 1909, "In Burbank's gardens the few tenderly cared for little potted plants or carefully grafted seedlings represent the surviving fittest, and the great bonfires of scores of thousands of uprooted others, the unfit, in this close mimicry of Darwin and [Herbert] Spencer's struggle and survival in nature."

New varieties and eventually new species are created by natural selection; likewise with artificial selection. If Darwin was the wizard of new species, then Burbank was the wizard of new varieties. Darwin had no explicit economic motive embedded in his evolutionary mechanism of natural selection. Burbank, on the other hand, was pragmatic above all in his application of Darwinian principles, and profit was his main motivation.

Burbank finally received the seeds in December 1906. From his office and residence in Santa Rosa, California, Burbank thanked Loomis and the Academy on December 8, 1906: "The exceedingly interesting cactus seeds from the Gallapagas [*sic*] Islands received and they will have the best possible care that I can give them. I highly appreciate this care by someone, I know not who, in collecting these for me." Burbank enclosed a check for $25, which was promptly returned by Loomis, who wrote, "The cactus seeds come in response to your request of a year and a half ago. I instructed the chief of the expedition [Rollo Beck] to make the collection and the Executive Committee wishes you to accept it with the expression of the desire of the Academy to do all in its power to further your investigations. I return your check." Burbank also sent a few books and publications to replace some of those lost in the fire of April 1906. Loomis and the Executive Committee held Burbank in high regard. Their respect for the great man of science meant they could not accept his money: they treated the Galápagos *Opuntia* seeds as a donation to the cause of plant breeding and a diplomatic gesture to further good relations.

Burbank also acquired seeds or cuttings of *Opuntia* cactus from collectors in Mexico, Central America, and mainland South America, North and South Africa, Australia, Japan, Hawaii, and the South Sea Islands, as well as eight kinds of partially thornless varieties from Sicily, Italy, France, and North Africa and a small collection of Mexican wild thorny varieties. To these varieties he crossbred "hardy wild species from Maine, Iowa, Missouri, Colorado, California, Arizona, New Mexico, Dakota, Texas and other States." He sought to distill from this genetic soup what soon emerged as the Burbank spineless cactus. The importance of each geographic source of seed material is lost to history, because Burbank kept scanty or chaotic records of which seeds were used in particular breeding crosses and experiments. Burbank's greenhouse techniques produced genetically modified organisms by employing artificial selection to mimic Darwin's natural selection. Both men envisioned new varieties and new species as the outcome of the selection process.

Burbank viewed himself as an extremely successful producer of new plant forms and received massive amounts of publicity and accolades during his lifetime. He variously called those new forms of plant life "varieties" or "horticultural productions" or "creations," and he was involved in "the business of selling the various new forms of plant life which I have evolved." Clearly, he saw himself as an active agent of evolution by means of artificial selection and as a creator of new life forms. In his view, he was not just the plant wizard of Santa Rosa, he was the plant "god" of Santa Rosa. His "creations" were new and important, just like those of God. David Starr Jordan wrote: "He has put into practical utility the teachings of his greatest master, Darwin, and he has enriched the world with thousands of fruits and flowers, useful and delightful, which but for him would have existed only among the conceivable possibilities of creation."

Concerning Burbank's processes of crossing and selection, Jordan wrote in 1909, "It is Darwin who first gave us the knowledge on which all this work rests. The origin of species demands variation, selection, segregation, and behind all this the law of heredity, the fact that 'like produces like' or nearly alike. Burbank is a creator of species. So is any man who applies these elements to animal or plant life. To call him a 'wizard,' as some men and some magazines do, is to injure him in reputation and to befog his great services with a trivial epithet."

Concerning his commercially available plant material, Burbank wrote, "Many hundreds of these productions, absolutely new to mankind and more useful and valuable than those now known, are already complete and await introduction." Burbank had a personal Noah's Ark overflowing with plants, and Burbank himself was ready to walk down the gangplank with these new creations, two by two like Noah, and introduce them to a waiting world that *needed* his newly evolved productions. Not only did Burbank improve plants, he improved the world.

Burbank realized that natural variation in spines and fruit in the cactus genus *Opuntia* rendered it a malleable subject for his subtle horticultural manipulations. Burbank would rapidly accomplish in his greenhouse and in his experimental plots in Sonoma County not only what nature *could* have done, but what nature *should* have done. Burbank judged nature and found it wanting. If natural selection as described by Charles Darwin was "neither merciful nor merciless," then Burbank was harshly judgmental about how varieties in nature should look. Burbank was correcting the "mistakes" of nature and improving on and augmenting them. Burbank wrote, "Systematic work for [*Opuntia*] cactus improvement has shown how pliable and readily moulded is this unique, hardy denizen of rocky, drought-cursed, windswept, sun-blistered districts, and how readily it adapts itself to more fertile soils and how rapidly it improves under cultivation and improved conditions."

The unassuming *Opuntia* cactus was an ideal subject for Burbank's outdoor laboratory of evolution, one that he could bend to his will with some certainty of success. Burbank changed what he found in nature and considered it *improving* on nature by emphasizing a desired physical trait or suppressing an undesirable attribute. Genetic traits were like musical notes, arranged by Burbank, varying by tone, pitch, strength, and octave. The outcome, as with a musical score, was a functionally coordinated variety of plant with the proportions Burbank desired, be they color, size, shape, spines, or flavor. The attributes of an organism were the notes in Burbank's symphony.

Hugo de Vries, the mutationist, visited Luther Burbank in Santa Rosa, a short distance north of San Francisco in the summers of 1904 and 1906, bracketing the Galápagos expedition. Burbank had been elected a an honorary life member of the California Academy of Sciences in 1905 and delivered an occasional lecture at Stanford University, where two members of the Academy expedition had attended college.

Even if Rollo Beck on the 1905–1906 expedition had acquired seeds for Burbank of the most thornless variety of *Opuntia* from the Galápagos Islands (the form on Tower [Genovesa] Island with hairlike spines), this might not have been good enough. Burbank realized how counterintuitive the breeding of cacti could turn out to be. Burbank wrote:

Many so-called thornless or partly thornless ones were obtained, but not one among the thousands from all these sources was free from thorns and spicules, and even worse, those which were the most promising in these respects often bore the poorest fruit, were the most unproductive of fruit or produced less fodder, or were less hardy than the wild thorny species and varieties.

Just as Burbank saw his popular Santa Rosa plums steadily replace the old varieties (read "inferior" in the Burbank sense of enhancing natural forms), so too would his new form of spineless cactus in the genus *Opuntia* be an improvement of natural varieties. Burbank placed himself in august company when he equated the spineless cactus described in his 1913 brochure as "only the beginnings of a great work with the Opuntias, which in importance may be classed with the discovery of a new continent." Burbank was a god, and Noah, and Christopher Columbus all wrapped into one horticulturalist living in Sonoma County, today an hour by freeway north of San Francisco and the Golden Gate Bridge.

NOTES

Page 221: *In Burbank's gardens*: Jordan (1909, 90).

Page 221: *I return your check*: Letters to and from L. Burbank and L. M. Loomis, 1906 correspondence file, Archives of the California Academy of Sciences.

Page 221: *Hardy wild species*: Burbank (1913, 10).

Page 222: *Burbank viewed himself*: Burbank (1913, 1).

Page 222: *Enriched the world*: Jordan (1909, x).

Page 222: *Burbank is a creator of species*: Jordan (1909, ix–x).

Page 222: *Many hundreds*: Burbank (1913, 6).

Page 222: *Cultivation and improved conditions*: Burbank (1913, 5).

Page 223: Burbank (1913, 4).

Our good ship labored, plunged, rose, trembled, plunged and rose again amidst the foaming billows, shaking off the feathery spray like a sea-lion, and rushing along her watery way with grandeur.

Titus Coan, 1834 voyage of the sailing merchant ship *Hellespont* in the Gulf Stream, bound for Honolulu.

14

The Voyage Home

SHIP'S LOG

ANY SUCCESSFUL SCIENTIFIC collecting expedition hinges on both getting the specimens in hand during the fieldwork and then getting them back safely to their museum repository. Neither the collecting nor the safety was guaranteed for the 1905–1906 expedition, and safety was in a precarious balance, teetering on failure, up until the very last hours as the men on the *Academy* reached San Francisco. This expedition could well have ended in failure.

A year and a day after landing in Galápagos on September 24, 1905, the men made their last landfall on September 25, 1906. Their visit to the final island of Culpepper, or Darwin Island as it is called today, was neither auspicious nor productive, despite the important symbolism of the island's modern name.

Leaving the penultimate island of Wenman, or Wolf Island as it is called today, at 3:00 a.m., they shaped course for Culpepper Island and arrived seven hours later. Culpepper is the northernmost and westernmost island in the archipelago and is rarely visited by ecotourists today, except for certain scuba divers seeking challenging diving conditions. The island is both physically and geologically removed from the main islands, reminiscent of the northwest Hawaiian Islands' removal from the eight main islands in that other famous archipelago. The men found Culpepper "a most uninviting spot," because it is merely a flat-topped volcanic remnant and considerably smaller than Wenman.

VAMPIRE FINCHES: BLOODY BUSINESS ON CULPEPPER ISLAND

Numberless seabirds circled high above Culpepper as the Academy men approached in the schooner. But at the top of Culpepper resided a fascinating evolutionary story involving finches and boobies that they would not learn on this trip, because the island was so steep and inaccessible. The story remained unknown to science until discovered, or possibly rediscovered, nearly sixty years later with the use of a helicopter.

For now, the men would do their best to collect specimens on this final island in the archipelago. Slevin, Beck, Ochsner, Hunter, and Williams put off in the skiff and attempted a landing that was difficult owing to the choppy sea and large boulders at the ledge at the base of the island where they had to jump ashore. Their last landing was a wet landing and a dangerous one at that, with several large sharks patrolling the landing site. A towering five-hundred-foot cliff loomed above them as they made the jump one at a time. They found the summit "quite inaccessible" and stayed only on the coastal ledge.

Rollo Beck had been to Culpepper before as a member of the Rothschild-funded Webster–Harris expedition in 1897–1898, which had similarly reported that there was "no chance whatever of reaching the top of the island." On that earlier trip the frigatebirds were very bold, and one swooped down and picked off Beck's cap. But the expedition members managed to collect finches on Culpepper and Wenman that proved to be new to science. At the top of Wenman, which could be easily scrambled up, they noted "some of the finches climbing on a booby's back and pecking in the feathers—probably in search of parasites. Saw three finches on one booby at a time." What they thought was pecking for parasites was actually pecking for blood. They had inadvertently discovered the existence of vampire finches on Wenman Island, and that same behavior likely existed then as it does now on Culpepper Island, but neither the 1897–1898 nor the 1905–1906 expedition could get to the top of Culpepper to make the observation.

Up on Culpepper and on Wenman, a subspecies or variety of the Darwin's finch group was engaged in the vampire activity. The vampire finch is known scientifically by its three-part subspecies name of *Geospiza difficilis septentrionalis*, and it was named in 1899 by Rollo Beck's wealthy English benefactor, Walter Rothschild, and his assistant, Ernst Hartert, at Rothschild's Tring Museum. They described the new species taxonomically from specimens Beck and colleagues collected while scrambling around on the same coastal ledge on Culpepper Island in late 1897, just as the men of the Academy expedition were now doing in late 1906. The bird's common name, sharp-beaked ground finch, seems appropriate for its blood-eating habit. Elsewhere in the archipelago the other two known subspecies of *difficilis* finches, the blood-eaters' closest relatives, have been found to lead rather conventional lives and eat the conventional finch food of seeds or insects. But at the top of Culpepper and Wenman they drink blood, booby blood.

The insurmountable island became surmountable in 1964, but only with the help of a military helicopter. The Galápagos International Scientific Project (GISP) was "the mother of all Galápagos expeditions" in terms of external manpower. The project had available two US Navy helicopters and made the first recorded summiting of Culpepper Island.

Gifford noted the propensity toward blood-eating, but did not quite put it all together when he observed on September 24, 1906, on Wenman that whenever he shot a sharp-beaked ground finch "several others would gather about it and pick at the blood." He clearly saw the vampire phenomenon but did not see it in action on a booby. Any specimens of seabirds for Loomis in late 1906 were only at the summit of Wenman, and that was where the interesting story remained unseen for so long.

SAILING HOME

The fieldwork of the 1905–1906 expedition lasted 366 days, a year and a day in the islands: from their arrival on September 24, 1905, using dead reckoning and the ship's compass to reach the islands without an accurate chronometer to determine their longitude, to September 25, 1906, when they shoved off from their last landfall, Culpepper (Darwin) Island, which they found "a most uninviting spot," and shaped their course for Mexico's Clarion Island. The men had made history, and they were firmly aware of their accomplishment. They knew of the extensive devastation of San Francisco by the earthquake and fires just five months ago and of the complete destruction of their home institution, the California Academy of Sciences, to which they remained fiercely loyal. Now it was time to go home.

With a fresh south-southeast wind directly behind them, they ran wing and wing, with the mainsail catching the wind over to one side of the vessel and the foresail over to the other side. To mark their progress, they put out the taffrail log, a torpedo-shaped spinning device at the end of a long line that works just like the speedometer on a car. Instead of terrestrial miles per hour, the taffrail log reports knots, or nautical miles per hour, with a nautical mile being only slightly longer than our more familiar land-based geographical mile. Knowing their speed in knots and how long they sailed each day enabled the crew of the schooner *Academy* to plot their progress home. On that 455th day of the voyage, the weather was cloudy with fog and there were rain squalls toward evening.

By the next morning the wind shifted to south-southwest, and the crew hauled in the foresail. With a heavy southerly swell running, the schooner made impressive speed at nine knots. This was real blue-water sailing. The strength of the wind and the height of the swell necessitated constant vigilance by the crew, now that their erstwhile navigator, Parker, had been left behind by his own choice at Puerto Vilamil on Albemarle Island. The eight young men were fully engaged as sailor-scientists, with the emphasis now on sailing. But the scientist part had not quite ended, with Slevin and King, with some assistance on the heaving vessel, skinning tortoises in order to clear up the decks before getting too far north. That night, September 26, there was a wind shift to the westward and heavy rains. By morning the heavy southerly swell continued and rain squalls pelted them until noon. Under the constant strain of sailing, a crucial part of the rigging broke, the main lift strop, and had to be replaced. Through the wind and rain and rough seas, Rollo Beck noted daily in his field notes any of the seabirds flying about the schooner. He waited with anticipation for his chance to collect some of them when the wind dropped and a skiff could be put out.

Making steady progress north, the *Academy* was rapidly approaching the doldrums, where the winds are notoriously fickle or absent. Without an engine on the schooner, the doldrums were a geographic latitude of concern. For now, on September 28, the day opened with a strong breeze from the southwest and heavy rain squalls. The heavy southerly swell continued, and there were occasional cross swells from the northwest. All this caused further damage to the running rigging that controlled the sails. At 5:00 a.m., the halyards for raising and lowering the forestaysail, a small sail near the bow, and the block and tackle for the gaff-topsail, way up at the top of the main mast, were "carried away," the sailing term for anything in the rigging that parts or breaks. The early-morning repair took the men an hour and a half, and they were able to raise the sails again at 6:30 a.m. and get back under way. A strong current was driving them eastward, as revealed by their noon sextant measurement. On deck, the men continued to make inroads on skinning the tortoises.

The next day, September 29, opened with strong southerly winds and heavy rain squalls. The schooner pitched heavily and shipped some green water over the bow. The phrase "shipping green water" is as descriptive as it is portentous. It is synonymous with heavy-weather sailing. Only when a vessel plunges deeply into the oncoming swells does it penetrate the sea surface to the extent that massive amounts green water pour over the bow and side rails. "Green water" distinguishes the spray and foam that commonly drive over a vessel under way from the really solid load of seawater on the deck caused by dramatic ship motions and rough seas. The pounding the vessel takes under such conditions puts tremendous strain on the hull and rigging, and the force of the green water can damage anything on deck. If a hatch is open or swept away, green water will pour into the vessel and could result in eventual sinking. The schooner *Academy* was in a fight for her life.

A pod of grampus or Risso's dolphins passed close by the schooner; Williams harpooned one, but it tore loose. Many flying fish took flight across the bow, as they often do in the open ocean. Heavy squalls continued all day, and the crew lowered the flying jib up in the bow and the gaff-topsail at the top of the main mast. Conditions were so rough that they stopped skinning tortoises. The entire next day, September 30, saw strong winds and heavy seas with squalls from the south-southwest. The schooner was clipping along at nine knots, and several sea turtles were sighted. The sea remained quite choppy, and the men were unable to work on the tortoises. The next day the wind subsided enough for them to continue work on the tortoises, and they took out a big fold in the mainsail, a reef, lifting the sail up to its full extent and increasing its surface area. In heavy winds, the mainsail is reefed to reduce its surface area, but in moderate winds they wanted the extra driving power of the full sail. The mate, Nelson, used his skills to put together, or rove, new halyards for raising and lowering some of the sails, to replace those that had been recently carried away. The reality was setting in that the aging schooner was showing her years. Built in 1875 and now thirty-one years old and having gone through hard service with the US Coast and Geodetic Survey and the US Navy, there was now an additional problem. The men came to realize that their long cruise from San Francisco to the Galápagos, and the long calms they had gone through in the archipelago for a year, had resulted in wear and tear on the running rigging. They now came to expect damage to the running rigging each time they hit heavy weather. Their safe return to San Francisco was not a sure thing.

The next day, October 2, saw more of the same, with strong winds and a heavy sea from the south-southwest. The crew dropped the flying jib and then the foresail, and then reefed in the mainsail, all to reduce the total surface area of sail in the increasing wind. Predictably, more of the running rigging gave way in the middle of the night; this time it was the main lift strop that serves to raise and lower the mainsail. Nelson got right on it and replaced it in the morning. Heavy winds and squalls continued the next day, October 3, so much so that the crew put a double reef in the mainsail. The sea appeared to be increasing and the wind was not quite so steady, which was a new and dangerous problem. Early that morning, the schooner listed heavily to leeward, spilling a jug of molasses and a crate of onions over the cabin floor, together with some of Stewart's plant specimens and all the cooking and eating gear on the galley table, breaking numerous dishes. The listing and crashing meant that no tortoises were skinned that day. All hands turned out to clean up the mess in the cabin. In addition, they pumped out considerable bilge water in the evening. The vessel was essentially slowly sinking, and the men were doing their best to keep her afloat.

Finally, the wind and sea went down the next day, October 4, and the morning opened with light winds and calms. This would also have a calming effect on the men's spirits after the chaos of the day before. They "shook out the reefs" in the foresail and mainsail, making them as large as possible. But in the early afternoon, the outer edge of the mainsail, called the leech, carried away and the sail was lowered for repairs. They decided to drop all the sails and let the schooner go with just the foresail, attached to the foremast, the mast nearest to the bow in a two-masted schooner. The day progressed and became a fine afternoon, although rather sultry. It was calm enough that Beck and Gifford went out in the skiff to shoot birds for two hours, getting quite a few specimens, including a wedge-tailed shearwater, ten Cook's petrels, two other species of petrel, a wedge-rumped storm petrel, a red-footed booby, and two additional species of boobies. Beck had perfected the technique of throwing out bits of food on the ocean surface to attract birds that he would then shoot. He got very good at this working for Leverett Mills Loomis and the Academy off the coast of Monterey and Pacific Grove in the 1890s. King continued working on deck on the tortoises, including a small one from Tagus Cove that had died the day before. Slevin used his sailing experience to help Nelson repair the mainsail.

During the light winds and calms on the morning of October 5, Beck went out alone in the skiff until noon shooting seabirds. When not taking turns at the helm, King and Slevin steadily put up all the tortoises. With his recent collecting activity, Beck was busy all day October 6 skinning birds (figure 14.1) and all hands were otherwise skinning tortoises. That day opened with light winds that fell to dead calm by 6:00 p.m., allowing them to lower the foresail for more repairs, then setting it again for sailing at 10:00 p.m. The weather was warm enough that some of the men slept on deck. October 7 was a decent day of calm sailing, the schooner proceeding steadily north. Then the wind dropped on the morning of October 8 to light airs and calms, so much so that the schooner could not be steered. This lack of "steerage-way" meant the schooner was adrift, but there was still considerable swell from the northwest causing the vessel to move about in all directions: up and down, left and right, and from bow to stern. These are the conditions that can easily produce seasickness. Beck took advantage of the light wind to head out again in the skiff to shoot birds, while the others skinned tortoises. The next day, October 9, saw no change in the weather. Beck was out in the skiff looking for birds, and the other men were taking advantage of the opportunity to get the decks cleared of tortoises, with all hands keeping busy with the skinning knives. Since the light wind did not allow northward progress, the schooner had drifted southward by about five miles in the past twenty-four hours.

On the morning of October 10, there was enough wind for the schooner to have some steerageway, and the weather was clear, but the considerable swell from the northwest continued. Beck set out in the skiff to shoot more birds, and the other crewmembers skinned tortoises. Nelson rigged a line called a preventer on the forestay, as the chain plates holding down the forestay had pulled out. Williams tried unsuccessfully to catch the couple of bonitos that swam around the bow. The next day, October 11, Williams continued fishing off the flying-jib boom up in the bow when a young booby lit on the spar within a couple of feet of him and was promptly grabbed by the neck and turned over to the ornithologists. The day otherwise saw light, unsteady winds and fine weather, with the occasional rain squall. The

FIGURE 14.1 **Rollo Beck Stuffing Birds.** Rollo Beck below deck on an earlier collecting expedition to illustrate how rapidly he worked and how exacting was his bird stuffing technique. *Archives, California Academy of Sciences.*

men busied themselves the next day, October 12, with skinning tortoises as fast as possible to take advantage of the fine weather, completing six that day. Some of the men took advantage of the spells of dead calm to go over the side for a swim, finding the water warm.

A tortoise milestone was reached on the morning of October 13, when the men skinned the last of the seventy tortoises on deck. They also kept ten tortoises alive to bring back to San Francisco. The light airs and calms continued, allowing Beck another excursion in the skiff for seabird shooting. Each skinned tortoise shell was placed in the pickle tubs to desiccate the remaining adhering flesh. These tubs contained a dehydrating solution of anhydrous aluminum sulfate similar to the material in a styptic pencil, which was previously a standard part of a shaving kit before the advent of safety razors and electric razors. After soaking in the pickle tubs for a couple of days, the tortoises were moved below deck and strapped down to prevent shifting as the vessel rocked and rolled. After the last tortoise was done in the pickle tubs in a couple of days, they could be knocked down and stored below in order to have a clear deck.

As a further example of the structure of the expedition, the cultural milieu of the time, and Loomis's instructions, Sunday, October 14, was a day of doing nothing except attending to the running the vessel. Sunday was a day off throughout the expedition. The schooner made steady progress of four knots with a fair breeze and moderate sea. A tone of irksome displeasure emerged among some of the party because they were unable to catch any fresh fish and were forced instead to stick to their monotonous diet of canned salmon and beans.

October 15 opened with a fresh breeze and moderate swell, with the schooner heading northwest at five to six knots. Navigational readings at noon each day with the sextant revealed that a strong westerly current was running, for which they had to compensate. King and Hunter knocked down the pickle tubs, affording a clear deck in anticipation of the heavy northwest winds they might expect off the California coast at that time of year.

With the tortoise work behind them, they spent the next few days running the schooner at a steady pace and shooting birds. This continued from October 16 to 20, with light to fresh breezes and a moderate swell. Some of the men returned to sleeping on deck at night and allowing time for maintenance, such as greasing down the masts, during the day. A new pest that they had brought with them suddenly emerged. Mosquitoes began hatching out of the tank of water they got at Chatham Island, becoming somewhat troublesome. On October 19, Williams finally caught a bonito, and on that day a landmark event occurred in San Francisco, with the arrival of the explorer Roald Amundsen in his seventy-foot former herring-fishing sloop *Gjøa*, the first vessel to sail through the Northwest Passage. By October 20, the men knew that Clarion Island was not far off but were unable to spot it until the next day, when it was sighted from the crosstrees of the mainmast about twenty-five miles away. Beck decided not to try to make a landfall at Clarion. The object in sighting it was to get a check on the chronometer and verify latitude and longitude readings. From here the crew set a confident course for San Francisco.

THE FINAL PUSH FOR HOME

Unfortunately, at Clarion Island on October 22, the men were not able to get a very good check on their chronometer, owing to hazy weather and uncertain latitude. Clarion was lost to their view at 2:00 p.m. as they continued north. The next day, October 23, a curious event happened. One of the tortoises they were keeping alive to bring back to San Francisco fell overboard, and they had to sail back to pick it up. At first, Slevin and Williams put off in the ship's boat (not the skiff Beck had been using to shoot birds) to rescue the tortoise, but the boat leaked so badly they had to bail vigorously and make for the schooner. They took the boat on deck for caulking and sailed the schooner after the tortoise, picking it up with a boat hook. A strong current continued to affect their progress, moving them sixteen miles to the west and southwest during the last twenty-four hours. Over the next four days, from October 24 to 27, with light winds and seas and fair progress of two to four knots, the men set about overhauling the ship's boat with caulk and paint, then scrubbing paintwork and getting ready to do some painting around the decks. Williams finally caught some bonitos in the morning of October 25, much to the delight of Hunter, who had grown to hate the sight of a can of salmon, to say nothing of its contents. On the twenty-seventh, the crew thought they sighted a sail to westward, but could not make it out well enough to be certain. With the arrival of Sunday the twenty-eighth, the men did no work outside of standing watches, with the schooner still making a good pace of four to five knots. On October 29, they had a fine sailing breeze, and painted and chipped rust. By the next day, October 30, they realized they had passed a geographic milestone, now being north of the Tropic of Cancer, which runs through the southern tip of the Baja California peninsula of Mexico. The weather was noticeably cooler and the wind stronger, and there was a growing suspicion among the men that they might be running into some bad weather ahead. At 6:00 p.m. they passed a three-masted, full-rigged ship bound southward. By the end of October on the thirty-first, they finished painting and turned to oiling down the decks. The wind died down that night, but the next day, November 1, cloudy weather and a heavy swell from the north-northwest continued. More running rigging damage also occurred during the night when the topsail sheet carried away, necessitating making the sail fast for the night. The crew repaired the damage in the morning and had all sails set just before noon. With the schooner spruced up with paint and oil and with all sails set before the wind, she would have made a proud sight on her triumphant voyage home from the Galápagos. The men felt the cold, and everyone started breaking out their long underwear. At 8:00 p.m. they sighted a sail to the windward.

November 2 opened with dead calm. Two black-footed albatrosses were sighted, and one was hooked by King while trolling a fishing line astern. The weather was officially cool, and sweaters and overcoats were in vogue on deck. The men were still able to marvel at natural wonders, including "a wonderful moon rise" as well as the surrounding ocean, which was "full of phosphorescent jelly-fish."

November 3 also opened with dead calm, so they took advantage of the weather to lower and repair the mainsail, some of the halyards for raising and lowering sails, and the staysail toward the bow. Animal life activity around them continued, with a large shark swimming by, and when Gifford shot an albatross from the schooner, they lowered the ship's boat to pick it up. After all the repairs, the ship's boat floated perfectly and was watertight and did not require the services of a bailer. The men spent a calm moonlit night waiting for the wind to rise. The dead calm continued on and off for two more days, November 4 and 5. On the fourth they picked up some wind from the north-northeast and sighted a sail to windward, but lost it when a heavy rainstorm passed over the schooner. They took in the flying jib way up in the bow to be repaired by Nelson and Slevin, and Williams "hooked a fine, large dolphin from the flying jib boom."

By November 6, a fresh breeze opened the day but became lighter toward noon, so sailing progress was only moderate. The next two days, November 7 and 8, also had light and unsteady winds, not at all what they wanted to drive them home. Tropicbirds and black-footed albatrosses were sighted on the seventh, and one of the latter was inadvertently hooked on a fishing line trailing astern. The flying-jib repairs were finally finished and the sail was "bent," or reattached, to the rigging by Nelson and Slevin. By 6:00 p.m. on the eighth, a large bark-rigged vessel passed across the schooner's bow, heading to the northeast. The breeze died down during the night and left both vessels becalmed in sight of each other.

Another milestone was reached on November 9, marking 500 days out from the June 28, 1905, departure from San Francisco. At daybreak, due west of Los Angeles and Santa Barbara and some 750 miles out to sea, the crew realized that the schooner had drifted overnight to within half a mile of the bark they had sighted the night before. At eight bells, or about 8:00 a.m., the men on the *Academy* showed their colors and sent a boat aboard the bark; in particular, Nelson wanted to get a check on its chronometer. The vessel proved to be the French bark *L'Hermite*, which was 125 days out from Dunkerque, France, bound for San Francisco with a load of cement. Nelson asked the captain to report on the schooner *Academy* when his crew got to San Francisco. *L'Hermite* arrived two weeks ahead of the *Academy*, and the captain reported having spoken with the expedition crew, giving their families and the museum director, Loomis, the first news of them in some months.

By sundown the following day, November 10, with light, unsteady breezes the entire day, the *L'Hermite* was still in sight to the west-northwest, and the men sighted another bark to the southeastward. The appearance of these sailing vessels off the California coast meant that the *Academy* crew had to keep a sharp lookout all night long, as they were now in the sailing ship track to and from San Francisco.

By November 23 the men were anxious for a shift of wind so they could head in toward the coast. They reached a maximum distance west of San Francisco of about 750 miles off-shore, at latitude 37° 55′ north and longitude 134° 37′ west. At 1:45 a.m. the schooner was put on the port tack toward the northeast, the last leg to San Francisco Bay. But again, the flapping edge, or leech, of the mainsail carried away and had to be lowered for repairs. The crew brought down the flying jib and staysail to make the schooner head up into the wind during the repairs. Toward evening the weather turned squally and they made fast the flying jib. On November 24, they took in the gaff-topsail and a cross swell began running, but not enough to make the vessel labor. November 25 saw danger and excitement on board when the foresail ripped out of the boltropes while the men were trying to reef it. The schooner was making heavy weather of it now and shipping green water over the bow. By sundown, they took two reefs in the mainsail to reduce its size considerably. On November 26 the wind and sea moderated, allowing Nelson and Slevin to repair the ripped foresail. The sky was overcast and it was "trying hard to rain."

November 27 opened with a moderate sea and northwest wind, so the men shook out the two reefs in the mainsail and set the flying jib. They were close to home and moving like horses back to the barn. A four-masted schooner passed to windward in the early afternoon. Repairs were finished on the foresail, which was set and raised with some difficulty owing to the fresh breeze. They had made considerable leeway, or sideways motion, over the past twenty-four hours on account of heavy seas and shortened sail. Then, at 3:00 a.m., they were hit by a "white squall," or what today we would call a "microburst." It carried away the martingale stay and the flying-jib boom up in the bow, and tore a hole in the mainsail. All hands turned out to clear away the wreckage. Even with the hole, the mainsail was holding out, and they watched for a chance to lower it. The wind died down toward nightfall, and the next day, November 28, there was only a light breeze to almost calm. The crew took advantage of the calm spell to lower the mainsail, and Nelson and Slevin make a hasty job of patching it. At 8:30 that morning they all got a thrill when the South Farallon Light was sighted by one of the men up the foremast in the fore crosstrees. The Farallon Islands were about eighteen miles away, and the light signaled the end of the expedition. However, all was far from safe, and events could still go badly for the men, the schooner, and the expedition as a whole. A sigh of relief would have been in order but premature at this point. By noon, South Farallon Island was one and a half miles away. But by nightfall the wind had dropped, and the men were becalmed within sight of their destination. Then things went from bad to worse.

On the final day of the expedition, sixty-five days out from Culpeper Island in the Galápagos and the 519th day of the voyage, the day opened with a light breeze from the north. It was Thursday, November 29, 1906, Thanksgiving Day. All on board were much elated over the prospects of getting in to port and having Thanksgiving dinner with their families. The men broke out razors, and some shaved for the first time in seventeen months, others for only the second time since leaving home. They threw their old clothes overboard in an attempt to prevent any delay by the San Francisco Board of Health. They washed in the remaining fresh water in the holding tanks, a practice unheard of when water was precious during the expedition. At noon, they set course for the lightship anchored well offshore, marking the entrance to the Golden Gate, the rocky headlands leading to San Francisco Bay. Making it up to the lightship by 3:00 p.m., they picked up their pilot, Captain George Kortz, who was well known to all mariners entering the Port of San Francisco. From there, they shaped course for the bridgeless Golden Gate. Passage into San Francisco Bay should have been a routine matter. But the wind dropped again and they lay becalmed. Because this was Thanksgiving, they could not get a tugboat, so instead of the turkey they all *longed* for, they had their daily dish of Alaska salmon. With the help of light breezes and a strong incoming tide, they drifted through the Golden Gate and narrowly missed colliding with the pilot boat *Lady Mine*, which was in the same predicament as the schooner *Academy*. By 9:00 p.m. they reached Lime Point, where today the northern anchorage of the Golden Gate Bridge is located in Marin County. Here, they almost met with disaster. The *Academy*, gravid with seventy-eight thousand biological specimens fresh from the Galápagos Islands and slated to be the saving grace of the destroyed California Academy of Sciences, nearly ran aground. It came so close to the rocks that the crew had to put out the ship's boat and tried to pull the schooner's head around, reminiscent of the frantic rowing in the 2003 film *Master and Commander: The Far Side of the World*. When this emergency maneuver failed, they hailed a passing crab fisherman in his fishing smack, *Louisa*. For the sum of $10, he agreed to tow them across the bay to the quarantine station, where they arrived at 10:15 p.m. The expedition was over. The men were home. All the specimens were saved (figure 14.2).

FIGURE 14.2 **Storage in San Francisco.** Temporary storage for the Galápagos collections in a walled-off section of the old Academy Museum building on Market Street complex, December 1906. Photo probably by Rollo Beck. *Archives, California Academy of Sciences.*

NOTES

Page 224: *Epigram*: Coan (1882).

Page 225: *No chance whatever*: Rothschild and Hartert (1899, 89, 107–108).

Page 225: *Three finches on one booby at a time*: Rothschild and Hartert (1899, 110).

Page 225: *The mother of all Galápagos expeditions*: Bowman (1966); Lowenstein (2003).

Page 225: *Propensity toward blood eating*: Gifford (1913, 242).

Page 226: *Rain squalls toward evening*: LSA, 150.

Page 228: *Tagus Cove tortoise*: RHB, 162.

Page 231: *Large dolphin*: LSA, 157. It is unclear if the dolphin was a marine mammal or if Slevin used the term "dolphin" to mean "dolphinfish," which is commonly known today as mahimahi and would have been a welcome addition to their monotonous diet.

Page 232: *White squall*: https://en.wikipedia.org/wiki/White_squall, accessed July 25, 2016.

Page 233: *Toward Lime Point*: For the Lime Point Lighthouse, see http://www.lighthousefriends. com/light.asp?ID=146, accessed June 30, 2010.

Page 233: *All the specimens were saved*: On the significance of the expedition, Academy Director Loomis wrote enthusiastically and insightfully from the Academy's temporary post-earthquake headquarters at 1812 Gough Street in San Franciso to naturalist and field collector Edward William Nelson (1855-1934) with the Bureau of Biological Survey in Washington, DC, who spent 1905-1906 on a collecting expedition in Baja California and who came aboard the schooner *Academy* in Ensenada, Mexico on July 5, 1905 (see LSA, page 11):

May 7, 1906 Loomis to Nelson: "The work accomplished by the Galapagos Expedition has exceeded our most sanguine expectations – among the treasures are a series of Darwin's Rail and tortoises from islands where they were long supposed to be extinct. The Galapagos collections will form the foundation of our new museum of the greater Academy."

June 23, 1906 Loomis to Nelson: "We have sufficient room in our temporary quarters to work up the Galapagos material. So next year we will be doing scientific work again. We are making special efforts to get the books together that will be required in working up the Galapagos collections."

July 16, 1906 Loomis to Nelson: "I wrote several letters [to museums in the U.S. and England] asking for their publications [to rebuild the Academy library]. We are trying to get together the literature necessary to work up the material from the Galapagos Expedition. I am now converted to the principle of working up material as soon as material comes to hand! The Galapagos Expedition has given particular attention to the Albatrosses and Petrels, and so far as numbers are concerned the Academy will be in the lead in this group. We mean to use our vessel off Monterey, and get together a greater local collection than we had before. In short, the work will be done on a larger scale. Out of all this disaster will come far greater results than would have been reached if we had meandered along the old path. Barriers have been burned away. A fault in many of our water birds in the old collection was that they were over stuffed. I finally got Beck and the others to the point where they made specimens natural size or a little less. The Galapagos specimens are from trained men and will be the best in the way of workmanship."

October 5, 1906 Loomis to Nelson: "Next month our Galapagos expedition returns. Now we are getting ready to properly install the collections. $500 of museum jars have just arrived for the reptiles, and three barrels of alcohol have been ordered. Twenty zinc cases have been ordered for the birds. These will be of the latest type with all the improvements. One hundred insect boxes of the type used for beetles have also been ordered. Dr. Van Denburgh, Mr. Anderson, and Dr. Van Dyke are full of zeal for their departments and eager to begin work on the Galapagos specimens. If it had not been for the long the long delay in outfitting the vessel [schooner Academy], which caused us to give up the trip to mainland [Ecuador] and the making of a shipment [back to San Francisco], many of these specimens would have been burned."

Source: http://sova.si.edu/details/Record%20Unit%207364#refd1e1019, accessed December 21, 2016; scans of letters courtesy Ellen Alers, Smithsonian Institution Archives.

15

Vindicating Darwin

IN 1909, A mere three years after the Academy's expedition to the Galápagos returned to California, scientific institutions around the world celebrated Darwin. The year marked both the fiftieth anniversary of the publication of Darwin's *On the Origin of Species* and the centennial of the great scientist's birth. The American Museum of Natural History in New York City, the museum the California Academy of Sciences sought to emulate, held an event celebrating the publication of *On the Origin of Species* on Darwin's birthday that year, February 12, 1909. In June, Cambridge University brought together 235 scientists from 167 different countries to fete evolution. The University of Chicago hosted an event on November 24, 1909, to honor the publication date of *Origin*, and a London publisher, Longmans, Green, released a landmark retrospective book on the *Origin*. But the index to that book contains no entry for "Darwin's finches," or even for "Galápagos finches," because those birds had not yet become the icons of evolution as we know them today.

And in fact, as of 1909, Darwinian evolution was still a problem in need of a solution. On November 30, 1906, the day after the schooner *Academy* successfully returned to the San Francisco docks, the Dutch botanist and geneticist Hugo de Vries received the Darwin Medal from the Royal Society at a gala dinner at its Whitehall rooms in London. De Vries was recognized that night for his experimental investigations into heredity and variation, particularly with the evening primrose. De Vries's ideas about genetics and heredity threatened Darwinian ideas, or at least served as a working alternative. Natural selection was *not* established fact at this point, but it soon would be.

Enter the specimens collected by the cadre of eight young field collectors Loomis sent to the Galápagos in 1905. Their remarkably varied collections, totaling seventy-eight thousand specimens in all, effectively provided an experimental sample base for working through the various theories of evolution. Natural history collections contain a wealth of information about the past and current biodiversity of our planet, but they also provide crucial information about species distribution and genetic variation. Museum specimens concretize the otherwise abstract concept of species. Spread out on a table, specimens can be carefully measured, photographed, and illustrated. Researchers can sample DNA from fragments of remaining tissue adhering to the inside of a bird skin, tortoise shell, or pressed plant leaf.

In the more than a century since the *Academy* returned to port, modern laboratory researchers have amplified the expedition's field efforts with any number of techniques that Loomis could not have imagined, including peering into the genetic material of species. From an analytical perspective, museum specimens yield more, and vitally different, information than living specimens observed from a distance or only momentarily captured for a blood sample before being released. One of their secrets has turned out to be that Darwin was right all along.

Far away from the Galápagos Islands, while the 1905–1906 expedition members pushed their way through unyielding vegetation, sweating profusely in the tropical sun and baking as they clambered over long-hardened lava flows, their bosses and backers awaited their return in cool, foggy San Francisco. Loomis had insured the schooner *Academy* and, most important, her valuable scientific contents against loss and made plans for how the specimens would advance the expanding post-earthquake professional agenda of the California Academy of Sciences. The expedition is remembered, through a quirk—or perhaps more accurately a fault—of history, as the solid foundation of the Academy's now-famous collections and research programs. The great earthquake and three days of firestorms in April 1906 are easily blamed for everything. The museum had been wiped out, and the expedition held the key to rebuilding it. Without those disastrous events, the expedition would have returned to an intact San Francisco and a flourishing California Academy of Sciences on Market Street. But these tragic events, particularly the "great conflagration" that swept across the city, ignoring economic and social boundaries as an equal opportunity destroyer, changed the significance of the Galápagos expedition. With the museum destroyed, the expedition constituted a veritable Noah's Arc of scientific specimens. The expedition far exceeded the ambitious goals of the expedition's organizational Noah, the museum director, Loomis. The sailor-scientists on the schooner *Academy* brought back not just an orderly two of every species from Darwin's remote archipelago, but dozens or hundreds or thousands of every species possible. These precious specimens jump-started the museum back into existence after its obliteration in the great conflagration. The timing of the earthquake could not have been worse, and the timing of the expedition could not have been better.

The finches might be the Galápagos's most famous example of evolution, but they were hardly the only example in the archipelago. The finches, the mockingbirds, the tortoises, even the sunflower-turned-tree called *Scalesia* are all reasonable candidates for best example of Galápagos evolution. But what about snails? It turns out that land snails in the Galápagos have produced more new species, in the sense of Darwin's origin of species, than all the finches and tortoises and *Scalesia* and iguanas and lava lizards combined. And within the remarkably varied world of land snails, the genus *Bulimulus* holds the record for variety, at eighty species (figure 15.1). There are "only" forty or so species of Darwin's finches and a similar number of giant tortoises.

Why have so few outside the field of malacology, the branch of zoology that is concerned with snails, heard of these marvels of evolution? It could be because they lack a backbone, or it could be their small size. And it doesn't help that they are difficult to locate deep in the

FIGURE 15.1 **Snail Evolution.** The story of Galápagos land snail evolution is arguably more dramatic than that of the tortoises, finches, or plants. Here a number of snails in a goat skull are found by Washington Henry Ochsner, the 1905–1906 expedition snail expert, who assembled the largest museum collection of these snails. *Archives, California Academy of Sciences.*

forest and that they are not nearly as photogenic as a finch or a tortoise or an iguana. Slow-moving snails have undergone rapid evolution, and evolution is all about diversity: How many different kinds are there? Those different kinds, however many there are, were produced by permanent genetic modification over time. That's evolution.

The 1905–1906 Galápagos expedition succeeded on many levels in specimen collection and returned with a nearly perfect safety record for the men involved. No deaths occurred, and no serious injuries or illnesses befell the eleven men living aboard the schooner *Academy* under primitive conditions and far from medical attention for the seventeen-month expedition. The expedition was conducted competently and safely and, above all, effectively. Other than a bruised ego or two and a minor scuffle, all the men returned to San Francisco healthy and ready to move on with their personal and professional lives. Only the navigator, Parker, did not return with the schooner, having left the vessel over accusations of incompetence following multiple groundings.

There was, however, more drama after the expedition than during it. Two episodes illustrate the values held by members of the expedition and by the museum curators of the California Academy of Sciences. The reason for the expedition was to boost the Academy's holdings in all the taxonomic areas to which the eight young men were assigned. Specimens would save a ruined museum. Specimens were worth fighting over. And two major fights ensued.

These two long-term personal and professional squabbles marred the otherwise resounding scientific success of the expedition. They started to brew not long after the *Academy* docked safely in San Francisco in late November 1906, in the shadow of the city ruined by the April 1906 earthquake and fires. In both cases, vital information about specimens was withheld by the collectors Stewart and Ochsner, who believed they had the moral and ethical high ground to publish on the material in the scientific literature. This illustrates how museum specimens, when removed from the context of locality information, become essentially useless. Information about where a specimen was found makes that specimen scientifically valuable, and the data must be as specific as possible, down to latitude and longitude GPS coordinates for today's collectors. Without detailed locality information, specimens are reduced to curios. This fact affords great leverage for anyone who feels wronged.

SHELLS OF CONTENTION

Two multi-year squabbles clouded the expedition's overall success. One involved the expedition's botanist, Alban Stewart. He battled the museum director, Leverett Mills Loomis, on the question of payment for services rendered while Stewart was working up and describing the expedition's plant specimens. Stewart eventually returned the plant material, but renumbered the specimens with a numbering system to which only he held the key, so that the specimens were not connected to specific island localities, and thus of little scientific value. Stewart only agreed to supply the decoding key after he was paid for his work. The other battle involved the expedition's geologist and conchologist, Washington Henry Ochsner, and the venerable Smithsonian scientist William Healey Dall. As a Stanford University undergraduate in geology, Ochsner took a leave of absence to participate in the expedition. He was charged with collecting modern seashells (conchology), rock specimens (geology), and fossils (paleontology, mostly marine invertebrates). In a complicated set of circumstances, he became embroiled for years with scientists at Stanford University and at the Smithsonian Institution over publication rights of the expedition's marine fossil specimens, despite his clear contractual rights with the Academy. Both battles lasted for years, and both damaged the quality and quantity of scientific information derived from the otherwise successful expedition.

Without a physical museum to adequately house the new specimens, the post-expedition research of the Academy scientists took nonstandard trajectories. Without the supporting infrastructure of a museum where expedition specimens could be unpacked, cataloged into an existing collection, and carefully studied, the museum management was obliged to allow some groups of specimens to be worked up at other locations, and out of their direct control. Normally, a specialist in a particular group of organisms would, over many months, spread out the new material on large tables in the museum for careful study. Day after day he or she would return to the task, making steady progress as insights accumulated.

Each biological specimen was processed on board the schooner in the Galápagos to preserve its scientific value. Anchored in the bays and inlets of the remote archipelago, the eight young men worked nights and weekends to "put up" everything they had collected. Daubs of glue and narrow strips of paper secured dried and brittle plant material to sheets of herbarium paper. Insects were labeled with nearly microscopic writing on a tiny piece of cardstock, then pierced squarely through the thorax with black pins topped by a small glass bead and laid out in orderly rows in wooden boxes the size of cigar boxes. Stuffed birds, big and small, whose innards were cut out in the field and replaced by a wad of excelsior or cotton, were laid out in wooden trays like so many identical cigarettes in a pack. Each bird was given a small, rectangular tag attached by string to the ankle with the species name, collector's name, and island and date where collected, not unlike a corpse in a morgue. Without this information, the specimens would have been virtually worthless.

Back in San Francisco, the vast material was prepared for further study, then compared with published works and other museum specimens. New specimens, and especially any new species, would be carefully drawn or photographed and written up in the coming months and years for publication in the scientific literature. This was, and still is, the activity of basic taxonomic description: documenting what species are present in the Galápagos Islands and how they compare with species found elsewhere in the world. Describing new species is the museum gold standard for post-expedition taxonomic research and writing. Ochsner and Stewart began these very tasks with the specimens they had personally collected and about which they possessed specialized knowledge.

The 1905–1906 Galápagos plant specimens were sent to the Gray Herbarium at Harvard University for analysis, and Alban Stewart worked on them there as part of his PhD dissertation. The mollusk material, including modern and fossil marine clams and snails and terrestrial snails, was sent first to Stanford University and later to the Smithsonian Institution. Perhaps it was no coincidence that the plants and mollusks were the only two groups removed from San Francisco for study, and Stewart and Ochsner were the only two who had become embroiled in rancorous episodes with their scientific colleagues during the expedition. Stewart, the tall and lanky botanist, and Ochsner, the short and stocky geologist, were also the two who had a documented fistfight in the Galápagos. Their separate battles of words in the years following the expedition would echo the muffled blows they landed on each other while in the islands.

Prior to leaving San Francisco in 1905, all the young men signed a contract with the California Academy of Sciences that prescribed and limited their activities before, during, and after the expedition. Ochsner's contract was dated June 15, 1905, just short of two weeks before the schooner set sail for the Galápagos. Each contract was in the form of a letter from Loomis, who signed the contract as "Agent and Attorney in Fact" on behalf of the Academy. Ochsner's contract stated:

This is to confirm previous understanding as to the character of your engagement on the expedition of the California Academy of Sciences to the Galápagos Islands in the capacity of conchologist, to which position you have been appointed by the undersigned.

1. You are to receive as salary twenty-five dollars a month and board, beginning June 15, 1905, and continuing till the return of the expedition to San Francisco, which is expected to be about December 1, 1906.
2. You are to have the opportunity of working up the conchological and geological material and having the results published by the Academy with ample illustrations. If it shall be necessary to remove specimens from the Academy museum, permission must be first obtained from the Council of the Academy.

3. It is expressly understood that all specimens, notes, photographs, or other material collected on this expedition are the exclusive property of the Academy, and it is further understood that you are not to engage in any kind of commercial transaction on your own account.

4. While on the vessel, you are to perform the duties of a sailor, as set forth in the shipping articles.

5. At no time during the expedition or after your return to San Francisco, are you to convey to newspapers directly or indirectly any information concerning the expedition.

If you consent to the above conditions, kindly indicate by signing your name below.

After agreeing to the terms and putting pen to paper below a line stating "I approve and consent to the above conditions," Ochsner was a full-fledged member of the expedition. For Ochsner, contract items 2 and 3 above, about publishing and about ownership of the specimens, would become bones of contention in the years to come.

HARD, DANGEROUS WORK

Ochsner's geological knowledge was important to the Academy. One of the first things Ochsner did after the schooner docked in San Francisco on Thanksgiving, November 29, 1906, was to visit the burned-out hulk of the museum on Market Street in response to a request by the Academy to inspect a pile of mineral specimens in the ruins. Ochsner and the Academy were still on very good terms. There was a sense of pulling together, which accompanied the general spirit of rebuilding that existed throughout San Francisco, to help the Academy overcome its disastrous loss. Ochsner found that "the greater part of it was a collection of common minerals of no economic value." After careful inspection, Ochsner placed the few specimens of value in a burlap sack to be picked up later and stored in temporary quarters with what few other natural history specimens were salvaged.

This bitter scientific and personal controversy was not resolved during the lifetime of either Ochsner or Dall. The men died within fifteen days of each other in 1927. Ironically for both men, they were united posthumously as coauthors of the fossil and land snail papers that kept them apart ideologically for some ten years. The enduring legacy of the 1905–1906 Galápagos expedition resides in its lasting contribution to furthering the Darwinian debate throughout the twentieth century and to the salvation of the Academy after the 1906 earthquake. But the delay in publication by the Ochsner–Oldroyd–Dall controversy severely impeded the malacological success of the expedition.

DARWIN VINDICATED BY BEETLES, SNAILS, AND FINCHES

A group of beetles that Darwin first collected in the Galápagos in 1835 and that were further collected by F. X. Williams on the 1905–1906 expedition are Darwin's darkling beetles, the

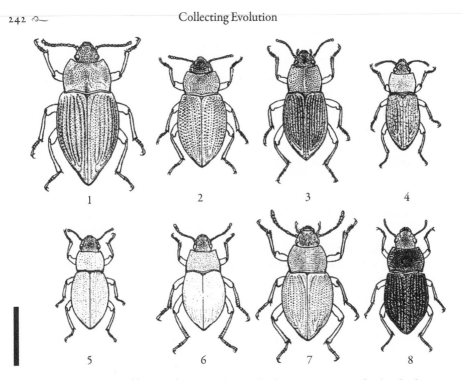

FIGURE 15.2 **Darwin's Darkling Beetles.** Not only was Charles Darwin interested in beetles from an early age, but a group of beetles from the Galápagos Island bear his name and exemplify the principles of speciation on remote oceanic islands. *Zoological Journal of the Linnean Society.*

best insect example of evolutionary radiation in the archipelago (figure 15.2). These little beetles are all members of the genus *Stomion*, and they, along with Darwin's finches and the native land snails, exemplify the diversification that occurs on remote oceanic islands. Van Dyke wrote in 1953 that beetle diversity aided by isolation, and thus reduced gene flow, "no matter how produced [by subsidence or by dispersal and divergence]," is an excellent example of Darwinian evolution.

Everyone who visits the Galápagos gets to see Darwin's finches. They flock around both main airports and are often the first birds, or even the first native organisms, one sees on arrival. These little Galápagos birds are a textbook example of evolution by natural selection and are the subject of the 1995 Pulitzer Prize–winning book, *The Beak of the Finch*, by Jonathan Weiner. Today, every student of biology from middle school through college learns about Darwin's finches. And yet, in 1905, when the schooner *Academy* set sail from San Francisco, hardly anyone called them "Darwin's finches." The collectors on the *Academy* saw these birds not as Darwin's finches, but simply as Galápagos finches.

In 1935, the English surgeon and ornithologist Percy R. Lowe (1870–1948) coined the phrase "Darwin's finches" to commemorate the hundredth anniversary of Darwin's visit to Galápagos on HMS *Beagle* in 1835. It took the work of the ornithologist David Lack to change how the finches were perceived. But even Lack didn't get it right the first time, or at least he changed his mind.

One of the great things about science is that you get to change your mind and no one holds it against you, because that's how science works. So long as there is a reasonable explanation for your change in interpretation—most commonly new data or new theories—it's fine. This is exactly what happened with David Lack. The story Galápagos visitors hear today from their naturalist guides or watch in numerous Galápagos natural history documentaries is different from Lack's original conclusion. The differences in interpretations are subtle, like the differences in the beaks of the species of Darwin's finches. Understanding what changed between Lack's first publication on the topic, in 1945, and his revised interpretation, in 1947, requires a detour into the history of evolutionary theory.

David Lack (1910–1973) was a British schoolmaster and evolutionary biologist who later became the director of the Edward Gray Institute of Field Ornithology at the University of Oxford. He undertook a four-month expedition to the Galápagos Islands at the urging of Julian Huxley, from late 1938 to April 1939. According to Lack's biographer, Ted Anderson, the expedition was not a complete success. It was, however, "the fulcrum on which the major change in David's life hinged." It resulted in Lack's having sufficient Galápagos experience to immediately expand his studies starting in April 1939 at the California Academy of Sciences, located since 1916 in Golden Gate Park after the debacle of April 1906.

The first version of Lack's research findings, submitted for publication in 1940, appeared in 1945 as one of the Occasional Papers of the California Academy of Sciences. This version differed greatly from his final, well-known version. In his first review, Lack attributed *no adaptive significance* to the different beak sizes and shapes, other than providing a visual cue for each species to distinguish members of its own group. The beaks were, in this scenario, a visual clue to identifying potential mating partners and species recognition signals. One can liken this to Freud's 1917 concept of the "narcissism of small differences," where minor variations in beak size and shape could have important implications. But Lack eventually abandoned that explanation; we no longer hear about the beaks as visual isolating mechanisms. Today, the beaks are all about food.

By the time Lack revised and rewrote the entire beak scenario for the well-known book version of his research, published by Cambridge University Press in 1947, he attributed the obvious beak differences, and thus species differences, to a differential *ability to utilize food*. He now interpreted the beaks as adaptations to specific food niches. I call this revised (and now-familiar) scenario "the Olympics and the Arms Race," as it involves simple measures of strength for opening seeds of variable toughness, grabbing insects, or extracting flower nectar. Lack wrote, "When two related bird species meet in the same region, they tend to compete, and both can persist there only if they are isolated ecologically either by habitat or food." Having beaks of different sizes and strengths would, in effect, isolate the birds by food source. In this version, birds are competing for limited food resources. The gold medal in this Olympics goes to the birds that get enough food to avoid death by starvation. In this competition, larger beaks are stronger beaks, enabling the birds to crack harder seeds for food. *Competition* between individuals and species for scarce food resources drives changes in beak shapes, creating differences in populations over time, ultimately creating different beak sizes, strengths, and shapes. Old species became new Darwinian species over time because they competed side by side in the evolutionary Olympics: big beaks for tough seeds, smaller beaks for smaller seeds, and delicate beaks for insects and flower nectar. The beaks are the keys to understanding these birds. Beaks are related to food, and food is related to reproduction. Lack summarized his views years later, stating, "In most birds, clutch size has evolved through natural selection to correspond with the largest number of young for which the parents can on the average find enough food." The ability of adult birds to adequately nourish their young made the food-gathering organ, the beak, of supreme importance. This might sound like a minor difference, species recognition based on beaks versus getting enough food with the beak, but to ecologists and evolutionary biologists it is huge. Recognizing the importance of ecological isolation in the process of making new species resulted in the recognition of Lack as the father of evolutionary ecology. Here the most Darwinian issue of all, the origin of species, boiled down to finding enough food.

Lack never could have conducted this kind of research without access to the California Academy of Science's vast collection of Galápagos finches. Indeed, Lack dedicated his 1947 book, *Darwin's Finches*, to the staff of the California Academy of Sciences.

The thirteen species of Darwin's finches are rightly well known throughout the world. But the fourteenth species is much less known and understood. Partly this brought back into the flock as a bird from Cocos Island in Costa Rica, and remains there as an important part, where the other thirteen species are thoroughly studied and form one of the iconic/emblematic groups of organisms in the archipelago. A scientific expedition similar to Darwin's found this rare bird, and its identity was obscured for many years by a simple mix-up of geographic locations. Eventually the fourteenth finch was brought back into the flock and remains there as an important part of an insightful evolutionary story of birds on Pacific islands.

In 1927, nearly six thousand stuffed Galápagos bird skins sat in neat rows in drawers at the California Academy of Sciences documented by rectangular paper tags attached to their ankles with thin string. The labels bore meticulously written locality information and the scientific name of each specimen in permanent India ink, along with the name of the field collector and the date of collection. Some of the labels had over time become saturated with oil from the birds' bodies and would need to be replaced, but never discarded. The new label would be attached with a new piece of string, like additional paper dog tags around the neck of a soldier or around the big toe of a corpse. The stuffed birds filled drawer after drawer in bug-proof museum cabinets where they resided for the twenty-one years after the 1905–1906 expedition returned to San Francisco. The specimens were professionally prepared during the expedition and accessioned into the Academy collections back in San Francisco. Today, they look as though they were prepared just yesterday.

Some of the more than eight thousand bird skins collected during the 1905–1906 expedition had already been studied, and the results published, by one of the expedition members, Edward Gifford, in 1913. And the oceangoing seabirds from the collection were documented by the museum director, Leverett Mills Loomis, in 1918, as they were his specialty. But it was not until 1927 that the Academy's curator of birds, Harry Swarth, began to examine the bulk of the Galápagos collection.

Swarth started an intellectual journey that has not ended. When he picked up the first bird in that impressively large Galápagos collection, he set the world of ornithology on a trajectory that would eventually link Charles Darwin with David Lack and ultimately with Princeton University evolutionary biologists Peter and Rosemary Grant. Swarth started with some changes in scientific names at the genus and species level, and continued with a change in thinking. Though few people knew about his work, Harry Swarth started the legend of Darwin's finches.

RECENT USE OF CALIFORNIA ACADEMY OF SCIENCES SPECIMENS

Biological materials collected during this expedition still have an impact on evolutionary studies of Galápagos organisms, and contribute to conservation efforts by researchers at the Charles Darwin Research Station and others throughout the world. For example, researchers take small tissue samples from the giant tortoises, no bigger than a fingernail clipping, to extract DNA so that they can determine how the tortoises from different islands are related. The results of this laboratory work would be very satisfying to Darwin. It was made possible only by the arduous fieldwork of the eight young men, an archipelago of personalities toiling in what would become known as Darwin's islands.

Another success made possible by the large number of specimens collected during the 1905–1906 expedition was the taxonomic revision of the giant tortoises by John Van Denburgh. With a large number of specimens at hand and reference to other material, Van Denburgh was able to create a "family portrait" of the giant tortoises for the first time (figure 15.3). His monograph on the group was the first treatment of all the extant and extinct species, and stands as a landmark achievement. Without a complete suite of specimens in one museum, Van Denburgh would have needed to travel to numerous museums before completing his taxonomic work. Having the Galápagos study series under one roof afforded him the luxury of knowing from which island most specimens were collected, and he photographed and figured them at standardized angles.

The morphology of shell and skin and bone that Van Denburgh worked with is today supplemented by genetic analysis. An active lab today, at Yale University, is run by Jeffrey Powell and Adalgisa Caccone, with collaborators around the world. They have accomplished a most Darwinian feat: determining that two species occur on Indefatigable (Santa Cruz) Island. The dominant species on the island was named by Walter Rothschild in 1903 from a specimen collected by Rollo Beck in 1902, *Chelonoidis porteri*, and was thought to be the only type of tortoise living there. Tissue taken from museum specimens at the University of Wisconsin, the Charles Darwin Research Station, and Rothschild's museum at Tring helped unravel the story. A new species was added to the Galápagos tortoise family tree, *Chelonoidis donfaustoi*, named for a former Galápagos National Park ranger, Fausto Llerena Sánchez, known to his friends and colleagues as Don Fausto.

As Henry Nicholls has pointed out, with climate change and a loss of biodiversity, natural history research and museums are of increasing relevance. Among the Galápagos organisms that one would expect not to change rapidly over time would be those organisms, especially birds, that freely migrate to and from the islands. Seabirds especially fall into this category, as by their very nature they take long flights that can result in "gene flow" from the archipelago to the continent or to other island groups. Gene flow introduces new genes into the population, potentially "diluting" an evolutionary change through natural selection. Imagine two buckets into which food coloring is being dripped at a steady rate. If the food coloring is red, the water in both buckets becomes redder and redder over time. Think of the red color

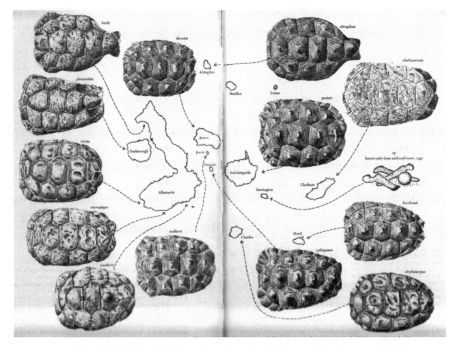

FIGURE 15.3 **Tortoises and Islands.** The notion that each island, or each volcano on the largest island of Albemarle (Isabela), should possess its own species of tortoise extends back to the expectations of Georg Baur and John Van Denburgh based on Baur's subsidence hypothesis of a former landmass that created the islands as they are today. Baur's geological expectations drove this view of biology, and Van Denburgh was an adherent of Baur's ideas. Source: Sherwin Carlquist, *Island Life* (1965), The Natural History Press, used with permission.

as genetic or evolutionary change. But if a garden hose with fresh water running from it is placed into one of the buckets , and the fresh water represents gene flow, the degree of change will be less in that bucket. In this analogy, gene flow is a "bad thing" that prevents genetic change (red color) from accumulating. Genetic change leads to evolutionary change. In 2010, Frank Hailer and six colleagues found that even the highly mobile magnificent frigatebirds, which could easily have gene flow with distant populations, are genetically distinguished in Galápagos from populations of the same species in places like Panama and into the Caribbean. A distinctive feature of Galápagos is the large proportion of endemic species, especially in terrestrial organisms. The magnificent frigatebirds have changed significantly over time, a Darwinian evolutionary change. Museum specimens for Hailer's study included eighteen specimens from the California Academy of Sciences, at least seven from the 1905–1906 expedition.

The specimens and observations from the 1905–1906 expedition have served as an invaluable baseline for science and conservation in Galápagos and have been used extensively. Examples include tissue samples collected from three tortoises on Abingdon (Pinta) Island being used to determine evolutionary relationships of taxa in the *Geochelone nigra* group. This information has guided the search for a potential mate for Lonesome George (see chapter 4). A single specimen of the now-extinct population of the Galápagos hawk (*Buteo galapagoensis*) collected on Chatham (San Cristóbal) Island was used to describe phylogenic relationships of these raptors and their colonization history. Specimens have also been used to increase our understanding of animal responses to environmental variables over time. On the basis of a mechanistic understanding of individual performances of marine iguanas (*Amblyrhynchus cristatus*), global warming was predicted to cause an evolutionary increase in maximum body size. The authors compared modern specimens with those collected during 1905 to validate their prediction.

Published observations from the expedition have also been used to reconstruct extinction events. The Grant's research group at Princeton showed that the warbler finch (*Certhidea fusca*) was uncommon in 1905–1906, and searches since 1979 have failed to detect their presence. Published observations and samples have been used to reconstruct the history of introduction of invasive species under current management, including feral donkeys (*Equus asinus*) and feral pigs (*Sus scrofa*). Back in the mid-1970s, Jim Patton's research group at the University of California, Berkeley, used allozyme analysis of a number of specimens, including those collected by the California Academy of Sciences, to show that the black rat (*Rattus rattus*) had been introduced to the archipelago in at least three separate events between 1600 and 1945. Patty Parker's research group at the University of Missouri-St. Louis used samples from the Academy and the Zoologisches Staatssammlung Muenchen to reconstruct the introduction of avipoxvirus into Galápagos. Using histopathology and viral genotyping, they concluded that this virus was introduced late in the 1890s and was dispersed among islands by a variety of mechanisms, including regular human movements among colonized islands.

A graduate student, Ore Carmi, working with Jack Dumbacher at the California Academy of Sciences has examined the genetics of the vermillion flycatcher, including the Galapágos populations and at least four specimens from the 1905–1906 expedition. Chatham Island populations of the vermillion flycatcher are genetically distinct and could tentatively be elevated to full species—they would thus represent the first bird species extinction in the Galápagos, if they are in fact extinct now.

Changes in the endemic land snail fauna, a more extensive diversification of species than Darwin's finches, have been documented by using specimens from the California Academy of Sciences collections. The specimens have died, but their legacy continues.

The plants of the Galápagos Islands also contain a detailed Darwinian story. One plant in particular, the genus *Scalesia*, embodies a story parallel to that of the tortoises, finches, land snails, and darkling beetles (figure 15.4). Darwin wrote that "Scalesia, a remarkable arborescent genus of the Compositae, is confined to the archipelago." And if found only there, it must have evolved there. The botanist on the 1905–1906 expedition, Alban Stewart, recognized seventeen species, which show a pattern of overall diversification following colonization and single-island endemics. Stewart also contributed to our ecological understanding of plant zonation in the islands, recognizing four zones: dry, transition, moist, and grassy.

The Galapagos Islands have proven intriguing for generations on so many scientific levels, especially those nearest to the topics investigated by Darwin: geology, botany, and zoology.

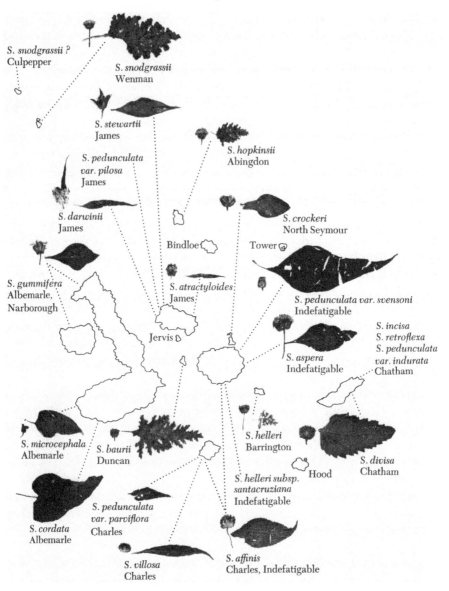

FIGURE 15.4 **Plant Evolution Vindicates Darwin.** The woody sunflower of the Galápagos has numerous species in the genus *Scalesia*. The plants show a range of growth habits and habitats, as well as variation in leaf size and in the nature of the flower heads. Most species are restricted to a single island, mirroring the evolution of other organisms in the archipelago. Source: Sherwin Carlquist, *Island Life* (1965), The Natural History Press, used with permission.

The archipelago also displays literary appeal through the writing of Herman Melville, who was himself influenced by Darwin. Perhaps no place in the world is so closely associated with one man and his ideas the way the Galápagos Islands are with Charles Darwin and evolution. That connection was made stronger, and continues to strengthen, through the specimens collected by the 1905–1906 scientific collecting expedition of the California Academy of Sciences.

NOTES

Page 236: *In June 1909*: Richmond (2006).

Page 236: *Landmark retrospective book*: Poulton (1909).

Page 236: *Icons of evolution*: Sulloway (1982a, b; 1983).

Page 236: *Natural history collections*: Wandeler, Hoeck, and Keller (2007).

Page 240: *Perhaps it was no coincidence*: RHB, January 22, 1906.

Page 240: *Prior to leaving San Francisco*: Contract letter, Washington Henry Ochsner archival material, Archives of the California Academy of Sciences.

Page 241: *No economic value*: Correspondence between Ochsner and CAS Executive Committee, December 1, 1906, Archives of the California Academy of Sciences.

Page 242: *Name "Darwin's finches" coined*: Lowe (1936). Lowe delivered this paper before the British Association meeting in 1935 in honor of the Darwin Centenary, and the printed version appeared in 1936.

Page 243: *Landmark work on Darwin's finches*: Lack (1953).

Page 243: *Fulcrum of life hinged*: Anderson (2013, 51).

Page 243: *Related bird species meet*: Lack (1947, 62).

Page 244: *Find enough food*: Anderson (2013, 64).

Page 243: *Fourteenth finch*: See: https://en.wikipedia.org/wiki/Cocos_finch, accessed January 6, 2016.

Page 244: *Studied by Gifford*: Gifford (1913).

Page 245: *Legend of Darwin's finches*: Sulloway (1982a).

Page 245: *Another success*: Jennings (1997).

Page 245: *Landmark achievement*: Pritchard (1996); Van Denburgh (1914).

Page 245: *Under one roof*: Pritchard (1996).

Page 245: *Increasing relevance*: Nicholls (2012; 2014, 133).

Page 246: *Mate for Lonesome George*: Caccone et al. (1999).

Page 247: *Now-extinct hawk population*: Bollmer et al. (2006).

Page 247: *Global warming*: Wikelski and Romero (2003).

Page 247: *Warbler finch*: Grant et al. (2005).

Page 247: *Invasive species*: (Carrion et al. 2007; Cruz et al. 2005).

Page 247: *Black rats*: Patton et al. (1975).

Page 247: *Introduction of avipoxvirus*: Parker et al. (2011).

Page 247: *Extinct vermillion flycatcher*: Carmi et al. (2016).

Page 247: *Land snail diversification*: Parent and Crespi (2006) and Parent et al. (2008).

Page 247: Scalesia *unique to islands*: Darwin (1845, 396).

Page 247: *Single-island endemics*: Stewart (1911, 1912b, 1915).

Page 247: *Plant zonation*: Stewart (1911, 206).

Page 248: *The archipelago also has literary appeal*: Blum (2008, 133–157).

Epilogue

Legacy

RECENT USE OF CALIFORNIA ACADEMY OF SCIENCES SPECIMENS

———————————————————————————

TODAY, WELL OVER a hundred years after the 1905–1906 Galápagos expedition, the scientific paradigms operating in the archipelago stress biological conservation, preservation, and restoration of rare species over collecting. An expedition like the one conducted by Rollo Beck and his young shipmates on the schooner *Academy* is unthinkable. Collecting is all but forbidden, as it should be, except by select scientists for very specific and limited reasons with the proper permits from the Galápagos National Park Service, the Ecuadorian government's dedicated and conscientious curator of this living outdoor museum of evolution. Circumstances, and scientists, have changed.

Beck and his men operated in a post-Darwinian, pre-conservation world that no longer exists. Their historical moment lacked a solid institutional framework for conservation, certainly not in the remote Galápagos archipelago of Ecuador. That would all come later. Likewise, while their unrestrained approach to field collecting seems foreign or even repulsive to us today, it was standard operating procedure for museums a century ago. Beck and his shipmates had a green light to collect as much as they possibly could and carry it back to San Francisco.

As in Rollo Beck's day, today the Galápagos Islands remain crucial to biologists, but now the emphasis is rightly focused on saving and studying. For the sake of science and collective knowledge, Beck and his collectors were willing to finish the work that eighteenth- and nineteenth-century whalers, pirates, and buccaneers had long ago started. But Beck's work actually resulted in new scientific knowledge, something that others throughout history could not necessarily claim. What Beck discovered was truly fantastic.

Collecting expeditions, such as this one, can have several objectives. One is to assemble in a museum an exceptional series of specimens of various species so that comparative, side-by-side studies can be performed at leisure after the expedition returns. These studies differ in objective depending on the zoological or botanical interests of the investigator. One very common type of museum study involves a researcher who lays out specimens in order to look at "variation within a species" and attempts to proscribe the limits of a species in terms of size, shape, or color. Here an investigator spreads out all the specimens available in order to get a sense, either through observation or measurement, of the natural variation present in a species. "Variation" is code for "evolution." Sometimes this involves visiting multiple museum collections over several years to get an inclusive sense of all the bumps and hooks and knobs that distinguish one species from another. In this type of study, the more specimens one has, the better. In this sense, specimens and their inclusive variation *are* evolution.

The desire to document variation within a species helps explain why just one or two specimens are inadequate. In an island group such as the Galápagos, where species were believed to be fast disappearing, there was no reason to show restraint in collecting. Either bring them back to be part of a documented series of specimens, or they will be lost to science forever. Extinction is forever, and at another level museum specimens are preserved forever. At the branching point in this dichotomy, researchers must choose one version of the infinite over the other.

Species that exhibit variation often require, after all, that a researcher examine tens or hundreds of specimens to get a handle on the range of variation. This gives some insight into why Rollo Beck and others collected such large suites of specimens in the Galápagos. This type of study focuses on differences between males and females of the same species, as well as the range of variation in size or shape or color from juvenile individuals up through adults. A common outcome of this type of "intra-specific," or within-species, analysis is a better understanding of the genus name and species name that should be applied. The taxonomy cannot be done comprehensively with just a few specimens at hand. Taxonomists often uncover redundant names, or synonyms, or previously undiscovered "new species" that are unnamed and totally new to science. These species often lurk in vast museum collections waiting to be discovered years later by a sharp-eyed taxonomist. The evolutionary divergence from ancestral populations occurred long ago and sits patiently waiting in museum drawers to be discovered. So the value and importance of a museum collection often derive from studies conducted long after the expedition returns. Loomis certainly knew this from his own research when he organized and dispatched this expedition. He based his personal research agenda on the new Academy collections until his death in 1928.

In another common type of specimen-based study, and some would argue the most interesting from an evolutionary perspective, one examines specimens to determine "who is related to whom." This type of study does not always support assumptions that organisms that look alike (in size, shape, color, etc.) are the most closely related. Two species can have distant ancestry and over time come to look alike through convergent evolution. An oversimplified example would be dolphins, ichthyosaurs, and sharks, with their similar streamlined bodies. The goal here is to figure out, despite outward appearance, which organisms are the most closely related to which other organisms and do not just superficially resemble each other. If done well and accurately, this brand of specimen study puts a distinctly Darwinian stamp, and confers a Darwinian history, on an otherwise lifeless museum collection.

Museum collections, no matter how complete, are not particularly good for addressing questions of ecology. The desire to know how organisms interact with each other and their environment, the textbook definition of "ecology," requires living animals, although paleontologists can investigate the paleoecology of long-dead species, at a different level of precision. Preferably these animals should be in the wild, not in captivity. Direct observations in field studies reveal who eats what and whom, and who lives where under what conditions. Thus, the ecologist working today in the Galápagos must be grateful that the 1905–1906 expedition left behind enough grasshoppers, tortoises, and land snails to conduct ecological studies.

New hope for using museum specimens preserved in formalin or formaldehyde has been announced. Formaldehyde, or embalming fluid, is excellent for preserving cellular structures and preventing decay of museum specimens, but for the double helix of DNA it serves as a sort of cellular paper shredder. Often in natural history fieldwork, specimens are preserved for a few days or weeks in formaldehyde and then transferred for long-term storage to ethyl alcohol. New work has uncovered a way to retrieve DNA in museum specimens and allow evolutionary studies.

More than a hundred years after the 1905–1906 Galápagos expedition, the specimens continue to yield their evolutionary secrets. The hard work of Rollo Beck and the other expedition members is still bearing fruit today.

Four of the eight men died during my lifetime: Joseph Richard Slevin in 1957, Edward Winslow Gifford in 1959, Francis Xavier Williams in 1967, and Joseph Slayton Hunter in 1972. Unfortunately, I did not become interested in the trajectories of their lives and their remarkable Galápagos expedition until 1982, and not in earnest until 1998, when I began researching this book. My years of living in Hawaii actually overlapped with those of the entomologist, Williams, who worked for many years for the Hawaiian Sugar Planters Association.

The expedition's ornithologist, Joe Hunter, died in 1972 in the San Francisco Bay Area when I was a junior in high school in Ewa Beach on the south shore of Oahu. In September 2000, I met Joe Hunter's eighty-five-year old son, Walter, and his wife at their home in Daly City, California, and quickly recovered my composure when I saw that Walter was the spitting image of his father, whom I only knew, though extensively, from photographs. At that moment, expedition history came alive.

Bob Bowman personally knew Slevin, Gifford, and Williams, and crossed paths briefly one day in 1950 with Rollo Howard Beck, the same year Beck died, in the Museum of Vertebrate Zoology on the University of California campus in Berkeley, where Bob was a beginning graduate student. During our many lengthy discussions, Bob was my living connection to history, making my research and writing so much more enjoyable and meaningful. And that connection goes all the way back to a book I read on evolution while growing up on the south shore of Oahu.

If Darwin had the best idea anyone has ever had, as asserted by the philosopher Daniel Dennett, then perhaps the Galápagos Islands are the best place in the world to see Darwin's idea in manifest form. How appropriate, is it not, that Darwin's monumental idea is celebrated daily in a monumentally beautiful archipelago? The combination of geology and organisms and ocean in the Galápagos is nearly perfect. Those factors combine to make the islands equal to the task of serving as the quintessential Darwinian location. Many other places in the world are rightly celebrated for their examples of biological evolution, the Hawaiian Islands, Madagascar, and Australia to name just three in a long list, but scientists have settled on one place to be the Greenwich of evolution. Direct observation by fieldwork is one way to vindicate Darwin. And another way is to conduct research on the many tens of thousands of specimens sitting in the dark in museum drawers, to be pulled out and studied by future generations of researchers. Evolution examines what happened in the past, how ancestor species diverged in the origin of descendant species. Museum specimens hold the promise of the future, when new techniques and new perspectives will make possible new discoveries that further vindicate the best idea ever.

NOTES

Page 250: *Their historical moment*: An impetus to practice conservation in the United States was the passage by Congress and President Theodore Roosevelt of the Antiquities Act in June 1906, in the last months of fieldwork of the 1905–1906 expedition. Conservation was not yet fully mainstream in 1905–1906. Yellowstone National Park was established in 1872, the Sierra Club was founded in 1892, and by 1905 there were three national parks in California.

Page 250: *Otherwise lifeless museum collection*: BioScience (1999) 49 (7): 511–512. http://bioscience.oxfordjournals.org/content/49/7/511.full. Accessed December 28, 2016.

Page 252: *New work*: Karmakar et al. (2015).

Page 252: *Best idea ever*: Dennett (1995).

CHRONOLOGY OF THE 1905–1906 GALÁPAGOS EXPEDITION

⌒ ───

1905	May 12	Schooner *Earnest* purchased from US Navy for $1,000 after being grounded and damaged on Yerba Buena Island, California. President Theodore Roosevelt approves sale and transfer of government property to civilian hands.
	June 25	Schooner *Earnest* is rechristened at Mission Street Pier 1 in San Francisco as the *Academy*; dignitaries and families of the eight young men aboard for celebration.
	June 26–27	Schooner rammed by the steamer *Argo* while docked at Mission Street bulkhead, owing to a mistake in the engine room signals. Cargo on the *Academy* is removed and the schooner listed over to repair damage below the waterline, delaying departure.
	June 28	Expedition departs from San Francisco through the bridgeless Golden Gate; most of the men become seasick immediately. The first obstacle for the schooner, and especially the eight young men, is the rough stretch of Pacific Ocean called the "potato patch" just outside the Golden Gate.
	July 3	Expedition reaches Ensenada, Mexico; clearance papers obtained for collecting on Mexican Islands.
	July 11	San Martin Island, Mexico.
	July 13	San Geronimo Island, Mexico.
	July 14–17	San Benito Island, Mexico.

July 18	Cerros Island, Mexico.
July 19	Natividad Island, Mexico.
July 26	San Benedicto Island, Mexico.
July 27–28	Socorro Island, Mexico.
August 10	Clipperton Island, Mexico.
September 3–13	Cocos Island, Costa Rica.
September 18–19	*Academy* crosses the equator, heading south; Schooner passes within a few miles of the city of Manta and Cape San Lorenzo, Ecuador, approx. 1°S, 81°W. The first of only two equatorial crossings of the trip occurs as the schooner passes close enough to the Ecuadorian mainland for the snow-capped Andes Mountains to be seen in the distance.
September 24	Expedition members arrive and drop anchor at Gardner Bay, Hood Island, Galápagos—88 days out of San Francisco; first Galápagos landfall.
October 3	Hunter loses shotgun while landing on Gardner Island near Charles.
October 4	Letter mailed by Gifford to his father in Alameda, California, from post office barrel at Post Office Bay, Charles Island, carried by British yacht *Deerhound*.
October 8	At Black Beach, expedition members nearly meet one of the men who participated in the killing of Manuel Cobos on Chatham Island in 1903.
October 12	At Black Beach, crewmembers receive first letters and news from home in 107 days; soldiers from Chatham dine aboard. Beck informed of necessity of visiting Chatham to collect clearance from governor.
October 15	Crewmembers meet General Plaza at El Progresso above Wreck Bay.
October 18	The men purchase anchor at Wreck Bay to replace one lost at Socorro Island; fluke broke off kedge anchor, keeping schooner off wharf.
October 25	First tortoise collected, Indefatigable Island.
October 30	Crewmembers meet Don Antonio Gil on Albemarle Island
November 5	Schooner strikes submerged reef on entering Academy Bay, Indefatigable Island, which crew names on this day in honor of the schooner.
November 9	Schooner touches bottom at low tide in Academy Bay.
November 23	While the men are on Daphne Island, small skiff drifts out to sea, capsizing in swells offshore. After crew rights and bails skiff, small sailboat is swamped by fancy "seamanship" of Parker in schooner.

November 27	Side trip in sailboat from Indefatigable to Daphne.
December 19	Keel of sailboat damaged on submerged reef off James Island.
December 25	Christmas Day, 181 days out of San Francisco; plumduff for dinner at James Bay, James Island.
December 26	Skiff swamped getting through surf at James Bay; Slevin loses field notebook.
December 31	New Year's Eve, anchored in James Bay.
1906 January 1	New Year ushered in with schooner's foghorn.
January 3	Aneroid barometer used to determine elevation of highest point on James Island, at 2,750 feet above sea level.
January 24	Men receive mail at Wreck Bay, Chatham Island.
February 9	Schooner anchored in Sappho Cove with the vessel *Josephine Cobos*.
February 11	Schooner touches bottom on reef outside Sappho Cove.
February 12	Camp in lava tube on Chatham Island; skeletons of 17 tortoises found in lava tube.
February 14	Schooner aground at low tide in Sappho Cove.
February 15	Schooner touches bottom departing from Sappho Cove; crew blames Parker.
February 17	Schooner nearly swept against Kicker Rock by strong current.
February 22	Schooner anchored at Wreck Bay along with the trading schooner *Manuel J. Cobos*; crew of *Academy* learns of revolution in Ecuador.
February 24	"Whiskey diplomacy" practiced by Slevin and Ochsner by concocting mixture of 25% whiskey and 75% specimen alcohol along with juice from stewed prunes; traded for bananas with crew of *Manuel J. Cobos*.
February 27	Flamingo for breakfast at Black Beach anchorage.
February 28	Expedition members spend night in "Pirate Caves" occupied by the Irishman Patrick Watkins in 1809.
March 4	Failure to reach Vilamil blamed on Parker. Schooner anchored in Turtle Cove with brigantine *Nellie*; men are told bottom of cove altered by recent earthquake.
March 7	Herr Brugermann, bookkeeper of Don Antonio Gil, aboard for dinner.
March 8	Hundreds of tortoise skeletons seen near water holes and on trail to Santo Tomás; Don Antonio Gil aboard for dinner.
March 12	In both skiffs, men pull down coast two miles and visit the Old Cobos Settlement; King becomes ill from drinking water out of a mud hole.
March 17	Small skiff smashed to pieces at Iguana Cove while two tortoises are loaded; men walk miles back to schooner, Beck barefooted.

April 3–4	Rollo Beck finds and skins only known tortoise from Narborough Island.
April 15	While men are camped at Banks Bay, sailboat drifts away and is saved by men in the skiff just before the sailboat hits the reef. Heroic action.
April 18	Great earthquake rocks San Francisco. Schooner anchored at Tagus Cove; all hands except Hunter, White, and Parker collecting on Narborough in sailboat, camping near Mangrove Bay.
April 18–21	Following the earthquake, fire ravages San Francisco, burning more than 80% of the city. California Academy of Sciences on Market Street destroyed in early afternoon of April 18; ruins later dynamited and carted away.
April 22–28	Schooner becalmed en route to Vilamil.
April 25	Men in skiff have narrow escape while rowing through heavy swells, just making the crest of one wave as it breaks.
April 30	Men receive first news of San Francisco and northern California earthquake, just 13 days after the event, from Guayaquil newspaper brought by Ecuadorian gunboat the *Cotopaxi*.
May 4–14	Schooner becalmed attempting passage from Vilamil to Wreck Bay; drifts 4° south of the equator.
May 14	Schooner anchored at Black Beach, Charles Island; men speak with captain of the brigantine *Nellie*, who spent 67 days unsuccessfully attempting passage to Guayaquil before returning to the Galápagos.
May 17	Tension with Parker increased over navigation, Parker accused Slevin of shifting helm on him while tacking, harsh words exchanged with several of the men, then Parker offered to go ashore and fight with knives; see August 9 and 20 and September 5.
May 18–23	Becalmed and drifting, schooner attempts passage to Chatham Island; drifts back dangerously close to Charles Island; kedge used to prevent grounding.
May 24–29	Dead calm conditions prevent departure from Black Beach Roads.
May 29–June 4	Schooner departs from Black Beach for Chatham Island but is forced by lack of wind to return there for seven days.
June 4	Crew sets sail for Hood Island; the *Tomasita* left anchorage, and the *Nellie* visible offshore sailing slowly and drifting.
June 4–23	Schooner becalmed and drifting for 19 days before reaching Hood Island; drifts as far as 200 miles south of the archipelago, unable to make headway.
June 28	One year out from San Francisco; schooner anchored at Hood Island.

July 3	While schooner anchored at Wreck Bay, expedition members receive additional news of San Francisco earthquake and fires from letters and newspapers; all family members and friends safe in San Francisco.
July 3–7	Ochsner attends to man at hacienda on Chatham Island who has accidental gunshot wound, with heavy ball lodged under his sternum.
July 13–19	Schooner becalmed at Academy Bay; unable to leave anchorage for 7 days.
July 20	Schooner under way, sailing north along Indefatigable coast; Slevin and Nelson take side trip in sailboat to Vilamil.
July 21	Slevin and Nelson attend baby christening at Vilamil, dancing and partying late into the night with Don Antonio Gil, Herr Brugermann, and residents of Vilamil.
August 7	At James Bay, fried flamingo for dinner, collected by Gifford.
August 9	Most serious grounding of schooner during trip: off coast of Albemarle Island; 3 feet of keel and kedge anchor lost; entire crew shaken, Parker blamed.
August 10	Entire crew signs ultimatum that Parker be relieved of duty. Beck suspends Parker as navigator; charts, chronometer, etc., removed from Parker's possession.
August 20	At Vilamil, Parker leaves schooner temporarily, crew protest his competence.
September 1	The brigantine *Goleta* and the sloop *Ballandra* at Vilamil; the former brought Don Antonio's father, forced to leave continent due to political troubles associated with Eloy Alfaro's coup d'état.
September 2	After keeping the Sabbath all day, all hands, except White and Slevin, attend evening farewell party with whiskey, wine, women, violins, and guitars until 2:00 a.m.; Hunter and Ochsner perform cakewalk.
September 5	Schooner goes aground off Albemarle Island (losing 3 feet of keel) under the helm of Navigator J. J. Parker. Crew members sign a complaint in the logbook; Parker is forced to resign, refuses all aid, removes effects from vessel, and leaves to find his own way back to California.
September 7	Final mail stop at Wreck Bay.
September 9–10	Governor of islands requests passage to Vilamil to interview father of Don Antonio Gil; schooner's chronometer found to be 39′44″slow on GMT at 4:00 p.m.

September 13	In foggy weather near midnight, schooner almost runs aground on Barrington Island; Slevin hears sound of surf in distance and saves vessel. Continuing north for Tower Island, schooner crosses equator for second and final time of expedition.
September 14	Tower Island reached, Culpepper Bay avoided due to swells and wind; schooner anchored on northeast side of island.
September 16	Schooner heading for Bindloe Island; men collect on island.
September 17	Crew sets sail for Abingdon Island. Four tortoises brought on board and "skinned as quickly as possible;" Lonesome George from this island.
September 23	Crew sets sail for distant Wenman Island, reaching it the following day.
September 24	Crew sets sail for Culpepper Island; 365 days in the Galápagos.
September 25	Brief landing on Culpepper before shaping course for San Francisco after "a year and a day" in the Galápagos.
November 29	Thanksgiving Day, schooner arrives at San Francisco after a day offshore in sight of the city, becalmed, 65 days out from Culpepper Island; narrowly misses colliding with pilot boat *Lady Mine*. Crab fisherman in fishing smack *Louisa* tows schooner to San Francisco quarantine station for $10; members of the expedition disembark at 10:15 p.m. Total duration of expedition: 519 days (17 months).
Birthdays	Beck, August 26, 1870 (35th and 36th birthdays)
	Stewart, January 14, 1875 (31st birthday)
	Hunter, August 9, 1879 (26th and 27th birthdays)
	Slevin, September 13, 1881 (25th and 26th birthdays)
	Ochsner, July 4, 1882 (23rd and 24th birthdays)
	Williams, August 6, 1882 (23rd and 24th birthdays)
	King, December 16, 1886 (19th birthday)
	Gifford, August 14, 1887 (18th and 19th birthdays)
	Nelson, April 7, 1877 (29th birthday)
	Parker, unknown
	White, unknown
1907 January	Parker returns to San Francisco, nearly two months after the schooner *Academy* and her crew, collects his back pay and is not heard from again.

BIBLIOGRAPHY

I agree this list could be helpful to the reader:

JSH: Joseph Slaton Huner field notes
LSA: Log of the Schooner "Academy," Slevin 1931
OOS: On the Origin of Species, Darwin 1839, 1845
RHB: Rollo Howard Beck field notes
RWE: Race with Extinction, Fritts and Fritts 1982
VOB: Voyage of the "Beagle," Darwin 1959

I. UNPUPLISHED PRIMARY SOURCE MATERIAL
Archives of the California Academy of Sciences, San Francisco, California

Beck field notes: Rollo Howard Beck, *Trip to Galapagos Islands*. Unpublished handwritten copy and transcribed copy of daily field notes from the 1905–1906 Galápagos expedition (RHB in notes).

Council correspondence, various years.

Executive Committee correspondence, various years.

Gifford bird notes: Edward Winslow Gifford, Unpublished journal documenting the 1905–1906 Galápagos expedition.

Gifford diary: Edward Winslow Gifford, Unpublished personal diary from the 1905–1906 Galápagos expedition. Courtesy of Gifford's daughter, Maureen Frederickson, Chico, CA.

Hunter field notes: Joseph Slayton Hunter, Unpublished typed field notes from the 1905–1906 Galápagos expedition (JSH in notes).

Ochsner geology: Washington Henry Ochsner, Unpublished field notes from the 1905–1906 Galápagos expedition.

Ochsner land snails: Washington Henry Ochsner, Unpublished field notes from the 1905–1906 Galápagos expedition.

Ochsner diary: Washington Henry Ochsner, Unpublished personal diary from the 1905–1906 Galápagos expedition. Courtesy of Vera Kirichenko, Oakland, CA.

Slevin log: Joseph Richard Slevin, *Log of the Schooner "Academy,"* 1931 (LSA in notes).

Slevin herpetological notes: Joseph Richard Slevin, *Race with Extinction: Herpetological Field Notes of J. R. Slevin's Journey to the Galapagos, 1905–1906*. Herpetological Monographs, no. 1, i–viii, 1–98 pp. Philadelphia: Fritts, T.H. & Fritts, P.R. 1982 (RWE in notes).

Williams 1: Francis Xavier Williams, Unpublished entomological manuscript.

Williams 2: Francis Xavier Williams, unpublished entomological manuscript.

Williams 3: Francis Xavier Williams, unpublished entomological manuscript.

Rollo Howard and Ida Menzies Beck papers.

Galápagos Islands collection.

Edward Winslow Gifford papers.

Edward Winslow Gifford photographs of Galápagos expedition.

David Lack papers.

Leverett Mills Loomis papers.

Washington Henry Ochsner papers.

Ornithology and Mammalogy Department field note collection, California Academy of Sciences Library.

Joseph Richard Slevin papers.

Alban Stewart papers.

John Van Denburgh papers.

Francis Xavier Williams papers.

Other Archives

Bancroft Library, University of California, Berkeley

Maryland Historical Society, Baltimore

Museum of Vertebrate Zoology, University of California, Berkeley

National Archives and Records Administration (NARA) branches located in

(1) Washington, DC (National Archives, downtown)

(2) College Park, Maryland (Archives II)

(3) San Carlos, CA

National Oceanic and Atmospheric Administration Central Library, Silver Spring, MD

Pacific Grove Museum of Natural History, Pacific Grove, CA

J. Porter Shaw Library, San Francisco Maritime National Historical Park, San Francisco, CA

Stanford University Archives

II. PUBLISHED SOURCES

Agassiz, A. 1892. "Reports on the Dredging Operations off the West Coast of Central America to the Galapagos, to the West Coast of Mexico, and in the Gulf of California, in Charge of Alexander Agassiz, Carried on by the U.S. Fish Commission Steamer 'Albatross,' Lieut. Commander Z. L. Tanner, U.S.N., Commanding. II. General Sketch of the Expedition the 'Albatross,' from February to May, 1891." *Bulletin of the Museum of Comparative Zoology* 23: 1–89.

Anderson, D. J. 1990a. "Evolution of Obligate Siblicide in Boobies. 1. A Test of the Insurance-Egg Hypothesis." *American Naturalist* 135(3): 334–350.

Anderson, D. J. 1990b. "Evolution of Obligate Siblicide in Boobies. 2: Food Limitation and Parent-Offspring Conflict." *Evolution* 44(8): 2069–2082.

Anderson, D. J. 1995. "The Role of Parents in Sibilicidal Brood Reduction of Two Booby Species." *Auk* 112(4): 860–869.

Anderson, T. R. 2013. *The Life of David Lack: Father of Evolutionary Ecology*. New York: Oxford University Press, 246 pp.

Anonymous, 1897. "Turtles from the Galapagos." *San Francisco Call* 83(5): 1.

Anonymous, 1912. "Two California Methodists Who Will Visit South America in Scientific Research." *California Christian Advocate*.

Barrow, M. V., Jr. 2000. "The Specimen Dealer: Entrepreneurial Natural History in America's Gilded Age." *Journal of the History of Biology* 33: 493–534.

Bateson, W. 1894. *Materials for the Study of Variation: Treated with Especial Regard to Discontinuity in the Origin of Species*. London: Macmillan, 630 pp.

Baur, G. 1891. "On the Origin of the Galapagos Islands." *American Naturalist* 25: 307–326.

Baur, G. 1892. "Professor Alexander Agassiz on the Origin of the Fauna and Flora of the Galapagos Islands." *Science* 19(477): 177.

Baur, G. 1895. "The Differentiation of Species on the Galápagos Islands and the Origin of the Group." Biological Lectures Delivered at the Marine Biological Laboratory of Woods Hole in the Summer of 1894, 67–78.

Beck, R. H. 1902. "In the Home of the Giant Tortoise." *Annual Report of the New York Zoological Society* 7: 160–174.

Beissinger, S. R. and Peery, M. Z. 2007. "Reconstructing the Historic Demography of an Endangered Seabird." *Ecology* 88(2): 296–305.

Bettelheim, M. 2006. "Left to Burn: The Recollection of a Scientific Collection." *Bay Nature*, July 1, 2006. http://baynature.org/article/left-to-burn/, accessed May 16, 2016.

Bishop, L. B. 1929. "In Memorium: Leverett Mills Loomis." *Auk* 46(1): 1–13.

Blum, H. 2008. *The View from the Masthead: Maritime Imagination and Antebellum American Sea Narrative*. Chapel Hill: University of North Carolina Press, 285 pp.

Bollmer, J. L., Kimball, R. T., and Whitman, N. K. 2006. "Phylogeography of the Galapagos Hawk (Buteo galapagoensis): A Recent Arrival to the Galapagos Islands." *Molecular Phylogenetics and Evolution* 39: 237–247.

Boss, K. J., Rosewater, J., and Ruhoff, F. A. 1968. "The Zoological Taxa of William Healey Dall." *United States National Museum Bulletin* 287: 1–427.

Bowler, P. J. 1984. *Evolution: The History of an Idea*. Berkeley: University of California Press, 412 pp.

Bowman, R. I. (ed.). 1966. *The Galápagos: Proceedings of the Symposia of the Galápagos International Scientific Project*. Berkeley: University of California Press, 318 pp.

Brinkley, D. 2009. *The Wilderness Warrior: Theodore Roosevelt and the Crusade for America*. New York: HarperCollins, 960 pp.

Burbank, L. 1913. *Luther Burbank's Spineless Cactus*. San Francisco, CA: The Luther Burbank, 40 pp.

Caccone, A., Gibbs, J. P., Ketmaier, V., Suatoni, E., and Powell, J. R. 1999. "Origin and Evolutionary Relationships of Giant Galapagos Tortoises." *Proceedings of the National Academy of Sciences* 96: 13223–13228.

Callahan, D. 2014. *A History of Birdwatching in 100 Objects*. London: Bloomsbury Publishing, 222 pp.

Carmi, O., Witt, C.C., Jaramillo, A., and Dumbacher, J. P. 2016. "Phylogeography of the Vermilion Flycatcher Species Complex: Multiple Speciation Events, Shifts in Migratory Behavior, and an Apparent Extinction of a Galápagos-endemic Bird Species." *Molecular Phylogenetics and Evolution* 102: 152–173.

Carrion, V., Donlan, C. J., Campbell, K., Lavoie, C., and Cruz, F. 2007. "Feral Donkey (Equus asinus) Eradications in the Galapagos." *Biodiversity and Conservation* 16: 437–445.

Coan, E. V. and Kellogg, M. G. 1990. "The Malacological Contributions of Ida Shepard Oldroyd and Tom Shaw Oldroyd." *Veliger* 33(2): 174–184.

Coan, T. 1882. *Life in Hawaii: An Autobiographic Sketch of Mission Life and Labors (1835–1881)*. New York: Anson D. F. Randolph & Co.

Cook L. M. 2003. "The Rise and Fall of the *Carbonaria* Form of the Peppered Moth." *Quarterly Review of Biology* 78: 399–417.

Coryn, S. G. P. 1905. "Why San Franciscans Need Not Fear Earthquakes." *San Francisco Chronicle*, January 22, 1905, 4.

Coues, E. 1894. *Key to North American Birds. Containing a Concise Account of Every Species of Living and Fossil Bird at Present Known from the Continent North of the Mexican and United States Boundary, Inclusive of Greenland and Lower California, with Which Are Incorporated General Ornithology: An Outline of the Structure and Classification of Birds; and Field Ornithology, a Manual of Collecting, Preparing, and Preserving Birds*. Boston: Estes and Lauriat, 895 pp.

Cruz, F., Donlan, C. J., Campbell, K., and Carrion, V. 2005. "Conservation Action in the Galapagos: Feral Pig (Sus scrofi) Eradication from Santiago Island." *Biological Conservation* 121: 473–478.

Cutright, P. R. and Brodhead, M. J. 1981. *Elliot Coues: Naturalist and Frontier Historian*. Chicago: University of Illinois Press, 513 pp.

Dall, W. H. 1900. "Additions to the Insular Land-Shell Faunas of the Pacific Coast, Especially of the Galapagos and Cocos Islands." *Proceedings of the Academy of Natural Sciences of Philadelphia*: 88–106, plate viii.

Dall, W. H. and Ochsner, W. H. 1928. "Tertiary and Pleistocene Mollusca from the Galapagos Islands." *Proceedings of the California Academy of Sciences*, 4th ser., 17(4): 89–138.

Dana, R. H. 1840. *Two Years Before the Mast*. New York: Modern Library edition 1936, 443 pp.

Darwin, C. 1839. *Journal of Researches into the Geology and Natural History of the Various Countries Visited by H. M. S. "Beagle."* London: Henry Colburn.

Darwin, C. 1859. *On the Origin of Species by Means of Natural Selection, or the Preservation of Favoured Races in the Struggle for Life*. London: Murray. (OOS in the notes and text.)

Darwin, C. R. 1845. *Journal of Researches into the Natural History and Geology of the Countries Visited During the Voyage of H.M.S. "Beagle" round the World, under the Command of Capt. Fitz Roy, R.N.* 2d ed. London: John Murray, 519 pp.

Darwin, C. R. 1958. *The Autobiography of Charles Darwin, 1809–1882—With Original Omissions Restored*. Edited with appendix and notes by his grand-daughter, Nora Barlow. London: Collins.

Darwin, C. R. 1959. *The Voyage of the "Beagle."* Introduction by H. Graham Cannon. Everyman's Library, New York: Dutton, 496 pp. Cited in the text as VOB.

Davies, A. R. 2012. *Saving San Francisco: Relief and Recovery after the 1906 Disaster.* Philadelphia: Temple University Press, 221 pp.

de Ford, M. A. 1941. "The Miser Who Brought the Stars to Earth: James Lick." In *They Were San Franciscans*, 53–76. Caldwell, ID: Claxton Printers, 321 pp.

Dennett, D. 1995. *Darwin's Dangerous Idea: Evolution and the Meanings of Life.* New York: Simon & Schuster, 592 pp.

Desrochers, A. 2010. "Morphological Response of Songbirds to 100 Years of Landscape Change in North America." *Ecology* 91(6): 1577–1582.

de Vries, H. 1905. *Species and Varieties, Their Origin by Mutation: Lectures Delivered at the University of California by Hugo De Vries.* Edited by D. T. MacDougal. Chicago: Open Court, 847 pp.

Dill, H. R. 1912. *Report of an Expedition to Laysan Island in 1911. Part I. Report on Conditions of the Hawaiian Bird Reservation with List of the Birds Found on Laysan.* US Department of Agriculture, Biological Survey, Bulletin no. 42, 30 pp.

Doyle, A. C. 1891. *A Scandal in Bohemia.* London: George Newnes, Ltd.

Dumbacher, J. P. and West, B. 2010. Collecting Galápagos and the Pacific: How Rollo Beck Shaped Our Understanding of Evolution. *Proceedings of the California Academy of Sciences*, 4th ser., 61, suppl. II, no. 13, 211–243.

Durham, W. H. 2012. "What Darwin Found Convincing in Galápagos." In *The Role of Science for Conservation*, edited by M. Wolff and M. Gardener, 3–15. Oxford: Routledge, 299 pp.

Ehrlich, P. R. and Raven, P. H. 1969. "Differentiation of Populations." *Science* 165: 1228–1232.

Evans, H. E. 1985. *The Pleasures of Entomology: Portraits of Insects and the People Who Study Them.* Washington, DC: Smithsonian Institution Press, 238 pp.

Finston T. L. and Peck, S. B. 1995. "Population Structure and Gene Flow in *Stomion*: A Species Swarm of Flightless Beetles in the Galápagos Islands." *Heredity* 75: 390–397.

Finston T. L. and Peck, S. B. 1997. "Genetic Differentiation and Speciation in *Stomion* (Coleoptera: Tenebrionidae): Flightless Beetles of the Galápagos Islands, Ecuador." *Biological Journal of the Linnean Society* 61: 183–200.

Finston, T. L. and Peck, S. B. 2004. "Speciation in Darwin's Darklings: Taxonomy and Evolution of *Stomion* Beetles in the Galápagos Islands, Ecuador (Insecta: Coleoptera: Tenebrionidae)." *Zoological Journal of the Linnean Society* 141: 135–152.

Finston, T. L., Peck S. B., and Perry R. B. 1997. "Population Density and Dispersal Ability in Darwin's Darklings: Flightless Beetles of the Galápagos Islands." *Pan-Pacific Entomologist* 73: 110–121.

Fradkin, P. L. 2005. *The Great Earthquake and Firestorms of 1906: How San Francisco Nearly Destroyed Itself.* Berkeley: University of California Press, 418 pp.

Fritts, T. H. & Fritts, P. R. 1982. *Race with Extinction: Herpetological Field Notes of J. R. Slevin's Journey to the Galapagos, 1905–1906.* Herpetological Monographs, no. 1, 98 pp. Philadelphia. Cited in text as RWE.

Fuller, E. 1987. *Extinct Birds.* New York: Facts on File Publications, 256 pp.

Gifford, E. W. 1913. "The Birds of the Galapagos Islands, with Observations on the Birds of Cocos and Clipperton Islands (Columbiformes to Pelecaniformes)." *Proceedings of the California Academy of Sciences*, 4th ser., vol. 2, 1–132.

Gould, S. J. 1981. *Ever since Darwin: Reflections in Natural History.* New York: W.W. Norton, 285 pp.

Grant, P. R. 1991. "Natural Selection in Darwin's Finches." *Scientific American* 256(4): 82–87.

Grant, P. R., Grant, B. R., Petren, K., and Keller, L. F. 2005. "Extinction Behind Our Backs: The Possible Fate of One of the Darwin's Finch Species on Isla Floreana, Galapagos." *Biological Conservation* 122(3): 499–503.

Grant, K. T. and Estes, G. B. 2009. *Darwin in Galápagos: Footsteps to a New World*. Princeton, NJ: Princeton University Press, 362 pp.

Hanna, G. D. 1932. "Barton Warren Evermann, 1853–1932." *Copeia* 4: 161–162.

Harris, M. 1982. *A Field Guide to the Birds of Galapagos*. London: Collins, 160 pp.

Hay, O. P. 1898. George Baur. *Science*, n.s., 8(135): 68–71.

Heinzel, H. and Hall, B. 2000. *Galápagos Diary*. Berkeley: University of California Press, 272 pp.

Heller, E. 1903. "Papers from the Hopkins Stanford Galapagos Expedition, 1898–1899, XIV, Reptiles." *Proceedings of the Washington Academy of Sciences* 5: 39–98.

Hickman, C. S. and Lipps, J. H. 1990. *History of Galapagos Geology: Oschner's [sic] Misplaced Galapagos Fossils*. Geological Society of America, Abstracts with Programs, A120.

Hickman, J. 1985. *The Enchanted Islands: The Galapagos Discovered*. Shropshire: Anthony Nelson, 169 pp.

Hittell, T. H. 1997. *The California Academy of Sciences, A Narrative History: 1853–1906*. Edited, revised, and enlarged by A. E. Leviton and M. L. Aldrich. San Francisco: California Academy of Sciences, 623 pp.

Hofkin, B. V. et al. 2004. "Ancient DNA Gives Green Light to Galápagos Land Iguana Repatriation." *Conservation Genetics* 4(1): 105–108.

Jackson, M. H. 1994. *Galapagos: A Natural History*, 2d ed. Calgary: University of Calgary Press, 336 pp.

James, M. J. 1999. "Rollo Beck: A Glimpse into the Life of an Endangered Species." *Proceedings of the American Association for the Advancement of Science, Pacific Division* 18(1): 60.

James, M. J. 2003. "Did Galápagos Experience an El Niño in 1878?" *Noticias de Galápagos* 62: 30–32.

James, M. J. 2010. "Collecting Evolution: The Vindication of Charles Darwin by the 1905–1906 Galápagos Expedition of the California Academy of Sciences." *Proceedings of the California Academy of Sciences*, 4th ser., vol. 61, suppl. II, no. 12, 197–210.

James, M. J. 2012. "The Boat, the Bay and the Museum: Significance of the 1905–1906 Galápagos Expedition of the California Academy of Sciences." In *The Role of Science for Conservation*, edited by M. Wolff and M. Gardener, 87–99. London: Routledge, 299 pp.

Jeffers, H. P. 2003. *Disaster by the Bay: The Great San Francisco Earthquake and Fire of 1906*. Guilford, Connecticut: The Lyons Press, 204 pp.

Jehl, J. R. 1971. "The Status of *Carpodacus mcgregori*." *Condor* 73: 375–376.

Jennings, M. R. 1997. "John Van Denburgh (1872–1924), Pioneer Herpetologist of the American West." In *Collection Building in Ichthyology and Herpetology*, edited by Theodore W. Pietsch and William D. Anderson, Jr., 323–350. Lawrence, Kansas: American Society of Ichthyologists and Herpetologists, Special Publication (3), 593 pp.

Jenyns, L. (1842). "Fish." In *The Zoology of the Voyage of H.M.S. "Beagle," under the Command of Captain Fitzroy, R.N., During the years 1832–1836*, edited by C. Darwin. London: Smith, Elder & Co. (in 4 parts): 1–32 (January 1840); 33–64 (June 1840); 65–96 (April 1841); 97–172 (1842).

Jenyns, L. (1862). *Memoir of the Rev. John Stevens Henslow, M.A., F.L.S., F.G.S., F.C.P.S. : Late Rector of Hitcham and Professor of Botany in the University of Cambridge* . London: John Van Voorst, 278 pp.

Jiménez-Uzcátegui, G. and Bettancourt, F. 2008. "Bird Mortality by Vehicles." In *Galápagos Report, 2007–2008*, Puerto Ayora, Galápagos, Ecuador, 103–106. See also http://www .galapagosdigital.com/2013/06/09/galapagos-roadkill-birds-threatened/, accessed July 30, 2014.

Johnston, A. K., Connor, R. D., Stephen, C. E., and Ceruzzi, P. E. 2015. *Time and Navigation: The Untold Story of Getting from Here to There*. Washington, DC: Smithsonian Books, 224 pp.

Jordan, D. S. 1909. *Some Experiments of Luther Burbank*, San Francisco: A. M. Robertson.

Jordan, D. S. and Kellogg, V. L. 1907. *Evolution and Animal Life: An Elementary Discussion of Facts, Processes, Laws and Theories Relating to the Life and Evolution of Animals*. New York: D. Appleton, 489 pp.

Karmakar, S. et al. 2015. "Organocatalytic Removal of Formaldehyde Adducts from RNA and DNA bases." *Nature Chemistry*. doi:10.1038/nchem.2307.

Kellogg, V. L. 1907. *Darwinism To-Day: A Discussion of Present-Day Scientific Criticism of the Darwinian Selection Theories, Together with a Brief Account of the Principal Other Proposed Auxiliary and Alternative Theories of Species-Forming*. New York: H. Holt and Company, 429 pp.

King, W. B. 1981. *Endangered Birds of the World: The ICBP Red Data Book*. Washington, DC: Smithsonian Institution Press, 800 pp.

Klett, M. and Lundgren, M. 2006. *After the Ruins 1906 and 2006: Rephotographing the San Francisco Earthquake and Fire*. Berkeley: University of California Press, (with the Fine Arts Museums of San Francisco), 134 pp.

Lack, D. L. 1945. *The Galapagos Finches (Geospizinae): A Study in Variation*. Occasional Papers of the California Academy of Sciences, no. 21, 152 pp.

Lack, D. 1947. *Darwin's Finches: An Essay on the General Biological Theory of Evolution*. Cambridge: Cambridge University Press. 208 pp.

Lack, D. 1953. "Darwin's Finches." *Scientific American* 188(4): 67–70.

Larson, E. J. 2001. *Evolution's Workshop: God and Science on the Galapagos Islands*. New York: Basic Books.

Latorre, O. 1990. *The Curse of the Giant Tortoise: Tragedies, Mysteries and Crimes in the Galapagos Islands*. Quito: Graficas Ortega, Quito, 237 pp.

Latorre, O. 1999. *El Hombre en las Islas Encantadas: La historia humana de Galápagos*. Quito, 446 pp.

Leviton, A. E., Aldrich, M. L., and Elsbernd, K. 2006. "The California Academy of Sciences, Grove Karl Gilbert, and Photographs of the 1906 Earthquake, Mostly from the Archives of the Academy." *Proceedings of the California Academy of Sciences* 4th ser., 57(1): 1–34.

Lick, R. 1967. *The Generous Miser*. Los Angeles: Ward Ritchie Press, 95 pp.

Lindberg, D. R. 1998. "William Healey Dall: A Neo-Lamarckian View of Molluscan Evolution." *Veliger* 41(3): 227–238.

London, J. 1906. "The Story of an Eyewitness." *Collier's Magazine*, May 5, 1906. http://london .sonoma.edu/writings/Journalism/sfearthquake.html, accessed June 2, 2015.

Lopez, Barry. 1998. *About This Life: Journeys on the Threshold of Memory.* New York: Alfred A. Knopf, 273 pp.

Lowe, P. R. 1936. "The Finches of the Galapagos in Relation to Darwin's Conception of Species." *Ibis* 78(2): 310–321.

Lowenstein, J. 2003. "Galápagos Adventure." *California Wild* 56(2): 48–50.

Manning, T. G. 1988. *U.S. Coast Survey vs. Naval Hydrographic Office: A 19th Century Rivalry in Science and Politics.* Tuscaloosa: University of Alabama Press, 202 pp.

Marble, C. C. 1900. "The Late Dr. Elliott Coues." *Birds and All Nature* 7(2): 3 pp.

Melville, H. 1851. *Moby Dick.* Oxford: Oxford University Press (Edition 2008).

Melville, H. 1856. *Billy Budd and the Piazza Tales.* New York: Dix & Edwards, 384 pp.

Mearns, B. and Mearns, R. 1998. *The Bird Collectors.* New York: Academic Press, 472 pp.

Mills, W. J. 2003. *Exploring Polar Frontiers: A Historical Encyclopedia,* vol. 1. Santa Barbara, CA: ABC-CLIO, 797 pp.

Morrell, B. 1832. *A Narrative of Four Voyages, to the South Sea, North and South Pacific Ocean, Ethiopic and Southern Atlantic Ocean, Indian and Antarctic Ocean, From the Year 1822 to 1831.* New York: J. & J. Harper, 492 pp.

Morus, I. R. 2016. "Invisible Technicians, Instrument-Makers and Artisans." In *A Companion to the History of Science,* edited by B. Lightman, Hoboken, NJ: John Wiley & Sons, 624 pp. doi: 10.1002/9781118620762.ch7.

Murphy, R. C. 1936. *Oceanic Birds of South America: A Study of Species of the Related Coasts and Seas, Including the American Quadrant of Antarctica, Based upon the Brewster–Sanford Collection in the American Museum of Natural History,* 2 vols. New York: Macmillan Co., and American Museum of Natural History, 1245 pp.

Murphy, R. C. 1947. *Logbook for Grace: Whaling Brig Daisy, 1912–1913.* New York: Time-Life Books, 371 pp.

Nicholls, H. 2006. *Lonesome George: The Life and Loves of a Conservation Icon.* London: Palgrave Macmillan, 256 pp.

Nicholls, H. 2012. A Natural Evolution. *Nature* 484: 36.

Nicholls, H. 2014. *The Galápagos: A Natural History.* New York: Basic Books, 195 pp.

Nichols, P. 2003. *Evolution's Captain: The Dark Fate of the Man Who Sailed Charles Darwin Around the World.* New York: Harper, 352 pp.

Parent, C. E., Caccone, A., and Petren, K. 2008. "Colonization and Diversification of Galapagos Terrestrial Fauna: A Phylogenetic and Biographical Synthesis." *Philosophical Transactions of the Royal Society,* B 363: 3347–3361.

Parent, C. E., and Crespi, B. J. 2006. "Sequential Colonization and Diversification of Galapagos Endemic Land Snail Genus Bulimulus (Gastropoda, Stylommatophora)." *Evolution* 60: 2311–2328.

Parker, P. G., Buckles, E. L., Farrington, H., Petren, K., Whiteman, N. K., Ricklefs, R. E., Bollmer, J. L., and Jiménez- Uzcátegui, G. 2011. "110 Years of Avipoxvirus in the Galapagos Islands." *PLoS ONE* 6:e 15989. doi: 10.137l/journal.pone.oo 15989.

Patton, J. L., Yang, S. Y., and Myers, P. 1975. "Genetic and Morphologic Divergence among Introduced Rat Populations (*Rattus rattus*) of the Galapagos Archipelago, Ecuador." *Systematic Zoology* 24: 269–310.

Peck, S. B. 2001. *Smaller Orders of Insects of the Galapagos Islands, Ecuador: Evolution, Ecology, and Diversity.* National Research Council of Canada Monographs, 281 pp.

Peck, S. B. 2006. *The Beetles of the Galapagos Islands, Ecuador: Evolution, Ecology, and Diversity (Insecta: Coleoptera)*. National Research Council of Canada Monographs, 315 pp.

Peck, S. B and Kukalová-Peck, J. 1990. "Origin and Biogeography of the Beetles (Coleoptera) of the Galápagos Archipelago, Ecuador." *Canadian Journal of Zoology* 68: 1617–1638.

Poulakakis, N., et al. 2015. "Description of a New Galapagos Giant Tortoise Species (Chelonoidis; Testudines: Testudinidae) from Cerro Fatal on Santa Cruz Island." *PLoS ONE.* doi:10.1371/journal.pone.0138779.

Poulton, E. B. 1909. *Charles Darwin and the Origin of Species: Addresses, Etc., in America and England in the Year of the Two Anniversaries*. London: Longmans, Green, 302 pp.

Pratt, H. D., Bruner, P. L., and Berrett, D. G. 1987. *A Field Guide to the Birds of Hawaii and the Tropical Pacific*. Princeton, NJ: Princeton University Press, 528 pp.

Pritchard, P. C. H. 1996. *The Galápagos Tortoises: Nomenclatural and Survival Status.* Chelonian Research Monographs, no. 1, 85 pp.

Quammen, D. 2009. "Darwin's First Clues." *National Geographic* 215(2): 34–53.

Quinn, W. H. and Neal, V. T. 1992. "The Historical Record of El Niño Events." In *Climate Since A.D. 1500*, edited by R. S. Bradley and P. D. Jones, 623–648. New York: Routledge, 706 pp.

Reynolds, R. P. and Marlow, R. W. 1983. "The Pinta Island tortoise (Geochelone elephantopus abingdoni): a case of limited alternatives." *Noticias de Galápagos* 37: 14–17.

Richmond, M. L. 2006. "The 1909 Darwin Celebration: Reexamining Evolution in the Light of Mendel, Mutation, and Meiosis." *Isis* 97: 447–484.

Ridley, M. 2009. "Modern Darwins." *National Geographic* 215(2): 56–73.

Ripley, S. D. 1977. *Rails of the World*. Boston: David R. Godine, 430 pp.

Roberts, J. H. 1988. *Darwinism and the Divine in America: Protestant Intellectuals and Organic Evolution, 1859–1900*. Madison: University of Wisconsin Press, 368 pp.

Ronson, J. 2016. The California Academy of Sciences Hero Versus the 1906 San Francisco Earthquake. April 18, 2016. https://www.inverse.com/article/14363-the-california-academy-of-sciences-hero-versus-the-1906-san-francisco-earthquake, accessed May 11, 2016.

Rothschild, W. and Hartert, E. 1899. "A Review of the Ornithology of the Galapagos Islands, The Diaries of Messrs. Harris and Drowne, and General Remarks about the Fauna of the Galapagos Islands." *Novitates Zoologicae* (Tring, England) 6: 85–205.

Schuchert, C. and LeVene, C. M. 1940. *O. C. Marsh: Pioneer in Paleontology*. New Haven, CT: Yale University Press.

Sclater, P. L. and Salvin, O. 1870. "Characters of New Species of Birds Collected by Dr. Habel in the Galapagos Islands." *Proceedings of the Zoological Society of London* (May 12): 322–327.

Sewell, J. S. 1907. "The Effects of the Earthquake and Fire on Buildings, Engineering Structures, and Structural Materials." *U.S. Geological Survey Bulletin* 324: 62–130.

Slevin, J. R. 1931. *Log of the Schooner "Academy": On a Voyage of Scientific Research to the Galapagos Islands, 1905–1906*. Occasional Papers of the California Academy of Sciences, no. 17, 162 pp.

Slevin, J. R. 1950. "Post Office in a Barrel." *Pacific Discovery* 3: 28–29.

Slevin, J. R. 1959. *The Galápagos Islands: A History of Their Exploration*. Occasional Papers of the California Academy of Sciences, no. 25, 150 pp.

Slocum, J. 1954. *Sailing Alone Around the World*. New York: Sheridan House, 294 pp.

Smith, A. E. 1900. A Great Ornithologist. *The Outlook* 64: 98.

Smocovitis, V. B. 1996. *Unifying Biology: The Evolutionary Synthesis and Evolutionary Biology*. Princeton, NJ: Princeton University Press, 230 pp.

Sobel, D. and Andrewes, W. J. H. 1998. *The Illustrated Longitude: The True Story of a Lone Genius Who Solved the Greatest Scientific Problem of His Time*. New York: Walker & Company, 216 pp.

Steinheimer, F. D. 2004. "Charles Darwin's Bird Collection and Ornithological Knowledge During the Voyage of H.M.S. *Beagle*, 1831–1836." *Journal of Ornithology* 145(4): 300–320.

Stewart, A. 1911. "Expedition of the California Academy of Sciences to the Galapagos Islands, 1905–1906. II. A Botanical Survey of the Galapagos Islands." *Proceedings of the California Academy of Sciences*, 4th ser., 1: 7–288.

Stewart, A. 1912a. "Expedition of the California Academy of Sciences to the Galapagos Islands, 1905–1906. V. Notes on the Botany of Cocos Island." *Proceedings of the California Academy of Sciences*, 4th ser., 1: 375–404.

Stewart, A. 1912b. "Expedition of the California Academy of Sciences to the Galapagos Islands, 1905–1906. VII. Notes on the Lichen of the Galapagos Islands." *Proceedings of the California Academy of Sciences*, 4th ser., 1: 431–446.

Stewart, A. 1915. "Some Observations Concerning the Botanical Conditions on the Galapagos Islands." *Transactions of the Wisconsin Academy of Science* 18: 272–340.

Sulloway, F. J. 1982a. "Darwin and His Finches: The Evolution of a Legend." *Journal of the History of Biology* 15(1): 1–53.

Sulloway, F. J. 1982b. "The *Beagle* Collections of Darwin's Finches (Geospizinae)." *Bulletin of the British Museum (Natural History) Zoology Series* 43(2).

Sulloway, F. J. 1982c. "Darwin's Conversion: The *Beagle* Voyage and Its Aftermath." *Journal of the History of Biology* 15(3): 325–396.

Sulloway, F. J. 1983. "The Legend of Darwin's Finches." *Nature* 303: 372.

Svanevik, M. and Burgett, S. 2002. "Alice's Trial by Quake and Fire." *San Francisco Examiner*, December 16, 2002.

Tebbich, S., Taborsky, M., Fessl, B., and Blomqvist, D. 2001. "Do Woodpecker Finches Acquire Tool-Use by Social Learning?" *Proceedings of the Royal Society B, Biological Sciences* 268(1482): 2189–2193.

Tebbich, S., Taborsky, M., Fessl, B., and Dvorak, M. 2002. "The Ecology of Tool Use in the Woodpecker Finch (*Cactospiza pallida*)." *Ecology Letters* 5: 656–664.

Thornton, I. 1971. *Darwin's Islands: A Natural History of the Galápagos*. New York: Natural History Press, 322 pp.

Tree, I. 1991. *The Ruling Passion of John Gould: A Biography of the British Audubon*. New York: Grove Weidenfeld, 248 pp.

Twain, M. 1897. *Following the Equator – A Journey Around the World*. Hartford, CT: American Publishing, 673 pp.

Van Denburgh, J. 1907. "Expedition of the California Academy of Sciences to the Galapagos Islands, 1905–1906. I. Preliminary Descriptions of Four New Races of Gigantic Land Tortoises from the Galapagos Islands." *Proceedings of the California Academy of Sciences*, 4th ser., 1: 1–6.

Van Denburgh, J. 1912a. "Expedition of the California Academy of Sciences to the Galapagos Islands, 1905–1906. IV. The Snakes of the Galapagos Islands." *Proceedings of the California Academy of Sciences*, 4th ser., 1: 323–374.

Van Denburgh, J. 1912b. "Expedition of the California Academy of Sciences to the Galapagos Islands, 1905–1906. VI. The Geckos of the Galapagos Islands." *Proceedings of the California Academy of Sciences*, 4th ser., 1: 405–430.

Van Denburgh, J. 1914. "The Gigantic Land Tortoises of the Galápagos Archipelago." *Proceedings of the California Academy of Sciences* 2, pt. 1, 203–374.

Van Denburgh, J. and Slevin, J. R. 1913. "Expedition of the California Academy of Sciences to the Galapagos Islands, 1905–1906. IX. The Galapagoan Lizards of the Genus Tropidurus; with Notes on the Iguanas of the Genera Conolophus and Amblyrhynchus." *Proceedings of the California Academy of Sciences*, 4th ser., vol. II, pt. I, 133–202.

Van Dyke, E. C. 1953. *The Coleoptera of the Galapagos Islands*. Occasional Papers of the California Academy of Sciences, no. 22, 181 pp.

van Wyhe, J. 2013. "'My Appointment Received the Sanction of the Admiralty': Why Charles Darwin Really Was the Naturalist on HMS *Beagle*." *Studies in History and Philosophy of Biological and Biomedical Science* 44: 316–326.

Victor, F. F. 1900. Dr. Elliott Coues. *The Quarterly of the Oregon Historical Society* 1: 192.

Wandeler, P., Hoeck, P. E. A., and Keller, L.F. 2007. "Back to the Future: Museum Specimens in Population Genetics." *Trends in Ecology and Evolution* 22 (12): 634–642.

Weidensaul, S. 2008. *Of a Feather: A Brief History of American Birding*. Boston: Houghton Mifflin Harcourt, 368 pp.

Weiner, J. 1994. *The Beak of the Finch: A Story of Evolution in Our Time*. New York: Knopf, 332 pp.

Wels, S., Farrington, G., McCosker, J., and Piano, R. 2008. *California Academy of Sciences: Architecture in Harmony with Nature*. San Francisco: Chronicle Books, 144 pp.

West, B. and James, M. J. 2001. "The Search for Alban Stewart's Field Notes." Personal correspondence. Archives of the California Academy of Sciences.

Wheeler, W. M. 1898. "George [*sic*] Baur's Life and Writings." *American Naturalist* 33(385): 15–30.

Wikelski M., and Romero, L. M. 2003. "Body Size, Performance and Fitness in Galapagos Marine Iguanas." *Integrative and Comparative Biology* 43: 376–386.

Williams, F. X. 1911. "Expedition of the California Academy of Sciences to the Galapagos Islands, 1905–06. III. The Butterflies and Hawk-Moths of the Galapagos Islands." *Proceedings of the California Academy of Sciences*, 4th ser., 1: 289–322.

Williams, F. X. 1926. "Expedition of the California Academy of Sciences to the Galapagos Islands, 1905–1906. The Bees and Aculeate Wasps of the Galapagos Islands." *Proceedings of the California Academy of Sciiences*, 4th ser., vol. II, pt. II, no. 18, 347–357.

Winchester, S. 2006. *A Crack in the Edge of the World: America and the Great California Earthquake of 1906*. New York: Harper Perennial, 463 pp.

Wong, K. M. 2003. "In Darwin's Wake: The *Academy* Sets Sail to the Galapagos." *California Wild* 56(2): 27–31.

Wylie, C. D. 2009. "Preparation in Action: Paleontological Skill and the Role of the Fossil Preparator." In *Methods in Fossil Preparation: Proceedings of the First Annual Fossil Preparation and Collections Symposium*, edited by M. A. Brown, J. F. Kane, and W. G. Parker, 3–12. Holbrook, Arizona: Petrified Forest National Park.

INDEX

Page references followed by an *f* indicate figures.